Electro-Mechanical Actuators for Safety-Critical Aerospace Applications

Electro-Mechanical Actuators for Safety-Critical Aerospace Applications

Editor

Gianpietro Di Rito

MDPI • Basel • Beijing • Wuhan • Barcelona • Belgrade • Manchester • Tokyo • Cluj • Tianjin

Editor
Gianpietro Di Rito
Dipartimento di Ingegneria
Civile ed Industriale–Sede Aerospaziale
Università di Pisa
Pisa
Italy

Editorial Office
MDPI
St. Alban-Anlage 66
4052 Basel, Switzerland

This is a reprint of articles from the Special Issue published online in the open access journal *Aerospace* (ISSN 2226-4310) (available at: www.mdpi.com/journal/aerospace/special_issues/Electro_mechanical_actuators).

For citation purposes, cite each article independently as indicated on the article page online and as indicated below:

LastName, A.A.; LastName, B.B.; LastName, C.C. Article Title. *Journal Name* **Year**, *Volume Number*, Page Range.

ISBN 978-3-0365-7933-7 (Hbk)
ISBN 978-3-0365-7932-0 (PDF)

© 2023 by the authors. Articles in this book are Open Access and distributed under the Creative Commons Attribution (CC BY) license, which allows users to download, copy and build upon published articles, as long as the author and publisher are properly credited, which ensures maximum dissemination and a wider impact of our publications.
The book as a whole is distributed by MDPI under the terms and conditions of the Creative Commons license CC BY-NC-ND.

Contents

About the Editor vii

Preface to "Electro-Mechanical Actuators for Safety-Critical Aerospace Applications" ix

Gianpietro Di Rito, Romain Kovel, Marco Nardeschi, Nicola Borgarelli and Benedetto Luciano
Minimisation of Failure Transients in a Fail-Safe Electro-Mechanical Actuator Employed for the Flap Movables of a High-Speed Helicopter-Plane
Reprinted from: *Aerospace* 2022, 9, 527, doi:10.3390/aerospace9090527 1

Jeremy Roussel, Marc Budinger and Laurent Ruet
Preliminary Sizing of the Electrical Motor and Housing of Electromechanical Actuators Applied on the Primary Flight Control System of Unmanned Helicopters
Reprinted from: *Aerospace* 2022, 9, 473, doi:10.3390/aerospace9090473 25

Shaojie Zhang, Han Zhang and Kun Ji
Incremental Nonlinear Dynamic Inversion Attitude Control for Helicopter with Actuator Delay and Saturation
Reprinted from: *Aerospace* 2023, 10, 521, doi:10.3390/aerospace10060521 55

Gaetano Quattrocchi, Pier C. Berri, Matteo D. L. Dalla Vedova and Paolo Maggiore
An Improved Fault Identification Method for Electromechanical Actuators
Reprinted from: *Aerospace* 2022, 9, 341, doi:10.3390/aerospace9070341 71

Zirui Liao, Shaoping Wang, Jian Shi, Dong Liu and Rentong Chen
Reliability-Oriented Configuration Optimization of More Electrical Control Systems
Reprinted from: *Aerospace* 2022, 9, 85, doi:10.3390/aerospace9020085 83

Aleksander Suti, Gianpietro Di Rito and Roberto Galatolo
Novel Approach to Fault-Tolerant Control of Inter-Turn Short Circuits in Permanent Magnet Synchronous Motors for UAV Propellers
Reprinted from: *Aerospace* 2022, 9, 401, doi:10.3390/aerospace9080401 105

Shuchang Liu, Zhong Yang, Zhao Zhang, Runqiang Jiang, Tongyang Ren and Yuan Jiang et al.
Application of Deep Reinforcement Learning in Reconfiguration Control of Aircraft Anti-Skid Braking System
Reprinted from: *Aerospace* 2022, 9, 555, doi:10.3390/aerospace9100555 129

Xiaoming Wang, Xinhan Hu, Chengbin Huang and Wenya Zhou
Multi-Mode Shape Control of Active Compliant Aerospace Structures Using Anisotropic Piezocomposite Materials in Antisymmetric Bimorph Configuration
Reprinted from: *Aerospace* 2022, 9, 195, doi:10.3390/aerospace9040195 155

Sami Arslan, Ires Iskender and Tuğba Selcen Navruz
Finite Element Method-Based Optimisation of Magnetic Coupler Design for Safe Operation of Hybrid UAVs
Reprinted from: *Aerospace* 2023, 10, 140, doi:10.3390/aerospace10020140 171

Aleksander Suti, Gianpietro Di Rito and Roberto Galatolo
Fault-Tolerant Control of a Dual-Stator PMSM for the Full-Electric Propulsion of a Lightweight Fixed-Wing UAV
Reprinted from: *Aerospace* 2022, 9, 337, doi:10.3390/aerospace9070337 193

Chenfei She, Ming Zhang, Yibo Ge, Liming Tang, Haifeng Yin and Gang Peng
Design and Simulation Analysis of an Electromagnetic Damper for Reducing Shimmy in Electrically Actuated Nose Wheel Steering Systems
Reprinted from: *Aerospace* **2022**, *9*, 113, doi:10.3390/aerospace9020113 . . . . **215**

Zhangbin Wu, Hongbai Bai, Guangming Xue and Zhiying Ren
Optimization of the Wire Diameter Based on the Analytical Model of the Mean Magnetic Field for a Magnetically Driven Actuator
Reprinted from: *Aerospace* **2023**, *10*, 270, doi:10.3390/aerospace10030270 . . . . **237**

Jean-Charles Maré
A Preliminary Top-Down Parametric Design of Electromechanical Actuator Position Control
Reprinted from: *Aerospace* **2022**, *9*, 314, doi:10.3390/aerospace9060314 . . . . **253**

About the Editor

Gianpietro Di Rito

Born in Massafra, Italy, October 2nd, 1974, he received his MSc Degree in Aerospace Engineering in 2001 and his PhD Degree in Aerospace Engineering in 2005, at the University of Pisa, Pisa, Italy. He was a Research Fellow at the Department of Aerospace Engineering of the University of Pisa from 2006 to 2011. From 2012 to 2021, he was an Assistant Professor of Aerospace Systems at the Department of Civil and Industrial Engineering of the University of Pisa, where in 2022 he became an Associate Professor of Aerospace Systems.

His main research interests are focused on the modelling, simulation, control, health-monitoring and testing of mechatronic systems for aerospace applications.

Preface to "Electro-Mechanical Actuators for Safety-Critical Aerospace Applications"

Aircraft electrification is one of the most important and strategic initiatives currently supporting the innovation of the aviation industry. This manifests in the well-known more-electric aircraft concept (with the ultimate aim of achieving the all-electric long-term target), which aims to gradually replace onboard systems based on mechanical, hydraulic, or pneumatic power sources with electrically powered ones to reduce the weight and costs, optimize energy, and increase the eco-compatibility and reliability of future aircrafts.

A key technological enabler for pursuing these challenging objectives is electro-mechanical actuation. The applicability of electro-mechanical actuators (EMAs) in aerospace has been proved in terms of dynamic performances, but it still entails several concerns in terms of reliability/safety and operation in a harsh environment. In civil aircrafts, EMAs are often avoided for safety-critical functions (flight controls, brakes, landing gears, and nose wheel steering), essentially because the statistical database on the components' fault modes is poor.

This Special Issue is thus focused on advancements and innovations in the design, modelling/simulation, architectural definition, reliability/safety analysis, control, condition-monitoring, and experimental testing of EMAs developed for safety-critical aerospace applications.

The research papers included in this Special Issue will undoubtedly contribute to progress toward the objective of more electric flights.

Gianpietro Di Rito
Editor

Article

Minimisation of Failure Transients in a Fail-Safe Electro-Mechanical Actuator Employed for the Flap Movables of a High-Speed Helicopter-Plane

Gianpietro Di Rito [1,*], Romain Kovel [2], Marco Nardeschi [3], Nicola Borgarelli [3] and Benedetto Luciano [4]

1 Department of Civil and Industrial Engineering, University of Pisa, Largo Lucio Lazzarino 2, 56122 Pisa, Italy
2 Airbus Helicopters, International Airport of Marseille-Provence, 13700 Marignane, France
3 Umbragroup Spa, Via Baldaccini 1, 06034 Foligno, Italy
4 AESIS srl, Via Carducci 62/H, San Giuliano Terme, 56017 Pisa, Italy
* Correspondence: gianpietro.di.rito@unipi.it; Tel.: +39-0502217211

Abstract: The work deals with the model-based characterization of the failure transients of a fail-safe rotary EMA developed by Umbragroup (Italy) for the flap movables of the RACER helicopter-plane by Airbus Helicopters (France). Since the reference application requires quasi-static position-tracking with high disturbance-rejection capability, the attention is focused on control hardover faults which determine an actuator runaway from the commanded setpoint. To perform the study, a high-fidelity nonlinear model of the EMA is developed from physical first principles and the main features of health-monitoring and closed-loop control functions (integrating the conventional nested loops architecture with a deformation feedback loop enhancing the actuator stiffness) are presented. The EMA model is then validated with experiments by identifying its parameters by ad-hoc tests. Simulation results are finally proposed to characterize the failure transients in worst case scenarios by highlighting the importance of using a specifically designed back-electromotive damper circuitry into the EMA power electronics to limit the position deviation after the fault detection.

Keywords: health monitoring; electro-mechanical actuators; modelling; simulation; testing; flight control; reliability; fault-tolerant systems; failure transient analysis

1. Introduction

The aircraft electrification is surely one of the most important and strategic initiatives currently supporting the innovation of the aviation industry [1,2]. In particular, the *more-electric aircraft* concept entails the gradual replacement of onboard systems based on mechanical, hydraulic, or pneumatic power sources with electrically powered ones, aiming to reduce weight and costs, to optimize energy and to increase the eco-compatibility of future aircrafts [3–6]. Electro-mechanical actuation clearly plays a key role for pursuing these challenging objectives. The applicability of Electro-Mechanical Actuators (EMAs) in aerospace is proven in terms of load and speed performances [7–12], but several reliability concerns still remain open [13–17]. The use of EMAs for safety-critical functions can be thus obtained only by fault-tolerant architectures, which apply hardware redundancies on electrical, electronic, or mechanical parts.

In general terms, depending on how the redundancy is applied, a fault-tolerant function can be maintained after a fault, or it can be lost while avoiding the extension of the fault effects to other functions, so that fail-operative or fail-safe functions are respectively obtained. With reference to flight control functions, this concept can be applied to both movables and actuators. To obtain a fail-operative flight control, different architectures can be used by applying load-level redundancy (splitting the movable into sub-movables and using a fail-safe EMA on each part), actuator-level redundancy (using multiple fail-safe EMAs on a single movable), or subsystem-level redundancy (using a single fail-operative EMA on a single movable). Any fault-tolerant system necessitates effective health-monitoring algorithms aiming to anticipate (i.e., avoid) or detect/isolate the fault, so that prognostic and diagnostic approaches are respectively defined. The prognostic solution, though potentially overwhelming [18–21], is nowadays far from being applicable to airworthy systems, and diagnostic approaches are typically preferred [22–24]. The diagnostic monitoring requires that the Fault Detection and Isolation (FDI) is implemented and executed in real time by onboard control electronics [25–28], so that, in case of fault, the redundant components or the isolation devices can be engaged [29,30].

The design and the validation of health-monitoring systems play a key role in this context. In a flight control EMA, the FDI output, including the consequent accommodation/compensation of the fault, must be provided with very small latency, and the failure transients must be adequately limited. The development of high-fidelity experimentally validated models of EMAs is of paramount importance for the validation of monitoring functions. Since nonlinearities, disturbances, environment, and loads can significantly affect the actuator response, an in-depth knowledge of both normal and faulty behaviours is required. The crucial problem entails the knowledge of faulty dynamics, especially in complex systems with a huge number of fault modes [31–33]. In the so-called data-driven techniques, this knowledge is achieved via experiments, by artificially injecting the major faults in the EMA and measuring its response [20,34–36]. This method provides accurate predictions, but rigging costs are often prohibitive. In addition, the FDI validation strongly depends on test conditions. As a relevant example, in [20], the mechanical degradation of the ball-screw elements of an aircraft EMA is investigated via a data-driven approach: the lifecycle of a rudder control actuator, including periodical maintenance checks, is simulated by testing a prototype EMA in laboratory environment with alternate endurance and monitoring trials. To accelerate the mechanical degradation, the prototype is intentionally modified with respect to the nominal design, by using a reduced number of recirculating paths in the ball-screw, by removing the anti-rotation device on the output shaft, by applying relevant radial loads, and by progressively removing the lubricant. Discrete-time and continuous-time fault symptoms are then computed by leveraging the EMA outputs via multivariate statistical methods (such as Hotelling's T^2 and Q techniques). The health monitoring demonstrates to be very effective, but the entire experimental activity required a specifically dedicated rig and took seven months. In addition, the experimental campaign did not take into account temperature effects.

In model-based techniques, the knowledge of the faulty dynamics is to a great extent obtained from mathematical models, capable of simulating the fault by physical first principles, and are experimentally validated with reference to normal and/or regime faulty conditions [26–28,37,38]. Oppositely to the data-driven case, this method generally provides less accurate predictions, but it is cost-effective, allows to verify the FDI functionalities in extreme conditions, and (above all) permits to generalize the validity of algorithms to similar equipment (i.e., governed by similar equations). As a relevant example, in [28], the major faults of a primary flight control movable driven by active-active EMAs are addressed via model-based approach: a set of monitoring algorithms are designed using a detailed nonlinear model of the system capable of fault simulation. Robust detection thresholds are determined taking into account parametric and input uncertainties, and the health-monitoring is verified through simulation, by injecting faults in an experimentally validated model of the system.

The basic objective of this work is to validate the monitoring algorithms of the fail-safe EMA developed by UmbraGroup (Italy) for the flap movables of the RACER (Rapid And Cost-Efficient Rotorcraft) helicopter-plane by Airbus Helicopters with reference to the model-based analysis of the failure transients related to the hardover of the control electronics (major EMA fault mode). The paper is articulated as follows: the first part is dedicated to the system description and to the EMA modelling; successively, the main features of the closed-loop control and health-monitoring functions are presented. Finally, an excerpt of simulation results is proposed by characterizing the EMA failure transients in selected worst-case scenarios. The results are finally discussed by highlighting the effectiveness and the most relevant criticalities of the proposed approach with suggestions of possible enhancements.

2. Materials and Methods

2.1. System Description

The reference EMA is used to control the six flap movables of the RACER helicopter-plane, an innovative, high-speed, more-electric air vehicle developed by Airbus Helicopters, Figure 1. The RACER helicopter-plane is designed to reach maximum cruise speed 50% faster than a conventional helicopter (the Velocity Never Exceed, VNE, is 115 m/s) and to consume 15% less fuel per distance at reference cruise speed (90 m/s) [39]. The aerodynamic concept essentially merges a conventional helicopter with a low aspect-ratio box-wing airplane; at cruise speed, the two wing propellers generate thrust and the box-wing contributes to lift, generating low induced drag and minimized interactions with the main rotor flow [40,41], so that the rotor can be slowed by up to 15%, preventing the blades from working with transonic local flow (which reduces performances). The electrical system is based on high-voltage direct current power generation, assuring a consistent weight reduction [42].

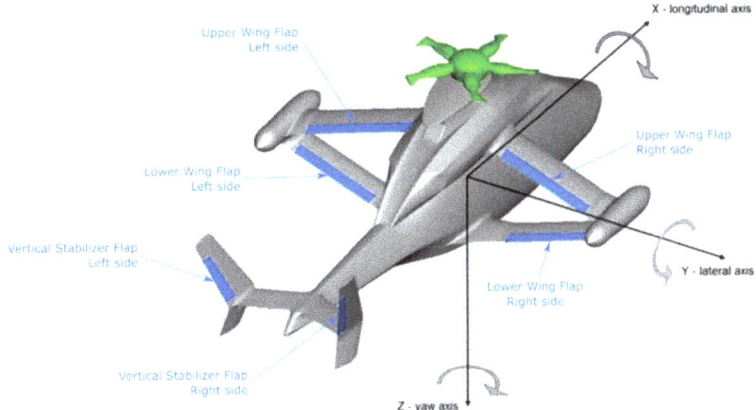

Figure 1. Flap movables on the RACER helicopter-plane by Airbus Helicopters.

The RACER flaps placed on both box-wing (four) and vertical stabilizers (two) are used to adapt the vehicle attitude, to enhance stability, to optimize the trim configuration, and to abate noise [43–45]. Depending on the vehicle weight, the airspeed, the altitude, and the rotors speed, the wing flaps are deflected to optimize the mean lift coefficient of the main rotor. On the other hand, the flaps on the vertical fins are used to eliminate the residual yawing torque generated by the propellers, by assuring that they only contribute to propulsion during cruise [43]. Given these basic flight control functions, the design of the closed-loop position control of the flap movables is mainly driven by disturbance rejection requirements (i.e., the capability to minimize the position deviation from the commanded setpoint under external disturbances).

It is worth noting that in the RACER helicopter-plane, the flaps are not used for manoeuvrability (trajectory control is managed through the cyclic stick, as for conventional helicopters), so that they are classified as secondary flight controls. The flap EMAs are thus designed to be fail-safe systems in such a way that, after a major fault, the actuator is still capable of maintaining the flap movable at a fixed deflection (last or neutral position, depending on the fault mode), providing an adequate torsional stiffness to avoid flutter concerns.

Each flap EMA is composed of two parts, Figure 2: an electromechanical rotary actuator (FLap Actuator, FLA) and a control electronic box (Actuator Control Electronics,

ACE). The equipment locations on the RACER helicopter-plane are shown in Figure 3 (note that the reported layout also depicts flaps on the horizontal tail since they have been initially included in the flight control system [41,43], but then eliminated from the final design).

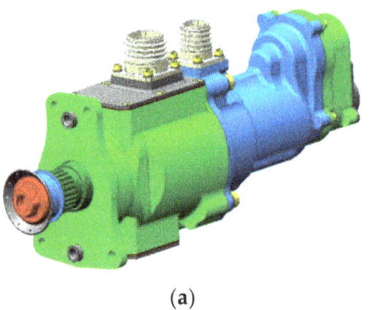
(a)

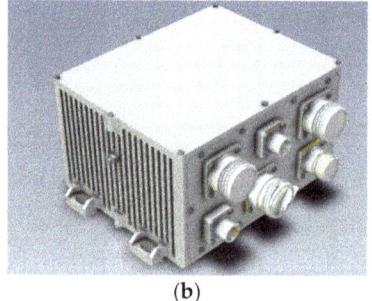
(b)

Figure 2. Flap EMA layout: (**a**) FLA; (**b**) ACE.

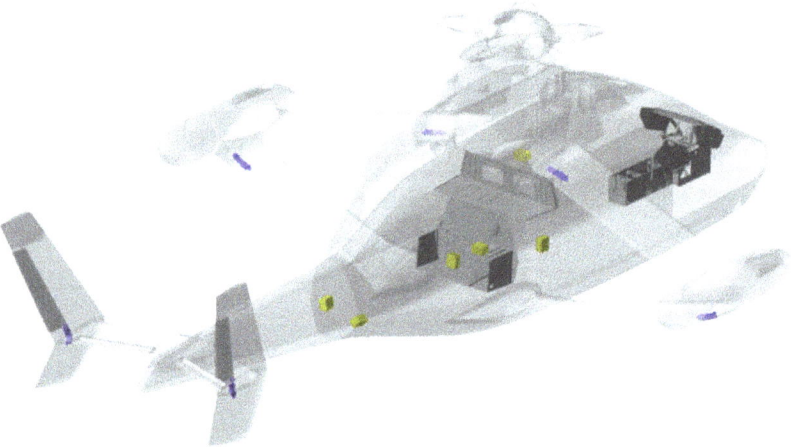

Figure 3. Locations of the EMA parts on the RACER helicopter-plane (yellow: ACEs; blue: FLAs).

The ACE includes three electronic boards, Figures 4 and 5:
- COMmand (COM) board, implementing the EMA closed-loop control functions and the control of one of the two motor brakes (BC_{COM} in Figure 4);
- MONitor (MON) board, implementing the EMA health-monitoring algorithms and the control of one of the two motor brakes (BC_{MON} in Figure 4);
- PoWeR (PWR) board, including the power supply regulation for all electrical components, the MOSFET bridge, the six currents sensors (three ones for the COM board and three ones for the MON board, $CF_{x\,COM}$ and $CF_{x\,MON}$ in Figure 4) and a BEMF (Back Electro-Motive Force) damper circuitry, Figure 5.

The activation of the BEMF damper circuitry in the PWR board is obtained by a logic signal named *system validity* (SV in Figures 4 and 5), which derives from an "AND" operator applied to the *local validity* signals provided by the two boards (LV_{COM} and LV_{MON} in Figures 4 and 5). When SV is true, the power bridge thyristors are opened, and the damper thyristor is closed (Figure 5), so that the motor phases are shorted to the ground, and an electromagnetic damping torque is developed and transmitted to the EMA output shaft. This strategy permits significantly limiting the failure transients related to major faults (e.g., the control electronics hardover) since the unavoidable delays needed to achieve the full engagement of the motor brakes can determine an excessive deviation from the commanded setpoint with potentially dangerous concerns due to the impact on mechanical stops.

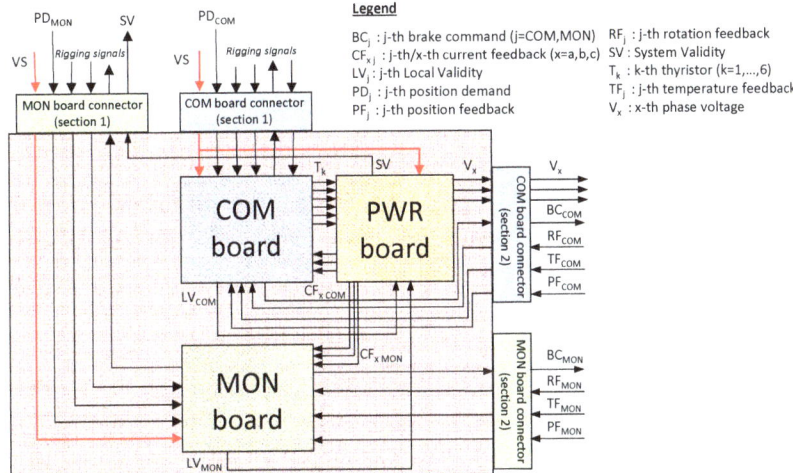

Figure 4. ACE: schematics of boards interfaces.

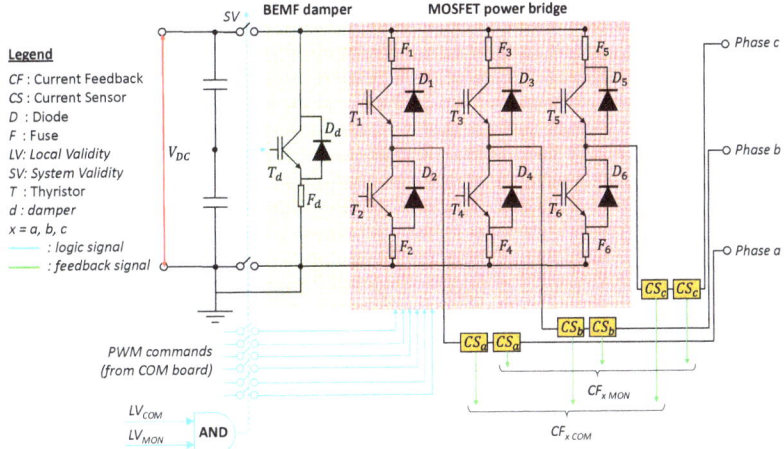

Figure 5. Schematics of the main sections of the PWR board.

Both COM and MON boards are controlled by Texas Instruments TMS570LC4357-EP ARM-based microcontrollers [46] using a 10 kHz sample rate for the digital signal processing. The FLA basically includes, Figure 6:

- a three-phase Permanent Magnet Synchronous Machine (PMSM) with surface-mounted magnets and sinusoidal back-electromotive forces, driven via Field-Oriented Control (FOC) technique;
- two motor rotation sensors: a resolver interfaced with the COM board and a magnetic encoder interfaced with the MON board (RF_{COM} and RF_{MON} in Figure 4);
- a dual magnetic encoder for the output shaft rotation sensing, interfaced to both COM and MON boards (PF_{COM} and PF_{MON} in Figure 4);
- two temperature sensors (TF_{COM} and TF_{MON} in Figure 4);

- two power-off electromagnetic brakes used to block in position the EMA after a major fault detection;
- an innovative Umbragroup-patented differential ball-screw mechanism implementing the mechanical power conversion from motor to output shaft, which, if compared with conventional gearboxes, assures a high gear ratio (more than 500) with minimum backlash (less than 0.05 deg) and superior efficiency (about 95%).

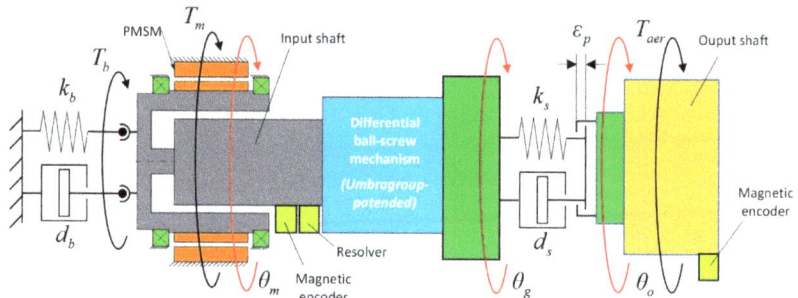

Figure 6. Schematics of the FLA electro-mechanical section.

Table 1 reports the main characteristics of the sensors used for the closed-loop control and health-monitoring functions of the actuator [47–51].

Table 1. EMA control sensors data.

Component	Model	Range	Accuracy
Current sensor	Allegro ACS723LLCTR-10AB-T	±10 A	0.1 A
Resolver	Tamagawa TS2610N171E64	±π rad	4×10^{-4} rad
Resolver analog-to-digital converter	Analog Devices AD2S1210	±π rad	2×10^{-4} rad
Magnetic encoder (motor)	Analog Devices ADA4571	±π rad	4×10^{-4} rad
Duplex magnetic encoder (output)	RLS AksIM-2	±0.157 rad	1.7×10^{-3} rad

2.2. Nonlinear Dynamic Modelling

The EMA health-monitoring algorithms have been designed with the essential support of the dynamic simulation by artificially injecting major system faults in a detailed nonlinear model of the actuator, developed via physical first principles and validated through experiments. This approach is of paramount importance, especially for the failure transient characterisation which is often unfeasible (or problematic) via testing.

The model of the RACER flaps EMA is essentially composed of:

- an electromechanical section, simulating
  - FOC current dynamics;
  - multi-harmonic modelling of PMSM torque disturbances (due to cogging effects [52–55] and/or BEMF waveform distortions);
  - 2-degree-of-freedom mechanical transmission with equations of motions related to motor and output rotations;
  - sliding friction on motors and output shaft, described via combined "Coulomb–tanh" model [56,57];
  - mechanical freeplay [21];
  - internal stiffness dependence on output shaft position;

- an electronic section simulating the control and health-monitoring algorithms implemented by the COM and MON boards, including
  - Clarke-Park transforms for the FOC technique implementation;
  - sensor errors and nonlinearities (bias, noise, resolution);
  - command nonlinearities (saturation, rate limiting);
  - digital signal processing at 10 kHz sampling rate;
  - control hardover fault simulation, implying that the voltage demands on both quadrature and direct axes suddenly assume and maintain random values, so that the EMA motion is out of control (as a worst case scenario, the quadrature voltage is set to saturation value, while the direct voltage is set to zero).

2.2.1. Electro-Mechanical Section of the Model

The electro-mechanical section of the model, schematically represented in Figure 6, is governed by Equations (1)–(10),

$$\mathbf{V_{abc}} = R\mathbf{I_{abc}} + L\dot{\mathbf{I}}_{abc} + \mathbf{e_{abc}}, \tag{1}$$

$$\mathbf{e_{abc}} = \lambda_m n_d \dot{\theta}_m \left[\sin(n_d\theta_m) \quad \sin\left(n_d\theta_m - \tfrac{2}{3}\pi\right) \quad \sin\left(n_d\theta_m + \tfrac{2}{3}\pi\right)\right]^T, \tag{2}$$

$$J_m \ddot{\theta}_m = T_m + T_b - T_{sfm}\tanh\left(\frac{\dot{\theta}_m}{\omega_{sfm}}\right) - d_{vfm}\dot{\theta}_m - \frac{d_s\dot{\theta}_s + k_s\theta_s}{\tau_g}, \tag{3}$$

$$J_o \ddot{\theta}_o = T_{aer} - T_{sfo}\tanh\left(\frac{\dot{\theta}_o}{\omega_{sfo}}\right) - d_{vfo}\dot{\theta}_o + d_s\dot{\theta}_s + k_s\theta_s, \tag{4}$$

$$T_m = \lambda_m n_d \left[I_a \sin(n_d\theta_m) + I_b \sin\left(n_d\theta_m - \tfrac{2}{3}\pi\right) + I_c \sin\left(n_d\theta_m + \tfrac{2}{3}\pi\right)\right] + \sum_{j=1}^{M} T_{hdj}\sin(n_{hdj}\theta_m), \tag{5}$$

$$k_s = k_{smin} + \gamma_k(\theta_o - \theta_{omax})^2, \tag{6}$$

$$T_b = \begin{cases} 0 & t < t_{FC} \\ -k_b[\theta_m - \theta_m(t_{FC})] - d_b[\dot{\theta}_m - \dot{\theta}_m(t_{FC})] & t \geq t_{FC} \end{cases}, \tag{7}$$

$$\dot{\theta}_s = \begin{cases} -\frac{k_s}{d_s}\theta_s & |\theta_g - \theta_o| < \varepsilon_p \\ \dot{\theta}_g - \dot{\theta}_o & |\theta_g - \theta_o| \geq \varepsilon_p \end{cases}, \tag{8}$$

$$\theta_s = \begin{cases} \int \dot{\theta}_s dt & |\theta_g - \theta_o| < \varepsilon_p \\ \theta_g - \theta_o - \varepsilon_p \mathrm{sgn}(\theta_g - \theta_o) & |\theta_g - \theta_o| \geq \varepsilon_p \end{cases}, \tag{9}$$

$$\dot{\theta}_g = \frac{\dot{\theta}_m}{\tau_g}, \tag{10}$$

where $\mathbf{V_{abc}} = [V_a, V_b, V_c]^T$ is the applied voltages vector, $\mathbf{I_{abc}} = [I_a, I_b, I_c]^T$ is the phase currents vector, $\mathbf{e_{abc}}$ is the back-electromotive forces vector (sinusoidal BEMF waveforms are assumed), λ_m is the magnet flux linkage, R and L are the resistance and the inductance of motor phases, respectively, n_d is the number of rotor pole pairs, θ_m is the motor rotation, θ_g is the theoretical rotation imposed by a rigid ball-screw drivetrain, θ_s is the torsional deformation referred to the first structural mode of the EMA, θ_o is the output rotation, J_m and J_o are the motor and output inertias, respectively, τ_g is the gear ratio of the differential ball-screw mechanism, ε_p is the internal freeplay, T_m is the motor torque, T_{hdj} and n_{hdj} are the amplitude and mechanical period indices, respectively, related to the j-th (j = 1, ..., M where M is an integer number) harmonic torque disturbance contribution, T_{aer} is the aerodynamic hinge moment, T_b is the brakes torque, k_b and d_b are the torsional stiffness and damping of the brakes, respectively, t_{FC} is the time at which the fault compensation occurs, k_s and d_s are the torsional stiffness and damping, respectively, referred to the first structural mode of the EMA, k_{smin} is the minimum internal stiffness, γ_k is the parameter defining the stiffness variation with respect to output position, d_{vfm} and d_{vfo} are the viscous friction coefficients related to the motor and output shafts, respectively, while T_{sfm}, ω_{sfm}, T_{sfo}, and ω_{sfo} are the parameters of the "Coulomb–tanh" models simulating the sliding friction on motor and output shaft, respectively.

Concerning the aerodynamic hinge moment applied on the EMA (T_{aer}, in Equation (4)), the requirements indicate that, apart from static loadings, two contributions of dynamic loads must be taken into account for the performance analysis: a deterministic one (related to the helicopter-plane motion, main rotor speed and angle, wing propeller speeds), in which harmonic loads of specific amplitudes and frequencies are superimposed, and a non-deterministic one, including gust loads and harmonic loads of constant amplitudes randomly applied along the the position-tracking frequency range. In this work, the study is focused on the vertical stabilizer flaps, since it represents the worst-case scenario for the EMAs employed in the RACER helicopter-plane, Table 2.

Table 2. Loads on vertical stabilizer FLA at VNE (derived from CFD analyses by Airbus Helicopters, worst-case scenario, all positions).

Static [Nm]	Harmonic Amplitude [Nm]	Harmonic Frequency [Hz]	Dynamic Load Definition
±100	2	15	Deterministic
	3	20	
	15	23	
	2	30	
	2	46	
	1.5	From 1 to 100	Non-deterministic

It is worth noting that the proposed model represents a balance between prediction accuracy, objectives of the study, and complexity of the model itself. More accurate simulations could include sophisticated friction models [56,57] and iron losses in the motor [58,59], but the inclusion of these features would entail minor effects for the examined application. In particular, the motor iron losses have been neglected because they depend on electrical frequency, which is relatively small in the position-tracking frequency range (<50 Hz, Table 2). On the other hand, more accurate friction models (including load and temperature dependence) could enhance the simulation, but a simplified approach has been preferred both for the lack of detailed information and to limit the number of model parameters.

The EMA model has been entirely developed in the Matlab–Simulink–Stateflow environment, and the numerical solution is obtained by a Runge–Kutta method with 10^{-6} s integration step. The choice of a fixed-step solver is not strictly related to the objectives of this work in which the model (once experimentally validated) is used for "off-line" simulations characterising the EMA failure transients, but it has been selected for the next steps of the project, when the algorithms for the closed-loop control and the health-monitoring will be implemented in the ACE boards via the automatic Matlab compiler and executed in "real-time".

The parameters of the electro-mechanical section of the model are given in Table 3.

Table 3. Parameters of the electro-mechanical section of the model.

Parameter	Meaning	Value	Unit	Identification Method (See Section 2.3)
L	Motor phase inductance	15×10^{-3}	H	Test 1, Test 2
R	Motor phase resistance	1.53	ohm	Test 1, Test 2
λ_m	Magnet flux linkage	0.014	N·m/A	Test 3
n_d	Motor pole pairs	10	–	Design

Table 3. Cont.

Parameter	Meaning	Value	Unit	Identification Method (See Section 2.3)
J_m	Motor inertia	4×10^{-5}	kg·m^2	Design, Test 5
T_{sfm}	Coulomb friction on motor shaft	0.015	N·m	Test 5
ω_{sfm}	Coulomb velocity on motor shaft	0.1	rad/s	Test 5
d_{vfm}	Viscous friction coefficient on motor shaft	10^{-4}	N·m s/rad	Test 5
τ_g	Differential ball-screw gear ratio	500	–	Design
$\theta_{o\max}$	Mechanical endstroke, from centred	0.14	rad	Design
$k_{s\min}$	Drivetrain torsional stiffness at $\theta_o = \theta_{o\max}$	1.15×10^4	N·m/rad	FEM analysis
γ_k	Parameter of the stiffness curve	1.3×10^5	N·m/rad^3	FEM analysis
d_s	Drivetrain damping (1st vibration mode)	2.6	N·m s/rad	FEM analysis
J_o	Output inertia, including flap movable	0.06	kg·m^2	Design, Test 5
T_{sfo}	Coulomb friction on output shaft	0.5	N·m	Test 5
ω_{sfo}	Coulomb velocity on output shaft	10^{-3}	rad/s	Test 5
d_{vfm}	Viscous friction coefficient on output shaft	0.1	N·m s/rad	Test 5
k_b	Brakes stiffness	150	N·m/rad	Test 4
d_b	Brakes damping	0.02	N·m s/rad	Test 4
ε_p	End-life internal freeplay	1.3×10^{-3}	rad	Design
M	Number of cogging torque harmonics	3	–	Test 5
T_{hd1}	Torque disturbance amplitude, 1st harmonic	0.001	N·m	Test 5
n_{hd1}	Torque disturbance period index, 1st harmonic	10	–	Test 5
T_{hd2}	Torque disturbance amplitude, 2nd harmonic	0.007	N·m	Test 5
n_{hd2}	Torque disturbance period index, 2nd harmonic	20	–	Test 5
T_{hd3}	Torque disturbance amplitude, 3rd harmonic	0.002	N·m	Test 5
n_{hd3}	Torque disturbance period index, 3rd harmonic	24	–	Test 5
V_{max}	DC voltage supply	28	V	Design
$I_{q\max}$	Maximum quadrature current	4	A	Design
$\omega_{m\ \max}$	Maximum motor speed	100	rad/s	Design

2.2.2. Electronic Section of the Model

The closed-loop control of the RACER flap EMAs is schematically represented in Figure 7. The position-tracking architecture integrates the conventional three nested loops on motor currents, motor speed, and output position [8,9] with a deformation feedback loop ("Stiffness Enhancement System, SES" block in Figure 7) and a model-based correction of voltage commands, aiming to decouple the currents dynamics from the motor motion ("electro-mechanical decoupler" block in Figure 7).

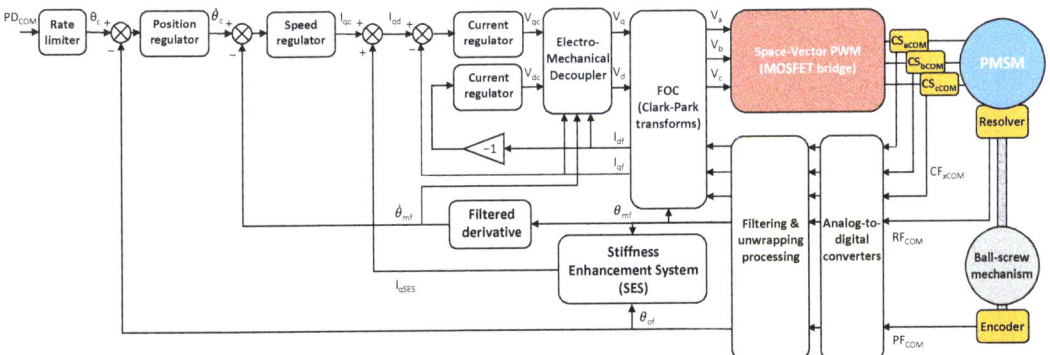

Figure 7. Closed-loop control scheme of the actuator.

The FOC technique implemented in the COM board applies the direct and inverse Clark–Park transforms [60] via Equations (11)–(13),

$$\mathbf{x}_{\alpha\beta\gamma} = \mathbf{T_C}\mathbf{x_{abc}} = \sqrt{\frac{2}{3}} \begin{bmatrix} 1 & -1/2 & -1/2 \\ 0 & \sqrt{3}/2 & -\sqrt{3}/2 \\ \sqrt{2}/2 & \sqrt{2}/2 & \sqrt{2}/2 \end{bmatrix} \mathbf{x_{abc}}, \quad (11)$$

$$\mathbf{x_{dqz}} = \mathbf{T_P}\mathbf{x}_{\alpha\beta\gamma} = \begin{bmatrix} \cos(n_d\theta_m) & \sin(n_d\theta_m) & 0 \\ -\sin(n_d\theta_m) & \cos(n_d\theta_m) & 0 \\ 0 & 0 & 1 \end{bmatrix} \mathbf{x}_{\alpha\beta\gamma}, \quad (12)$$

$$\mathbf{x_{dqz}} = \mathbf{T_P T_C}\mathbf{x_{abc}} \Leftrightarrow \mathbf{x_{abc}} = (\mathbf{T_P T_C})^T \mathbf{x_{dqz}}, \quad (13)$$

where $\mathbf{x}_{\alpha\beta\gamma} = [x_\alpha, x_\beta, x_\gamma]^T$, $\mathbf{x_{abc}} = [x_a, x_b, x_c]^T$, and $\mathbf{x_{dqz}} = [x_d, x_q, x_z]^T$ are generic three-phase vectors in the Clarke, Park, and stator reference frames, respectively, while $\mathbf{T_C}$ and $\mathbf{T_P}$ are the Clarke and Park transforms.

The digital regulators on position, speed, and currents implement proportional/integral actions on tracking error signals, plus anti-windup functions with back-calculation technique [61] to compensate for commands saturation. Each j-th (with $j = \theta$, ω, and I indicating the position, speed, and currents loops, respectively) digital regulator is governed by Equations (14) and (15):

$$y_{PI}^{(j)} = k_P^{(j)} \varepsilon^{(j)} + \frac{k_I^{(j)} T_s}{z-1} \left[\varepsilon^{(j)} + k_{AW}^{(j)} \left(y^{(j)} - y_{PI}^{(j)} \right) \right] \quad (14)$$

$$y^{(j)} = \begin{cases} y_{PI}^{(j)} & \left| y_{PI}^{(j)} \right| < y_{sat}^{(j)} \\ y_{sat}^{(j)} \mathrm{sgn}\left(y_{PI}^{(j)}\right) & \left| y_{PI}^{(j)} \right| \geq y_{sat}^{(j)} \end{cases} \quad (15)$$

where z is the discrete-time operator, $\varepsilon^{(j)}$ is the regulator input (tracking error), $y^{(j)}$ is the regulator output, $y_{PI}^{(j)}$ is the saturator block input (proportional–integral with respect to error, if no saturation is present), while $k_P^{(j)}$ and $k_I^{(j)}$ are the proportional and integral gains, $k_{AW}^{(j)}$ is the back-calculation anti-windup gain, $y_{sat}^{(j)}$ is the saturation limit, and T_s is the sampling time.

Concerning the SES loop, its basic objective is to enhance the loads disturbance rejection in the frequency range where the first resonant pulsation of the ball-screw mechanism is located (according to FEM analyses performed by Umbragroup, from 70 to 90 Hz, depending on output shaft position, Equation (6)). The control task is achieved by superimposing to the current demand generated by the speed regulator (I_{qc}, in Figure 7) an additional one (I_{qSES}, in Figure 7) that depends on the torsional deformation (δ_f) reconstructed by the motion feedbacks (θ_{mf} and θ_{of} in Figure 7), Equations (16)–(18).

$$I_{qd} = I_{qc} + I_{qSES}, \quad (16)$$

$$\ddot{I}_{qSES} = -a_{SES} \dot{I}_{qSES} - b_{SES} I_{qSES} - k_{SES} \dot{\delta}_f, \quad (17)$$

$$\delta_f = \frac{\theta_{mf}}{\tau_g} - \theta_{of}, \quad (18)$$

The structure of the current demand regulator (a second-order system responding to deformation rate input, Equations (17) and (18)) is defined by pursuing the following requirements:

- the loop shall not affect the EMA low-frequency behaviour (maxima loads, position tracking, etc.);
- the loop shall generate demands only in the frequency range where the first resonant pulsation of the ball-screw mechanism is located, and the compensation shall imply

an increase of EMA stiffness, enhancing the disturbance rejection capabilities related to external loads.

To fulfill these objectives, the positive-defined parameters, k_{SES}, a_{SES}, and b_{SES} in Equation (17) are set in such a way that

- by tuning a_{SES} and b_{SES}, the phase response of the SES current demand (I_{qSES}) with respect to torsional deformation is about $-180°$ from 70 to 90 Hz;
- by tuning k_{SES}, the SES current demand (I_{qSES}) implies an effective compensation without affecting the control stability.

Figure 8 shows the Bode diagram of the transfer function defined in Equation (19), which relates in the Laplace domain (i.e., s represents the complex variable) the SES current demand with the reconstructed deformation feedback,

$$\frac{I_{qSES}(s)}{\delta_f(s)} = -\frac{k_{SES}s}{s^2 + a_{SES}s + b_{SES}}, \tag{19}$$

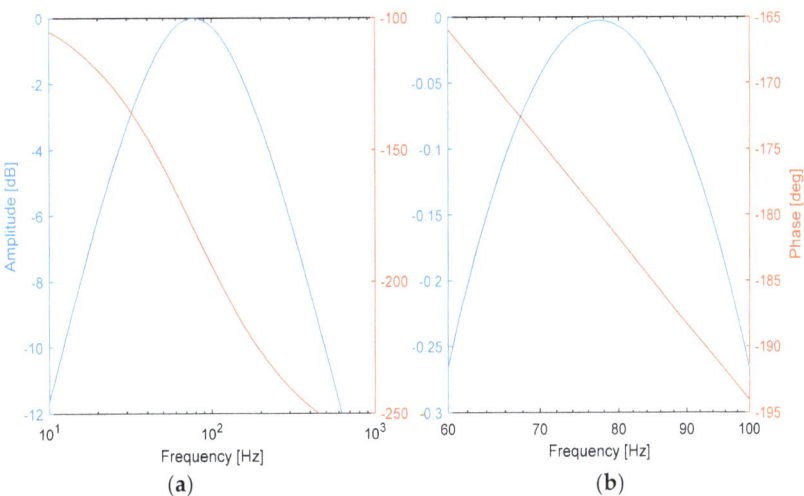

Figure 8. Bode diagram of the SES regulator (0 dB = 1 A/m): (**a**) behaviour from low to high frequencies; (**b**) detail in the frequency range where the first resonant pulsation of the ball-screw drivetrain is located.

It can be noted that the design implies that from 70 to 90 Hz the phase response ranges from $-175°$ to $-190°$, and the regulator gain achieves its maximum, while it tends to be negligible at both low and high frequencies.

Regarding the currents-motion decoupling, it is obtained via Equations (20) and (21):

$$V_d = V_{dc} - L_q n_d I_{qf} \dot{\theta}_{mf}, \tag{20}$$

$$V_q = V_{qc} + \left(\sqrt{\frac{3}{2}}\lambda_m + L_d I_{df}\right) n_d \dot{\theta}_{mf}, \tag{21}$$

where V_{dc} and V_{qc} are the direct and quadrature voltage demands generated by the current regulators, and L_d and L_q are the inductances on the direct and quadrature axes (in the reference PMSM, having surface-mounted magnets, $L_d = L_q = L$). The currents dynamics imposed by the FOC technique implies that in the PMSM rotor frame (Equation (13)),

$$L\dot{I}_d = V_d - RI_d + Ln_d I_q \dot{\theta}_m, \tag{22}$$

$$L\dot{I}_q = V_q - RI_q - \left(\sqrt{\frac{3}{2}}\lambda_m + LI_d\right)n_d\dot{\theta}_m, \tag{23}$$

Thus, by substituting Equations (20) and (21) into Equations (22) and (23), we have

$$L\dot{I}_d = V_{dc} - RI_d + Ln_d\left(I_q\dot{\theta}_m - I_{qf}\dot{\theta}_{mf}\right), \tag{24}$$

$$L\dot{I}_q = V_{qc} - RI_q - \sqrt{\frac{3}{2}}\lambda_m n_d\left(\dot{\theta}_m - \dot{\theta}_{mf}\right) - Ln_d\left(I_d\dot{\theta}_m - I_{df}\dot{\theta}_{mf}\right), \tag{25}$$

Now, if the sensor dynamics imply minor phase delays and/or attenuations ($\dot{\theta}_{mf} \approx \dot{\theta}_m$, $I_{qf} \approx I_q$, $I_{df} \approx I_d$), the residuals terms at second hands in Equations (24) and (25) can be neglected, so that the currents dynamics on both direct and quadrature axes behave independently and are decoupled from the rotor motion.

To protect the system from major faults and to permit its reversion into a fail-safe configuration (EMA with engaged brakes, maintaining the flap at fixed deflection), the following set of health-monitoring algorithms are executed by the MON board:

- over-temperature monitor, checking that the motor stator temperature does not exceed a pre-defined threshold;
- over-current monitor, checking that the quadrature current does not exceed a pre-defined threshold;
- Over-Speed Monitor (OSM), checking that the motor speed does not exceed a pre-defined threshold;
- currents consistency monitor, checking that the sum of the phase currents is lower than a pre-defined threshold;
- mechanical consistency monitor, checking that the EMA torsional deformation is lower than a pre-defined threshold;
- position deviation monitor, checking that the deviation of the output position feedback from the commanded setpoint is lower than a pre-defined threshold.

For the examined application, the most feared EMA failure is the control hardover, i.e., an electronic fault for which the COM board applies and maintains random voltage demands on both quadrature and direct axes, so that the actuator motion is out of control. The coverage of this failure is here provided by the OSM, whose working flow chart is reported in Figure 9. The OSM fault flag (F_{mon}) is generated by elaborating as fault symptom the amplitude of the speed feedback signal (ω_{mon}) at the k-th monitoring sample (processed at 10 kHz rate): if the fault symptom is greater than a pre-defined threshold (ω_{th}), a fault counter (c_{mon}) is increased by 2; if the threshold is not exceeded, the fault counter is decreased by 1 if it is positive at the previous step, otherwise it is held at 0. The fault is thus detected when the fault counter exceeds a pre-defined value ($c_{mon\ max}$, which basically defines the OSM FDI latency).

The parameters of the electronic section of the model are given in Table 4.

Table 4. Parameters of the electronic section of the model.

Parameter	Meaning	Value	Unit
T_s	Digital control sample time (all regulators)	10^{-4}	s
$k_P^{(\theta)}$	Proportional gain of the position regulator	1.58×10^4	1/s
$k_I^{(\theta)}$	Integral gain of the position regulator	1.1×10^5	$1/s^2$
$k_{AW}^{(\theta)}$	Anti-windup gain of the position regulator	0.69	s
$y_{sat}^{(\theta)}$	Saturation limit of the position regulator	100	rad/s
$k_P^{(\omega)}$	Proportional gain of the speed regulator	0.07	A s/rad
$k_I^{(\omega)}$	Integral gain of the speed regulator	2	A/rad
$k_{AW}^{(\omega)}$	Anti-windup gain of the speed regulator	0.28	rad/(A s)

Table 4. Cont.

Parameter	Meaning	Value	Unit
$y_{sat}^{(\omega)}$	Saturation limit of the speed regulator	4	A
$k_P^{(I)}$	Proportional gain of the current regulators	2.78	V/A
$k_I^{(I)}$	Integral gain of the current regulators	4.1×10^3	V/(A s)
$k_{AW}^{(I)}$	Anti-windup gain of the current regulators	150	A/V
$y_{sat}^{(I)}$	Saturation limit of the current regulators	28	V
a_{SES}	SES regulator parameter 1	1.02×10^3	rad/s
b_{SES}	SES regulator parameter 2	2.37×10^5	rad^2/s^2
k_{SES}	SES regulator gain	10^3	A/(m s)
ω_{th}	OSM fault symptom threshold	0.0175	rad/s
$c_{mon\,max}$	OSM fault counter threshold	250	–

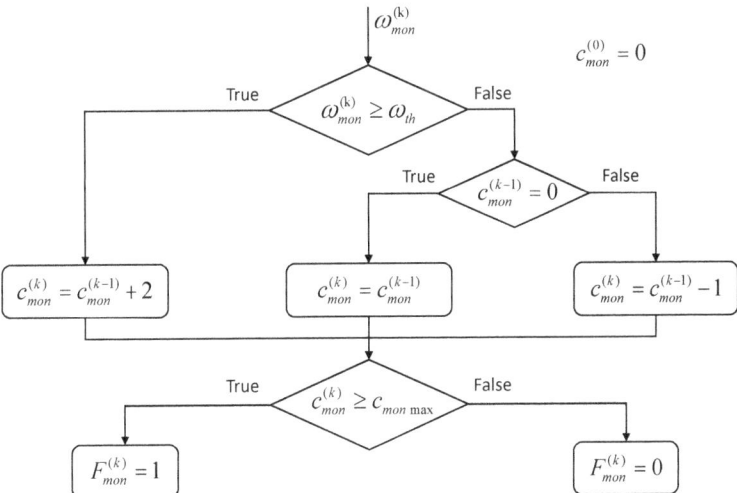

Figure 9. Fault detection logics of the OSM.

2.2.3. Fault Simulation

As previously mentioned, the basic objective of the work is to validate the EMA health-monitoring algorithms with reference to the control hardover fault (worst-case failure), by particularly focusing on the failure transient characterisation. The model has been developed as a finite-state machine by using Matlab-Simulink-Stateflow charts and logics so that the simulations of both hardover fault and the subsequent activation of the back-electromotive circuitry are integrated in the EMA simulator.

The hardover fault is simulated by Equation (26), so that when the fault is injected ($t = t_{FI}$), the direct and quadrature voltages are switched from the values demanded by the EMA control laws (Equations (20) and (21)) to zero and saturation values, respectively:

$$V_d = \begin{cases} V_{dc} - Ln_d I_{qf}\dot{\theta}_{mf} & t < t_{FI} \\ 0 & t \geq t_{FI} \end{cases} ; V_q = \begin{cases} V_{qc} + \left(\sqrt{\frac{3}{2}}\lambda_m + LI_{df}\right)n_d\dot{\theta}_{mf} & t < t_{FI} \\ V_{\max} & t \geq t_{FI} \end{cases}, \quad (26)$$

As described in Section 2.1, the PWR board of the EMA includes a BEMF damper circuitry, which, in case of a detected fault, imposes that the motor phases are shorted to the ground (so that an electromagnetic damping torque is developed and transmitted to the EMA output shaft). In the model, the BEMF damper activation is simulated via Equation (27), where t_{FD} is the time at which the fault is detected by the OSM.

$$\mathbf{V_{abc}} = \begin{cases} (\mathbf{T_P T_C})^T \mathbf{V_{dqz}} & t < t_{FD} \\ 0 & t \geq t_{FD} \end{cases}, \qquad (27)$$

2.3. Experimental Test Campaign for the Model Validation

To substantiate the failure transient analysis presented and discussed in Section 3, the EMA model has been experimentally validated through a specific test campaign carried out at the Umbragroup facilities. In particular, the following tests have been performed, aiming to identify the model parameters reported in Table 3:

- Unloaded, open-loop tests
    - Test 1 (blocked motor with engaged brakes): chirp wave inputs are given as direct voltage demand, while the quadrature voltage is set to zero, aiming to identify motor phase resistance and inductance (R and L). The test is repeated at a different position of the PMSM rotor to verify that the phase inductance does not significantly depend on motor angle (assumption of the model);
    - Test 2 (blocked motor with engaged brakes): step inputs of different amplitudes are given to the quadrature voltage demand, while the direct voltage is set to zero, aiming to confirm the values of motor phase resistance and inductance. The test is repeated at different position of the PMSM rotor;
    - Test 3 (free-wheeling motor with disengaged brakes and open phases): the PMSM rotor is dragged by an external motor at different speed amplitudes and the phase-to-phase BEMF is measured, aiming to identify the motor flux linkage (λ_m) and to eventually highlight higher harmonic components in the BEMF waveform;
- Unloaded, closed-loop tests
    - Test 4 (blocked motor with engaged brakes): current loop tracking is tested by providing square-wave inputs of different amplitudes as quadrature current demand, while the direct current is set to zero, aiming to identify the damping and stiffness of the brakes (d_b and k_b);
    - Test 5 (disengaged brakes): speed loop tracking is tested, by providing square-wave inputs of different amplitudes as speed demand, aiming to identify the torque disturbance parameters (M, T_{hd1}, n_{hd1}, T_{hd2}, n_{hd2}, T_{hd3} and n_{hd3}), the viscous damping coefficients (d_{vfm} and d_{vfo}), the parameters of the sliding friction models (T_{sfm}, T_{sfo} ω_{sfm} and ω_{sfo}), and the actuator inertias (J_m and J_o).

At the current stage of the campaign, loaded tests and position-loop tests have not been carried out, but they have been planned for the future steps of the research, mainly to confirm the predictions of the resonant frequency of the ball-screw drivetrain, currently estimated through FEM analyses.

3. Results

3.1. Experimental Validation of the Model

An excerpt of the results obtained during the model validation campaign is reported from Figures 10–13.

Figures 10 and 11 are devoted to the identification of the electrical parameters of the motor phases (i.e., resistance and inductance), and it can be noted that the model succeeds in predicting the hardware response in both steady-state and dynamic operations. The repetition of tests at different positions of the PMSM rotor provided essentially identical results, thus confirming that the position-dependence of electrical parameters is negligible (basic assumption of the model).

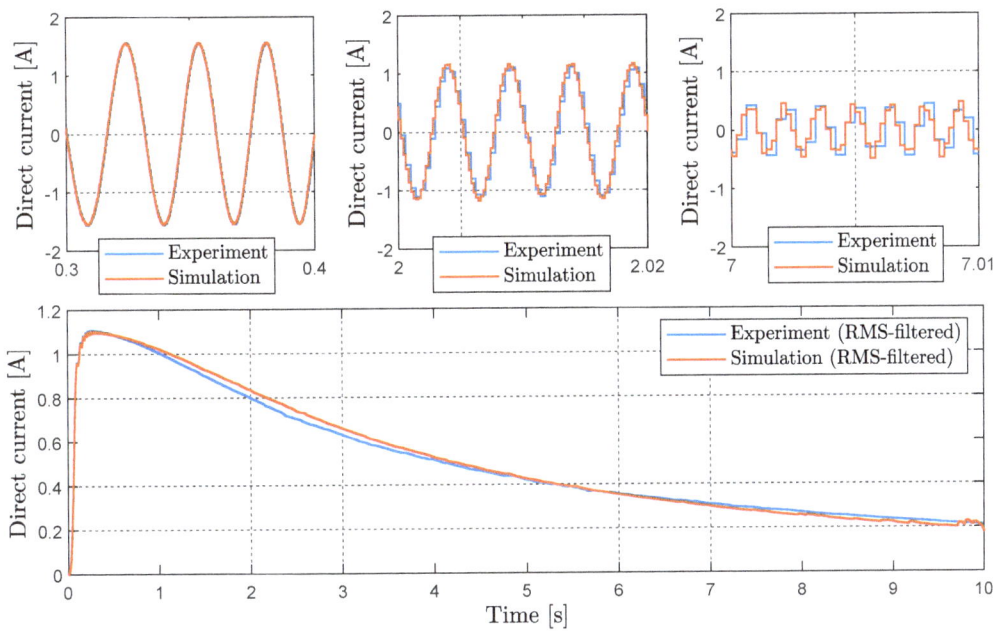

Figure 10. Response to Test 1 (open-loop, engaged brakes, chirp wave input applied on direct voltage demand, ±2 V ranging from 10 Hz, at t = 0 s, to 10 kHz, at t = 10 s).

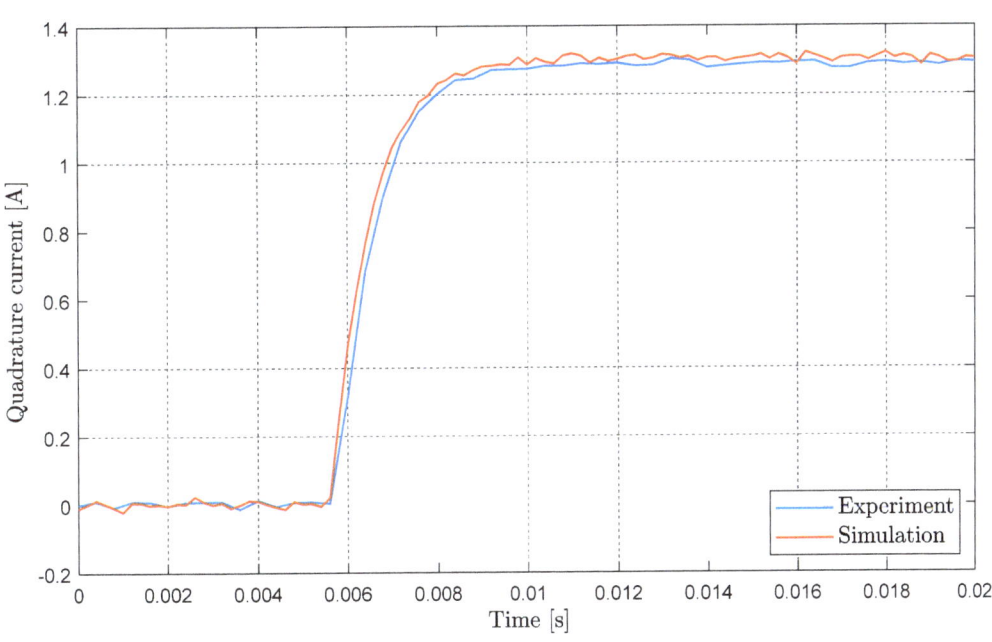

Figure 11. Response to Test 2 (open loop, engaged brakes, 2 V step input applied on quadrature voltage demand).

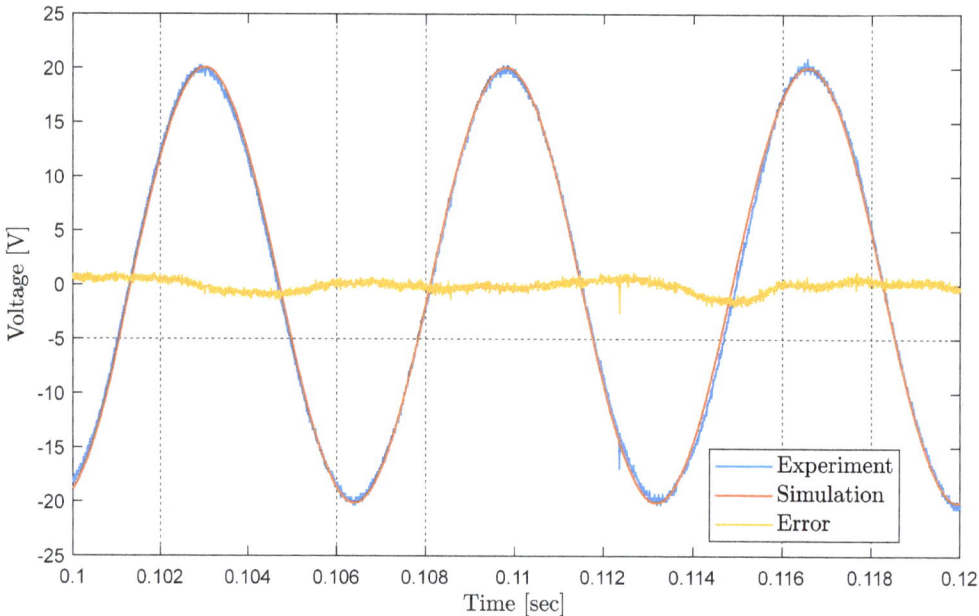

Figure 12. Response to Test 3 (open loop, open phases, disengaged brakes, motor dragged at 80 rad/s, phase-to-phase BEMF measured).

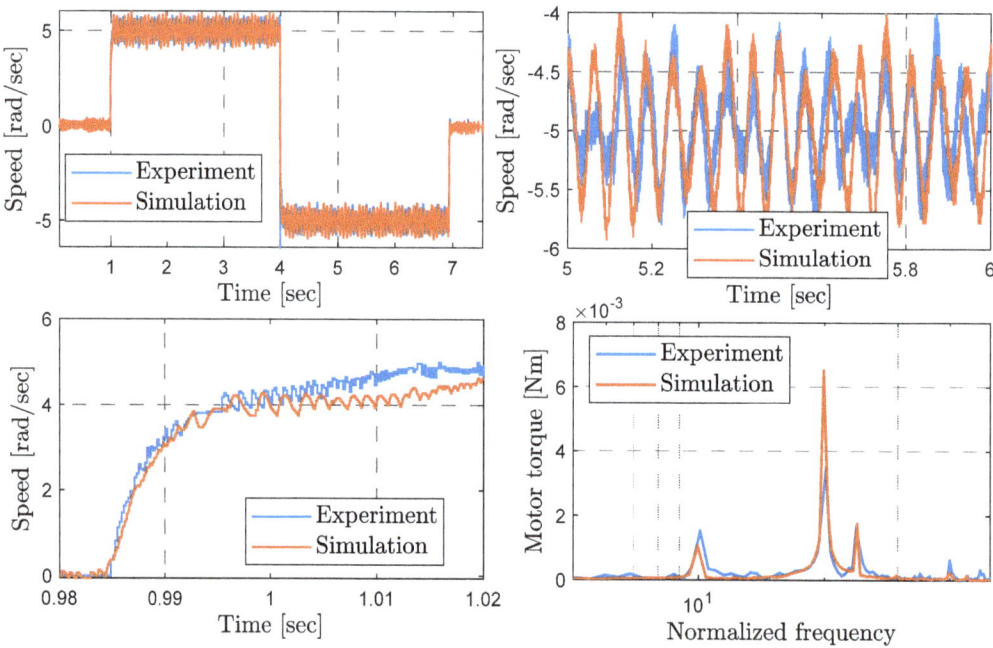

Figure 13. Response to Test 5 (closed loop on speed and currents, disengaged brakes, square-wave speed demand ±5 rad/sec). FFT analysis is performed on quadrature current signal (proportional to torque) and frequencies are normalized with the steady-state mechanical frequency (0.8 Hz).

Figure 12 is instead relevant for the identification of the magnet flux linkage as well as for the characterisation of the BEMF waveform with respect to the motor angle. Again, the model succeeds in predicting the hardware response in terms of both magnet flux linkage and BEMF waveform, which is essentially sinusoidal for the reference PMSM (basic assumption of the model).

Figure 13 finally reports a closed-loop speed-tracking response, which is essentially relevant for the identification of system inertias (lower left plot in Figure 13), torque disturbance parameters (upper and lower right plots in Figure 13), and friction parameters. It is interesting to note that the FFT analysis performed on quadrature current signal clearly highlights the presence of three harmonic disturbances. The first two harmonic components are multiples of the fundamental electrical frequency, being 10 and 20 times the mechanical frequency (n_d = 10, Table 3), and they derive from small deviations of the BEMF from the sinusoidal waveform. The resting harmonic component is instead located at 24 times the mechanical frequency, and the disturbance can be interpreted as an effect of cogging torque, due to assembly tolerances and/or magnet imperfections. As discussed in [52–55], these irregularities generate torque harmonics at frequencies that are multiple of the stator slots, which in the reference PMSM is 12.

3.2. Loads Disturbance Rejection Capability

Before performing the failure transient analysis (Section 3.3), the control system performances have been verified using the experimentally validated model of the EMA by characterizing the disturbance rejection capability against external loads.

The performance specification actually requires that under any of the load conditions reported in Table 2, the EMA position deviation shall be lower than the position sensor accuracy (0.1 deg). To demonstrate the effectiveness of the developed control system with special reference to the application of the SES deformation loop, an extensive simulation campaign has been carried out, and a summary of the results is given in Figure 14.

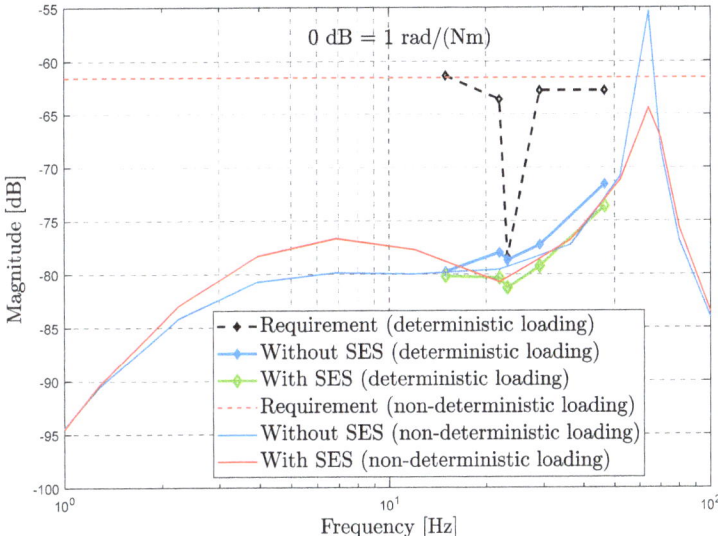

Figure 14. EMA dynamic compliance response: simulation results obtained with the experimentally validated model of the actuator (position setpoint corresponding to minimum stiffness).

The results clearly highlight that the use of the SES deformation loop implies a relevant enhancement of performances. The compliance response under deterministic loading profiles is marginal without using the SES loop, while it significantly diminishes if the

SES loop is present. Further, the compliance under non-deterministic loads exceeds the requirement limit in the resonant pulsation region (70 Hz) if the SES loop is not used, while it becomes adequate if the SES loop is applied.

3.3. Failure Transient Analysis

The experimentally validated model of the EMA has been finally used to characterise the failure transients related to a control hardover fault, as simulated in Equation (26). The following worst-case scenario has been simulated:

- the maximum static load plus the deterministic dynamic loads defined in Table 2 are applied to the output shaft;
- the EMA is demanded to move to the maximum positive deflection (minimum stiffness);
- the control hardover fault occurs immediately after the EMA reaches the position setpoint ($t = t_{FI} = 0$ s);
- the brakes activation occurs with a predefined delay from the fault detection ($t_{FC} - t_{FD} = 51$ ms, Umbragroup information).

To evaluate the effectiveness of the BEMF damper circuitry, the simulation is performed by activating or not the system, obtaining the results in Figures 15 and 16.

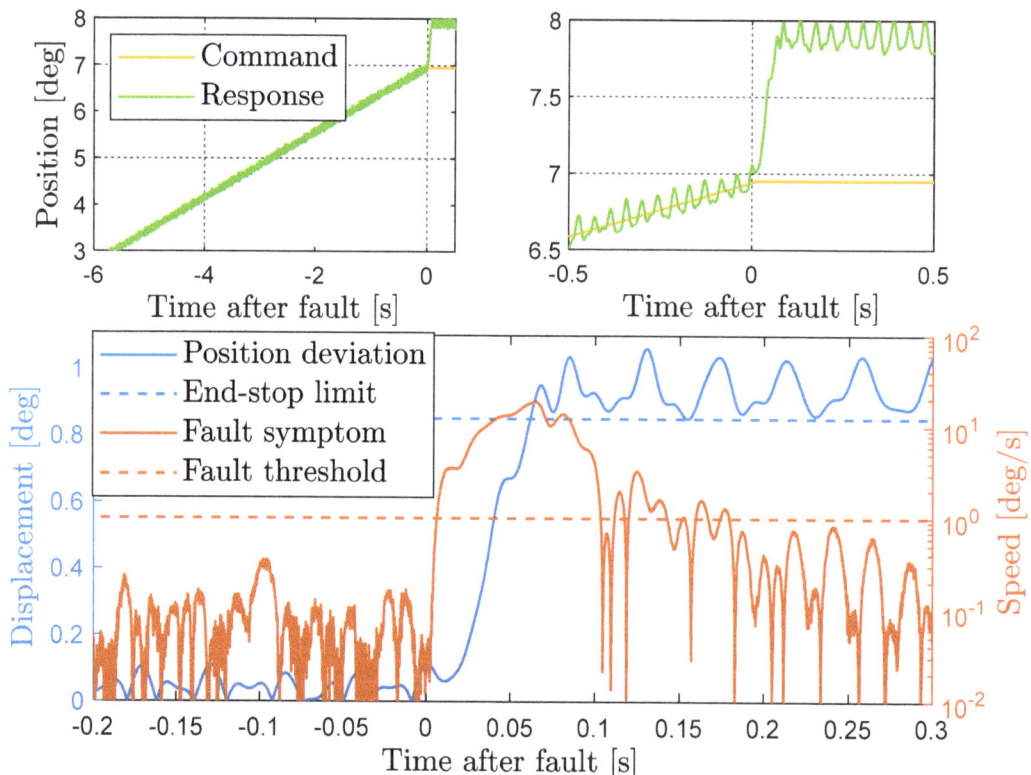

Figure 15. EMA failure transient with a control hardover fault without BEMF damper circuitry: simulation obtained with the experimentally validated model of the actuator ($t_{FI} = 0$ s; $t_{FD} = 13.4$ ms; $t_{FC} = 64.4$ ms).

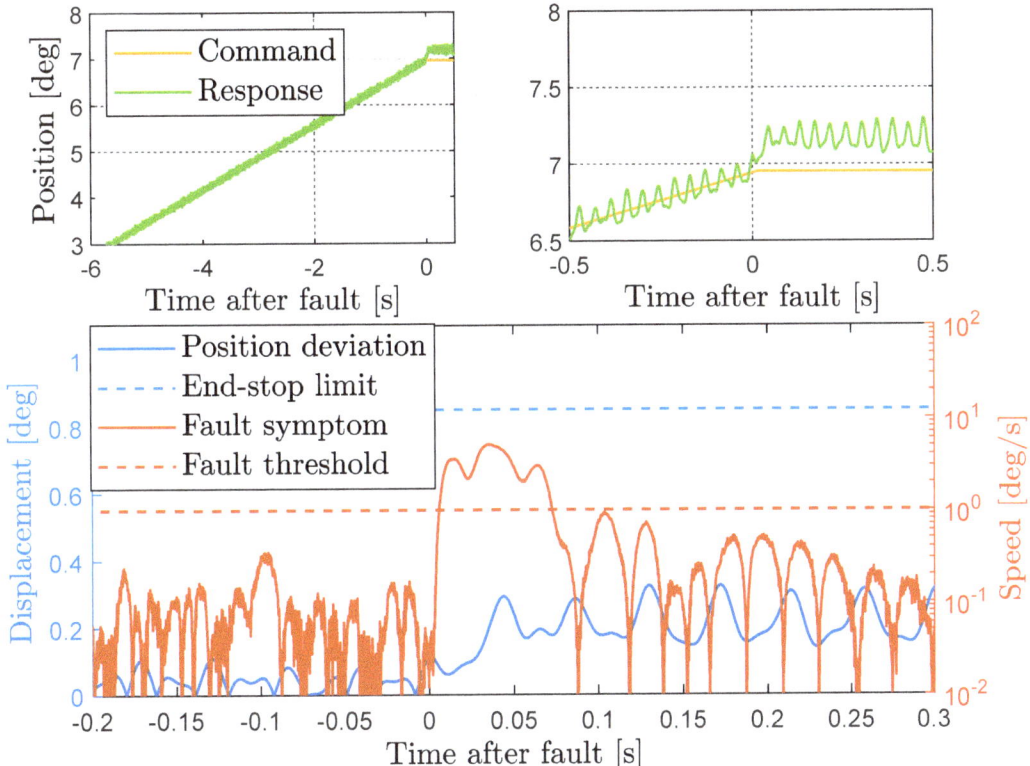

Figure 16. EMA failure transient with a control hardover fault with BEMF damper circuitry: simulation obtained with the experimentally validated model of the actuator (t_{FI} = 0 s; t_{FD} = 13.4 ms; t_{FC} = 64.4 ms).

It is interesting to note that, though the fault detection latency is extremely small (in both simulations, t_{FD} = 13.4 ms), the position deviation without BEMF damper is excessive and the EMA reaches the mechanical endstroke at high speed, possibly implying permanent damages (Figure 15). On the other hand, the use of the BEMF damper permits to strongly limit the position deviation during the failure transient, and the EMA can be blocked by the brakes in safety.

4. Discussion

As highlighted by the results in Section 3.3, the control hardover fault can determine EMA damages and potentially unsafe operation for the FLA movables. The failure transient analysis conducted using the experimentally validated model of the EMA highlights that, in some special operating conditions, even if the fault is detected with extremely small latency (less than 15 ms), the actuator can reach the mechanical endstroke impacting at high speed, thus causing permanent damages. To counteract this adverse situation, essentially caused by unavoidable delays in the activation of EMA brakes, the PWR board of the RACER flaps EMAs includes a specifically designed BEMF damper circuitry, which, immediately after the fault detection, opens the power bridge thyristors and connects all the motor phases to the ground, thus generating an electromagnetic damping torque.

5. Conclusions

The failure transients related to control hardover fault in the EMA employed for the flap movables of the RACER helicopter-plane are characterised using an experimentally validated model of the system, which includes multi-harmonic torque disturbance simulation, "Coulomb–tanh" friction, mechanical freeplay, and position-dependant stiffness of the ball-screw drivetrain. The failure transient characterization performed in a worst-case scenario in terms of external loads and position setpoint demonstrates that, though the fault detection is executed with extremely small latency (less than 15 ms), a potentially dangerous actuator runaway can occur, causing high-speed impacts on the EMA mechanical endstroke, caused by the activation delay of the EMA brakes (about 50 ms). Simulation is thus used to point out that an effective solution can be obtained by including a BEMF damper circuitry in the EMA power electronics, which, immediately after the fault detection, opens the power bridge thyristors and connects all the motor phases to the ground, thus generating an electromagnetic damping torque.

The future developments of the research will be focused on:

- extension of the model validation with loaded position-loop tests, aiming to:
    - verify the actual location of the resonant pulsation of the ball-screw drivetrain (currently estimated via FEM analyses);
    - characterise the actual disturbance rejection of external loads;
- model enhancement, by including a friction model that takes into account dependence on applied loads and temperature;
- robustness analysis of the health-monitoring performances against model parameters uncertainties.

Author Contributions: Conceptualization, G.D.R.; methodology, G.D.R.; software, B.L.; validation, R.K., M.N. and N.B.; formal analysis, R.K., M.N. and N.B.; investigation, G.D.R. and B.L.; resources, R.K., M.N. and N.B.; data curation, B.L.; writing—original draft preparation, G.D.R.; writing—review and editing, G.D.R. and B.L.; visualization, R.K., M.N. and N.B.; supervision, R.K., M.N. and N.B.; project administration, M.N. and N.B.; funding acquisition, M.N. and N.B. All authors have read and agreed to the published version of the manuscript.

Funding: This research received no external funding.

Institutional Review Board Statement: Not applicable.

Informed Consent Statement: Not applicable.

Data Availability Statement: Not applicable.

Conflicts of Interest: The authors declare no conflict of interest.

References

1. Flightpath 2050: Europe's Vision for Aviation. Available online: https://op.europa.eu/en/publication-detail/-/publication/7d834950-1f5e-480f-ab70-ab96e4a0a0ad/language-en (accessed on 30 August 2022).
2. Schäfer, A.W.; Barrett, S.R.H.; Doyme, K.; Dray, L.; Gnadt, A.R.; Self, R.; O'Sullivan, A.; Synodinos, A.; Torija, A.J. Technological, economic and environmental prospects of all-electric aircraft. *Nat. Energy* **2019**, *4*, 160–166. [CrossRef]
3. Howse, M. All-electric aircraft. *Power Eng. J.* **2003**, *17*, 35–37. [CrossRef]
4. Rosero, J.A.; Ortega, J.A.; Aldabas, E.; Romeral, L. Moving towards a more electric aircraft. *IEEE Aerosp. Electron. Syst. Mag.* **2007**, *22*, 3–9. [CrossRef]
5. Roboam, X.; Sareni, B.; De Andrade, A. More electricity in the air: Toward optimized electrical networks embedded in more-electrical aircraft. *IEEE Ind. Electron. Mag.* **2012**, *6*, 6–17. [CrossRef]
6. Madonna, V.; Giangrande, P.; Galea, M. Electrical power generation in aircraft: Review, challenges, and opportunities. *IEEE Trans. Transp. Electrif.* **2018**, *4*, 646–659. [CrossRef]
7. Maré, J.-C.; Fu, J. Review on signal-by-wire and power-by-wire actuation for more electric aircraft. *Chin. J. Aeronaut.* **2017**, *30*, 857–870. [CrossRef]
8. Mazzoleni, M.; Di Rito, G.; Previdi, F. Introduction. In *Electro-Mechanical Actuators for the More Electric Aircraft; Advances in Industrial Control*; Springer: Cham, Switzerland, 2021; Volume 2, pp. 1–44. [CrossRef]

9. Maré, J.-C. *Aerospace Actuators: Signal-by-Wire and Power-by-Wire*; Bourrières, J.-P., Ed.; ISTE Ltd.: Washington, DC, USA; John Wiley & Sons: Hoboken, NJ, USA, 2017; Volume 2, pp. 171–217. [CrossRef]
10. Di Rito, G.; Galatolo, R.; Schettini, F. Experimental and simulation study of the dynamics of an electro-mechanical landing gear actuator. In Proceedings of the 30th Congress of the International Council of the Aeronautical Sciences (ICAS), Daejeon, Korea, 25–30 September 2016.
11. Giangrande, P.; Al-Timimy, A.; Galassini, A.; Papadopoulos, S.; Degano, M.; Galea, M. Design of PMSM for EMA employed in secondary flight control systems. In Proceedings of the 2018 IEEE International Conference on Electrical Systems for Aircraft, Railway, Ship Propulsion and Road Vehicles & International Transportation Electrification Conference (ESARS-ITEC), Nottingham, UK, 7–9 November 2018; pp. 1–6. [CrossRef]
12. Qiao, G.; Liu, G.; Shi, Z.; Wang, Y.; Ma, S.; Lim, T.C. A review of electromechanical actuators for More/All Electric aircraft systems. *Proc. Inst. Mech. Eng. Part C J. Mech. Eng. Sci.* 2018, *232*, 4128–4151. [CrossRef]
13. Mazzoleni, M.; Di Rito, G.; Previdi, F. Reliability and safety of electro-mechanical actuators for aircraft applications. In *Electro-Mechanical Actuators for the More Electric Aircraft*; Advances in Industrial Control; Springer: Cham, Switzerland, 2021; Volume 2, pp. 45–85. [CrossRef]
14. Balaban, E.; Bansal, P.; Stoelting, P.; Saxena, A.; Goebel, K.F.; Curran, S. A diagnostic approach for electro-mechanical actuators in aerospace systems. In Proceedings of the 2009 IEEE Aerospace Conference, Big Sky, MT, USA, 7–14 March 2009; pp. 1–13. [CrossRef]
15. Ossmann, D.; Van der Linden, F.-L.-J. Advanced sensor fault detection and isolation for electro-mechanical flight actuators. In Proceedings of the NASA/ESA Conference on Adaptive Hardware and Systems, Montreal, QC, Canada, 15–18 June 2015; pp. 1–8.
16. Ismail, M.A.A.; Balaban, E.; Spangenberg, H. Fault detection and classification for flight control electromechanical actuators. In Proceedings of the 2016 IEEE Aerospace Conference, Big Sky, MT, USA, 5–12 March 2016; pp. 1–10. [CrossRef]
17. Ismail, M.A.A.; Windelberg, J. Fault detection of bearing defects for ballscrew based electro-mechanical actuators. In Proceedings of the World Congress on Condition Monitoring (WCCM), London, UK, 13–16 June 2017; pp. 1–12.
18. Byington, C.; Stoelting, P.; Watson, M.; Edwards, D. A model-based approach to prognostics and health management for flight control actuators. In Proceedings of the 2004 IEEE Aerospace Conference Proceedings, Big Sky, MT, USA, 6–13 March 2004; pp. 3551–3562. [CrossRef]
19. Mazzoleni, M.; Maccarana, Y.; Previdi, F.; Pispola, G.; Nardi, M.; Perni, F.; Toro, S. Development of a reliable electro-mechanical actuator for primary control surfaces in small aircrafts. In Proceedings of the 2017 IEEE International Con-ference on Advanced Intelligent Mechatronics (AIM), Munich, Germany, 3–7 July 2017; pp. 1142–1147. [CrossRef]
20. Mazzoleni, M.; Previdi, F.; Scandella, M.; Pispola, G. Experimental development of a health monitoring method for electro-mechanical actuators of flight control primary surfaces in more electric aircrafts. *IEEE Access* 2019, *7*, 153618–153634. [CrossRef]
21. Di Rito, G.; Schettini, F.; Galatolo, R. Model-based prognostic health-management algorithms for the freeplay identification in electromechanical flight control actuators. In Proceedings of the 2018 IEEE International Workshop on Metrology for AeroSpace, Rome, Italy, 20–22 June 2018; pp. 340–345. [CrossRef]
22. Todeschi, M.; Baxerres, L. Health Monitoring for the Flight Control EMAs. *IFAC-PapersOnLine* 2015, *48*, 186–193. [CrossRef]
23. Blanke, M.; Kinnaert, M.; Lunze, J.; Staroswiecki, M.; Schröder, J. *Diagnosis and Fault-tolerant Control*; Springer: Berlin, Germany, 2016. [CrossRef]
24. Di Rito, G.; Galatolo, R.; Schettini, F. Self-monitoring electro-mechanical actuator for medium altitude long endurance unmanned aerial vehicle flight controls. *Adv. Mech. Eng.* 2016, *8*, 1–12. [CrossRef]
25. Di Rito, G.; Schettini, F. Health monitoring of electromechanical flight actuators via position-tracking predictive models. *Adv. Mech. Eng.* 2018, *10*, 1–12. [CrossRef]
26. Ferranti, L.; Wan, Y.; Keviczky, T. Fault-tolerant reference generation for model predictive control with active diagnosis of elevator jamming faults. *Int. J. Robust Nonlinear Control* 2018, *29*, 5412–5428. [CrossRef]
27. Ferranti, L.; Wan, Y.; Keviczky, T. Predictive flight control with active diagnosis and reconfiguration for actuator jamming. *IFAC-PapersOnLine* 2015, *48*, 166–171. [CrossRef]
28. Arriola, D.; Thielecke, F. Model-based design and experimental verification of a monitoring concept for an active-active electromechanical aileron actuation system. *Mech. Syst. Signal Process.* 2017, *94*, 322–345. [CrossRef]
29. Annaz, F.Y. Fundamental design concepts in multi-lane smart electromechanical actuators. *Smart Mater. Struct.* 2005, *14*, 1227–1238. [CrossRef]
30. Yu, Z.Y.; Niu, T.; Dong, H.L. A jam-tolerant electromechanical system. In Proceedings of the ACTUATOR 2018: 16th International Conference on New Actuators, Bremen, Germany, 25–27 June 2018; pp. 551–554.
31. Gao, Z.; Cecati, C.; Ding, S.X. A survey of fault diagnosis and fault-tolerant techniques—Part I: Fault diagnosis with model-based and signal-based approaches. *IEEE Trans. Ind. Electron.* 2015, *62*, 3757–3767. [CrossRef]
32. Gao, Z.; Cecati, C.; Ding, S.X. A survey of fault diagnosis and fault-tolerant techniques—Part II: Fault diagnosis with knowledge-based and hybrid/active approaches. *IEEE Trans. Ind. Electron.* 2015, *62*, 3768–3774. [CrossRef]
33. Mazzoleni, M.; Di Rito, G.; Previdi, F. Fault diagnosis and condition monitoring approaches. In *Electro-Mechanical Actuators for the More Electric Aircraft*; Advances in Industrial Control; Springer: Cham, Switzerland, 2021; Volume 3, pp. 87–117. [CrossRef]

34. Smith, M.J.; Byington, C.S.; Watson, M.J.; Bharadwaj, S.; Swerdon, G.; Goebel, K.; Balaban, E. Experimental and analytical development of health management for electro-mechanical actuators. In Proceedings of the 2009 IEEE Aerospace conference, Big Sky, MT, USA, 7–14 March 2009; pp. 1–14. [CrossRef]
35. Chirico, A.J.; Kolodziej, J.R. A data-driven methodology for fault detection in electromechanical actuators. *J. Dyn. Syst. Meas. Control* **2014**, *136*, 041025. [CrossRef]
36. Bodden, D.S.; Clements, N.S.; Schley, B.; Jenney, G. Seeded failure testing and analysis of an electro-mechanical actuator. In Proceedings of the 2007 IEEE Aerospace Conference, Big Sky, MT, USA, 7–14 March 2007; pp. 1–8. [CrossRef]
37. Di Rito, G.; Luciano, B.; Borgarelli, N.; Nardeschi, M. Model-based condition-monitoring and jamming-tolerant control of an electro-mechanical flight actuator with differential ball screws. *Actuators* **2021**, *10*, 230. [CrossRef]
38. Mazzoleni, M.; Di Rito, G.; Previdi, F. Fault diagnosis and condition monitoring of aircraft electro-mechanical actuators. In *Electro-Mechanical Actuators for the More Electric Aircraft*; Advances in Industrial Control; Springer: Cham, Switzerland, 2021; Volume 4, pp. 119–224. [CrossRef]
39. Airbus Racer on Pace for 2022 First Flight. Available online: https://www.ainonline.com/aviation-news/business-aviation/2022-03-07/airbus-racer-pace-2022-first-flight (accessed on 30 August 2022).
40. Stokkermans, T.; Veldhuis, L.; Soemarwoto, B.; Fukari, R.; Eglin, P. Breakdown of aerodynamic interactions for the lateral rotors on a compound helicopter. *Aerosp. Sci. Technol.* **2020**, *101*, 105845. [CrossRef]
41. Thiemeier, J.; Öhrle, C.; Frey, F.; Keßler, M.; Krämer, E. Aerodynamics and flight mechanics analysis of Airbus Helicopters compound helicopter RACER in hover under crosswind conditions. *CEAS Aeronaut. J.* **2020**, *11*, 49–66. [CrossRef]
42. Airbus Helicopters Reveals RACER High-Speed Demonstrator Configuration. Available online: https://www.airbus.com/en/newsroom/press-releases/2017-06-airbus-helicopters-reveals-racer-high-speed-demonstrator (accessed on 30 August 2022).
43. Huot, R.; Eglin, P. Flight Mechanics of the RACER compound H/C. In Proceedings of the 76th Vertical Flight Society's Annual Forum & Technology Display, Virginia Beach, VA, USA, 6–8 October 2020.
44. Advanced Assembly Solutions for the Airbus RACER Joined-Wing Configuration. Available online: https://www.mobilityengineeringtech.com/component/content/article/adt/pub/features/articles/37105 (accessed on 30 August 2022).
45. Airbus Begins Assembly of Racer Compound Helicopter. Available online: https://www.ainonline.com/aviation-news/general-aviation/2021-03-25/airbus-begins-assembly-racer-compound-helicopter (accessed on 30 August 2022).
46. Texas Instruments. TMS570LC4357-EP ARM-Based Microcontroller. Available online: https://www.ti.com/product/TMS570LC4357-EP (accessed on 30 August 2022).
47. Allegro Microsystems. Automotive Grade, Fully Integrated, Hall-Effect-Based Linear Current Sensor IC with 2.1 kVRMS Voltage Isolation and Low-Resistance Current Conductor. Available online: https://www.allegromicro.com/en/products/sense (accessed on 30 August 2022).
48. Tamagawa. Brushless Resolvers (Smartsyn). Available online: https://www.tamagawa-seiki.com/products/resolver-synchro/brushless-resolver-smartsyn.html (accessed on 30 August 2022).
49. Analog Devices. AD2S1210 Variable Resolution, 10-Bit to 16-Bit R/D Converter with Reference Oscillator. Available online: https://www.analog.com/en/products/ad2s1210.html (accessed on 30 August 2022).
50. Analog Devices. ADA4571 Integrated AMR Angle Sensor and Signal Conditioner. Available online: https://www.analog.com/en/products/ada4571.html (accessed on 30 August 2022).
51. RLS. AksIM-2™ Off-Axis Rotary Absolute Magnetic Encoder Module. Available online: https://www.rls.si/eng/aksim-2-off-axis-rotary-absolute-encoder (accessed on 30 August 2022).
52. Islam, R.; Husain, I.; Fardoun, A.; Mc Laughlin, K. Permanent magnet synchronous motor magnet designs with skewing for torque ripple and cogging torque reduction. In Proceedings of the 2007 IEEE Industry Applications Annual Meeting, New Orleans, LA, USA, 23–27 September 2007; pp. 1552–1559. [CrossRef]
53. Gasparin, L.; Cernigoj, A.; Markic, S.; Fiser, R. Additional cogging torque components in permanent-magnet motors due to manufacturing imperfections. *IEEE Trans. Magn.* **2009**, *45*, 1210–1213. [CrossRef]
54. Dajaku, G.; Gerling, D. New methods for reducing the cogging torque and torque ripples of PMSM. In Proceedings of the 4th International Electric Drives Production Conference (EDPC), Nuremberg, Germany, 30 September–1 October 2014; pp. 1–7. [CrossRef]
55. Gao, J.; Wang, G.; Liu, X.; Zhang, W.; Huang, S.; Li, H. Cogging torque reduction by elementary-cogging-unit shift for permanent magnet machines. *IEEE Trans. Magn.* **2017**, *53*, 1–5. [CrossRef]
56. Olsson, H.; Åström, K.; Canudas-De-Wit, C.; Gäfvert, M.; Lischinsky, P. Friction Models and Friction Compensation. *Eur. J. Control* **1998**, *4*, 176–195. [CrossRef]
57. Andersson, S.; Söderberg, A.; Björklund, S. Friction models for sliding dry, boundary and mixed lubricated contacts. *Tribol. Int.* **2007**, *40*, 580–587. [CrossRef]
58. Dutta, C.; Tripathi, S.M. Comparison between conventional and loss d-q model of PMSM. In Proceedings of the International Conference on Emerging Trends in Electrical Electronics & Sustainable Energy Systems (ICETEESES), Sultanpur, India, 11–12 March 2016; pp. 256–260. [CrossRef]

59. Zhang, C.; Tian, Z.; Dong, Y.; Zhang, S. Analysis of losses and thermal model in a surface-mounted permanent-magnet synchronous machine over a wide-voltage range of rated output power operation. In Proceedings of the IEEE Conference and Expo Transportation Electrification Asia-Pacific (ITEC Asia-Pacific), Beijing, China, 31 August–3 September 2014; pp. 1–5. [CrossRef]
60. Fitzgerald, A.E.; Kingsley, C., Jr.; Umans, S.D. *Electric Machinery*, 6th ed.; Mc Graw-Hill: New York, NY, USA, 2003.
61. Åström, K.J.; Hägglund, T. *Advanced PID Control*; ISA, The Instrumentation, Systems, and Automation Society: Pittsburgh, PA, USA, 2006.

Article

Preliminary Sizing of the Electrical Motor and Housing of Electromechanical Actuators Applied on the Primary Flight Control System of Unmanned Helicopters

Jeremy Roussel [1,2,*], Marc Budinger [1] and Laurent Ruet [2]

1 National Institute of Applied Sciences of Toulouse (INSA), Clement Ader's Institute (ICA), 31077 Toulouse, France
2 Airbus Helicopters, 13700 Marignane, France
* Correspondence: jroussel@insa-toulouse.fr

Abstract: Helicopter dronization is expanding, for example, the VSR700 project. This leads to the integration of electromechanical actuators (EMAs) into the primary flight control system (PFCS). The PFCS is in charge of controlling the helicopter flight over its four axes (roll, pitch, yaw, and vertical). It controls the blade pitch thanks to mechanical kinematics and actuators. For more than 60 years, the actuators have been conventionally using the hydraulic technology. The EMA technology introduction involves the reconsideration of the design practices. Indeed, an EMA is multidisciplinary. Each of its components introduces new design drivers and new inherent technological imperfections (friction, inertia, and losses). This paper presents a methodology to specify and pre-design critical EMAs. The description will be focused on two components: the electrical motor and the housing. This includes a data-driven specification, scaling laws for motor losses estimation, and surrogate modeling for the housing vibratory sizing. The tools are finally applied to two study cases. The first case considers two potential redundant topologies of actuation. The housing sizing shows that one prevails on the other. The second case considers the actuators of helicopter rotors. The electrical motor sizing highlights the importance of designing two separate actuators.

Keywords: specification; flight analysis; dimensional analysis; vibration; multidisciplinary optimization

1. Introduction

1.1. Context

1.1.1. Helicopter Dronization

Today, we observe a fast increase in the number of projects about OPVs (Optional Pilot vehicles), UASs (Unmanned Aerial Systems), UAVs (Unmanned Aerial Vehicles), UAM (Urban Air Mobility), and VTOL (Vertical Take-Off Landing). For instance, the Ehang 184, the Vahana, the CityAirbus, and Boeing's self-piloted passenger drone can be seen on the civil range; the aerial fighter Northrop Grumman X-47B, the Airbus VSR700 (Figure 1), and the Leonardo AWHero can be seen on the military range. Furthermore, the market for aerial delivery already grows, with vehicles carrying parcels weighing less than 10 kg (DPD France drones in rural areas) up to 100 kg (Kawasaki K-Racer X1 drone prototype). It is clear that flying vehicles and future ones are already required to develop new functionalities to be more autonomous and to be safer. Reducing pilots' working load is part of the current helicopter development roadmaps for enhanced safety. Thus, today's market trend is globally facing a technological watershed toward more electrical solutions. Aircraft makers aim for the More Electric Aircraft (MEA) achievement [1]. The drone concept comes progressively by the implementation of new electrical solutions on already existing vehicles [2].

Figure 1. VSR700, the multi-mission naval UAS (©Airbus).

1.1.2. Primary Flight Control Systems: From Hydraulic to Electromechanical Technologies

The primary flight control system (PFCS) is in charge of controlling the helicopter flight over its four axes (roll, pitch, yaw, and vertical) by the control of the blade attack angles. The paper [3] describes the PFCS with pictures. More details can be found in the book [4] and the handbook [5]. The automatic pilot function is ensured through EMAs located in series and in parallel with the mechanical kinematic (Figure 2a).

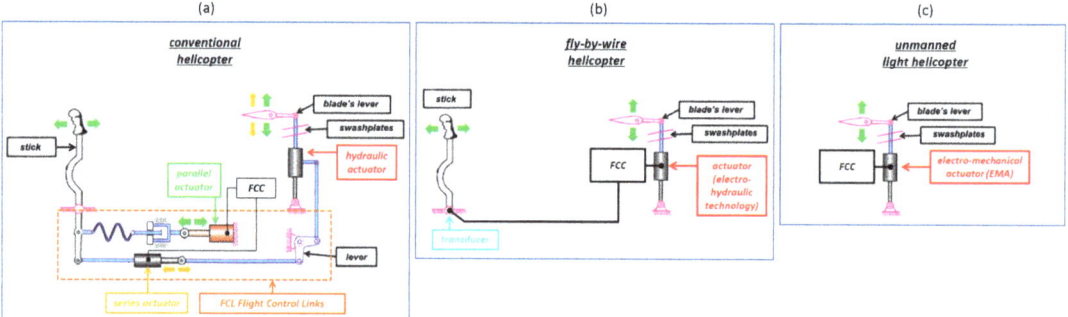

Figure 2. PFCS architectures: sketch of principle. From (**a**–**c**), this figure shows basically the evolution of PFCS architectures as it applies to autonomous helicopters. This evolution clearly shows that the helicopter mass can be significantly reduced with the reduction in part numbers. As long as the actuation control loop is concerned, this part number reduction decreases the response delay sources. The quid pro quo for it is the increase in the actuator critical level because it gathers nearly all piloting functions.

Only one helicopter in the Airbus fleet (NH90) is fly-by-wire. The hydraulic actuators (Direct Drive Valve, DDV) are commanded directly by four electrical torque-motors connected to the FCC (Flight Control Computer), as shown in Figure 2b.

The hydraulic technology has been conventionally used in actuators for more than 60 years [1,6]. A new trend uses EMAs as substitutes for hydraulic actuators in the PFCS in actual helicopters or as part of fly-by-wire PFCSs in new autonomous helicopters (Figure 2c). This requires the reconsideration of the design practices right at the preliminary design phase. The VSR700 (Figure 1) is a use case. It is an already proved light helicopter (Cabri G2) turned into a drone by the integration of electrical components. These components include four EMAs in the PFCS.

1.1.3. Business Need

Today's business need is to perform EMA preliminary studies and learn more about this technology. Indeed, further to an application in the primary flight control system (main rotor and tail rotor), the EMA technology finds interest in many other helicopter applications, such as the automatic pilot actuators or the secondary flight control system

(SFCS). This includes the actuation of landing gears and wing flaps and the propellers' pitch control (for high-speed helicopters, such as the RACER helicopter).

1.2. Helicopter Specificities

The helicopter PFCS is a specific application for actuators. Linked to safety-critical surfaces, the actuator failure is qualified as "catastrophic" as it leads to the helicopter's crash. The actuation unit must comply with fail safe characteristics. Therefore, the actuation unit must have a redundant topology by force or position summing. The loading spectrum coming from the rotor spinning contains high-frequency content (fundamental at 20 hz minimum). Moreover, the operational pilot demand scenarios are difficult to predict. The required actuation performance is demanding as it combines a significant bandwidth response, reduced plays, high stiffness, and a low mass. The on-board environment is severe with temperature variations within $[-40; 85]$ °C, humidity, a salty atmosphere, and vibrations (6 to 20 g, [7]). Finally, concerning the component lifespan, the time between overhauls is required to be from 3000 flight h onwards. This set of specificities highlights the critical function occupied by the actuator in terms of aviation safety.

1.3. Electromechanical Actuators

An electromechanical actuator (EMA) includes components from multiple disciplines. There are mechanical parts (rod ends, bearings and rotary/linear conversion mechanisms, and a clutch), electrical parts (a motor, brake, and clutch), and electronic parts for power and control. The paper [3] presents an example of an EMA in detail and the different possible EMA architectures.

Except for low-power and/or less safety-critical applications (flaps, slats, spoilers, and a trim horizontal stabilizer), the EMA technology is not mature enough yet for primary flight controls [6]. This is essentially because of their lack of accumulated return of experience. The statistical database on components fault modes is poor [8]. EMAs entail some concerns in terms of reliability, risk of failures due to the jamming in the mechanical transmission components, health monitoring (HM) and assessment, and thermal management. The EMA applicability in aerospace has been proved in terms of dynamic performances [8]. In addition, EMAs offer interesting perspectives in terms of maintenance, integration, reconfiguration in case of failure, ease of operation, harsh running environment ($[-50, 125]$ °C), and management of power [1,6].

1.4. Preliminary Sizing Method

The preliminary design methodologies can be divided into two phases. The first phase is the system architecture choice, commonly guided by reliability studies, such as those presented in [9–11]. At this level, there are difficulties in taking into account the entire set of design criteria, important to evaluate an architecture. A study with a higher level of details is necessary, especially in the case of the EMA as it includes many constraints and interdisciplinary couplings between components. This is up to the second phase: the preliminary sizing. This phase is mainly based on multidisciplinary optimizations. The models used are usually analytical models or response surface models (RSM) to facilitate the design space exploration and the design optimization.

Some already existing preliminary sizing methodologies can be cited. The references [12,13] present a methodology to select the motor and gearhead of the actuators in the automotive field. The methodology includes a selection based on scaling laws. It outputs graphs showing all feasible motor/gear ratio combinations. In the aeronautic field and regarding the secondary flight control actuators, the paper [14] presents a methodology for the preliminary design of mechanical transmission systems. It is formalized as a constraint satisfaction problem (CSP) with an automated consistency checking and a pruning of the solution space. The mechanical components are modeled by scaling laws. In addition, the paper [10] presents a simulation and an optimization strategy to evaluate two concepts of actuation systems: the conventional hydraulic actuators and the electro hydrostatic actuators (EHAs). Moreover, the paper [15] describes a preliminary design method of EHAs based on multi-objective optimization (MOO) with Pareto dominance. Two objectives are set: the minimization of the mass and the maximization of the efficiency. The weight prediction is achieved using scaling laws, and the efficiency is calculated by a static energy loss model. The method outputs the design parameter, leading to a Pareto front in mass and efficiency.

The scaling laws are good candidates for preliminary design. They are an example of analytical models. They are interesting because they require few data to build them and validate them (existing industrial product ranges). In aerospace, the scaling laws are broadly used in the conceptual design of aircraft, especially regarding aerodynamics, propulsion, structure, and mass [16]. Moreover, the propellers are often described by scaling laws [17–19]. Furthermore, the scaling laws can be used in the field of robotic actuators where low speeds and high torques are usually required [20].

This paper suggests a preliminary sizing methodology applied in the aeronautic field, on helicopters, and on critical EMAs. It contains similar concepts as found in the previously cited literature. Indeed, it includes scaling laws. It includes an optimization where the component selection must satisfy a specification and design constraints while minimizing the actuator total mass. This paper adds its value by the introduction of two elements not considered in the literature yet. The first element is the motor heating based on dynamic criteria extracted from an equivalent representation applied on complex real mission profiles. The helicopter application is more dynamic than the aircraft application, as highlighted in the paper [3]. The second element is the vibratory environment through the actuator housing sizing.

1.5. Objective and Outline

The objective is to set up a design methodology supported by tools estimating the actuator component characteristics for any helicopter PFCS application. The considered EMA architecture is a direct drive in-line EMA. Details regarding this architecture are provided in Section 7.

To address this topic, firstly, this paper briefly presents the proposed global methodology. Secondly, it briefly recalls the tools established in [3] to better understand the actuation need and develops the actuator specification from the given flight data records. Thirdly, the modeling is detailed, limited to the electrical motor and the housing. Finally, the toolbox is applied to two real use cases. One compares two potential redundant topologies of actuation in terms of housing. The other compares the electrical motor characteristics obtained for the specifications of a given helicopter (main rotor and tail rotor).

2. Global Methodology

To answer the preliminary design need, this paper offers the global methodology presented in Figure 3. It consists of three main steps:

1. The actuation need must be understood. Flight mission profiles are analyzed. An equivalent actuator specification is synthesized, with key design values corresponding to component specificities.
2. Each actuator component is modeled according to the available inputs.
3. All models are set up into a sizing loop. An optimization algorithm is implemented with the objective of mass minimization.

This methodology takes flight data and high-level project requirements as inputs. It outputs the characteristics of each component necessary to answer the actuation need with a minimized mass.

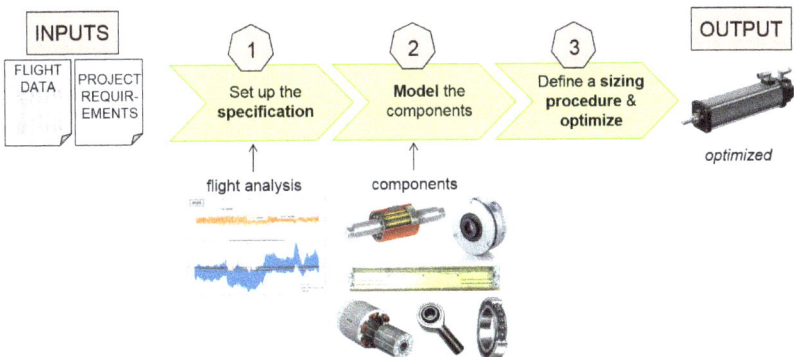

Figure 3. Proposed methodology for actuator sizing.

This paper addresses step 1 based on already published work in [3]. It details step 2 focusing only on the tools set up to model two components: the electrical motor and the housing. Step 3 is not explained. The mass minimization is a multidisciplinary design optimization (MDO) problem with numerous interdisciplinary couplings [21–23]. The MDO architecture is chosen to be with a single optimizer. Some variables and constraints are added into the sequencing. This optimization settings correspond to a hybrid individual disciplinary feasible (IDF) architecture and were inspired from [23–25].

3. Mission Profile Analysis

To reach the "first time right" objective of a design office, the requirements of any application must be well understood. In that way, the specification is fully representative, and the design is well guided. The EMA application in a PFCS of a helicopter combines three types of difficulties which impede easily writing down its specification. These difficulties are: a set of multidisciplinary design drivers due to the different technologies of the components; an external loading spectrum, coming from the rotor spinning in air, difficult to model; and operational piloting scenarios difficult to predict.

At the time of the project development of unmanned vehicles, a reduced number of short flight-test records with an interesting sampling rate can be available. These data are called mission profiles. They are the most trustful inputs representative of the application requirements. The paper [3] offers to set up an EMA specification based on the analysis of these mission profiles. The offered methodology is briefly recalled in the following subsection.

3.1. Methodology

The methodology offered by [3] has the objective to build the specification whilst simplifying the data analysis for the engineer and keeping them as the decision maker. The methodology is summed up in Figure 4. It is inspired from the work developed on railway trains and aircraft by [26,27]. It contains 4 main steps:

1. Considering the actuator architecture and components to extract a list of key parameters driving the design, called key design drivers.
2. Preparing the data to be analyzed (mission profiles) by filtering and transforming within the temporal or frequency domains.
3. Linking the data to the component key parameters, setting up the mathematical indicators to estimate over mission profiles. Each design driver has its own representative indicator(s).
4. The evaluation of indicators over mission profiles to develop the final EMA specification.

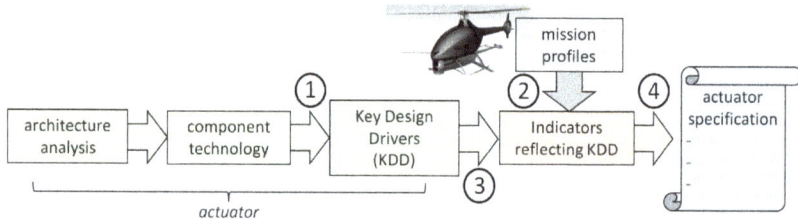

Figure 4. Proposed methodology for specification setup.

3.2. Estimation over Real Mission Profiles

The paper [3] illustrates the specification methodology, applying it on real flight data, and emphasizes the specificity of a rotorcraft PFCS for EMA application. It compares the specification of a main rotor actuator (MRA) to the specification of a tail rotor actuator (TRA). The MRA and TRA integration is illustrated in Figure 5. The analysis of the data coming from a real helicopter flight (VSR700, Figure 1) over three flight phases (take-off, cruise, and landing) provides the specifications of the MRA and TRA. In the paper [3], both specifications are compared using ratios elaborated for comparisons. In this paper, both specifications are presented in Table 1. For confidentiality reasons, the values cannot be displayed. Therefore, Table 1 presents ratios such as X_{MRA}/X_{TRA}, where X is the concerned specification indicator. These specifications will be the inputs in Section 7.3.

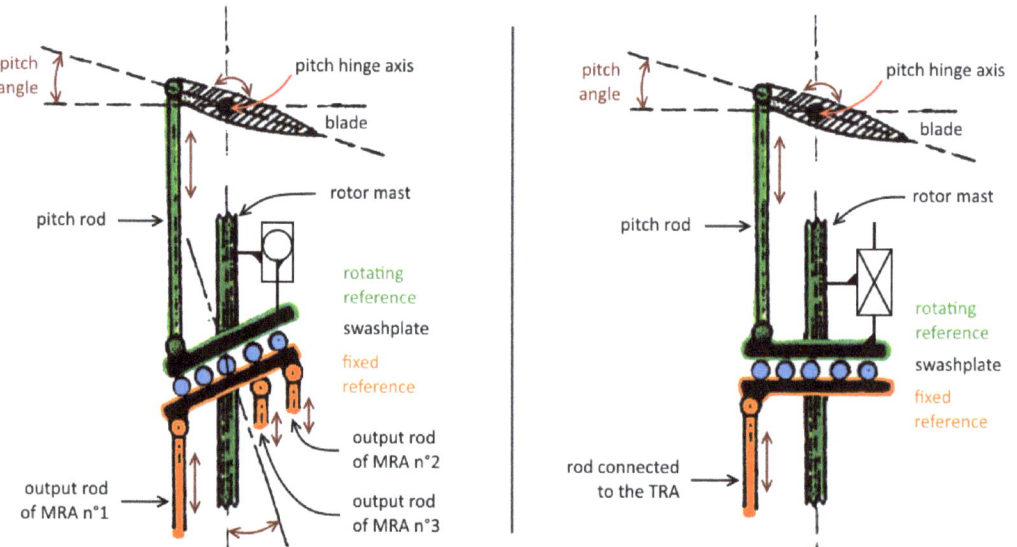

Figure 5. Sketch of principle of the main rotor (**left**) and the tail rotor (**right**) on helicopters. The main rotor is responsible for lifting the helicopter weight. The tail rotor ensures the control of the helicopter yaw axis.

At first glance, Table 1 clearly shows that the MRA faces loads roughly 10 times higher than the TRA and moves with an equivalent continuous speed and acceleration roughly 40% lower than the TRA. In the paper [3], the comparison between both applications is emphasized regarding the continuous and maximum loading (from the rotor inertia and external load), the rolling fatigue of the mechanical components, the rotor inertia impact compared to the external load, and the motor loss identification in terms of the continuous motor torque contribution. This paper adds an additional indicator regarding the load-pitting analysis.

Table 1. VSR700 specification for MRA and TRA. a is the acceleration, F is the load; they come from measures on the VSR700 helicopter PFCS. $PR = F(t) \cdot a(t)$ is the power rate. The mean power rate $\overline{PR} = | <F(t) \cdot a(t)> |$ takes an absolute value to be conservative. v_{iron} is a mean value of speed representative of motor iron losses. F_{RMC} is an equivalent rolling fatigue load based on 10^6 cycles. (details in [3]).

Measures	Indicator	Ratio MRA/TRA [-]
position	stroke max s_{max}	1.1
	equivalent distance travelled L_{eq}	0.5
	speed max v_{max}	1.1
	speed iron v_{iron}	0.5
	acceleration max a_{max}	0.6
	acceleration rms a_{rms}	0.7
load	load max F_{max}	7.0
	load rms F_{rms}	15.1
	load rmc F_{rmc}	13.6
position & load	power rate mean $\overline{PR}$	4.6
	power rate max PR_{max}	3.0
	pair $(a_{PRmax}; F_{PRmax})$	(0.6; 4.7)

The pitting load refers to the fluctuating load when the actuator is motionless. This paper suggests estimating the averaged contribution of the load upper frequencies. This can be performed compared to the entire load frequency content. Therefore, the load is separated from its low frequencies using a filter with a cut-off frequency set at the actuator bandwidth value. This resulting load is averaged by the RMS to estimate an equivalent continuous load. This result is normalized by the equivalent continuous load, including all load frequency content. The indicator is R_p:

$$R_p = \frac{[butter_{HP}[F(t)]]_{rms}}{[F(t)]_{rms}} \quad (1)$$

The indicator R_p is evaluated on a nontransient phase of the VSR700 mission profile (Table 2). The indicator shows a higher level of the pitting load in the MRA. This result is understood considering the helicopter rotor architectures (Figure 5). As a piece of information, the MRA selectively controls the blade attack angle during one rotor azimuth. Meanwhile, the TRA simultaneously controls the attack angle of all the blades whatever the azimuth.

Table 2. Comparison of the indicator R_p for MRA and TRA applications on VSR700 mission profiles. The indicator is evaluated over steady state flight phases (cruise).

Domain	Indicator	Unit	MRA	TRA
pitting fatigue	R_p	-	28	5

4. Models for the Preliminary Sizing of EMA

4.1. Modeling Overview

The EMA includes multidisciplinary components. Each of them has multiple and different key design drivers and operational scenarios. The selected components should comply with the actuator specification and ensure a minimized total mass. To answer the need of component selection, a knowledge-based process around component modeling must be set. It covers two levels [28]:

1. The component level: It deals with the determination of component characteristics from a reduced number of parameters to facilitate the optimization. The models involved in it are called the *estimation models* (Figure 6).

2. The system level: It deals with interactions between components, operational scenarios, and component operational limits. The models involved in it are called *simulation models* and *evaluation models* (Figure 6).

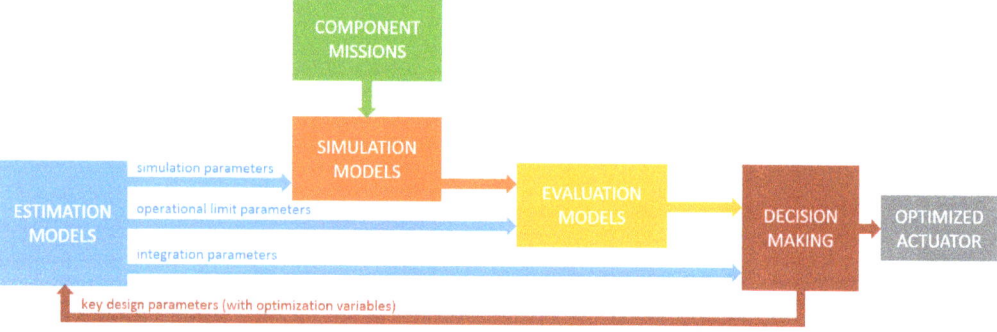

Figure 6. Structure of model-based preliminary design.

This paper proposes to focus only on the *estimation models*. To cover the tools setup, only two components are presented. They are those at stake within two onboard specificities: the thermal and vibratory environments. Therefore, the following sections deal with the modeling proposed for the electrical motor and the housing characteristics.

4.2. Need of Estimation Models and Approach

As shown in Figure 6, to start a first iteration in the sizing loop, the main characteristics of each component have to be identified from a reduced set of key parameters. The *estimation models* are introduced for this purpose. Per component, they directly link the primary characteristics, which define the component functionally, to the secondary characteristics, which can be seen as the dimensions and features of the imperfections. Thus, the *estimation models* provide the necessary parameters for the integration study, *simulation models*, and *evaluation models*.

Generally, at the component level, the models link the physical dimensions and characteristics of in-use materials to the primary and secondary characteristics. The design at the component level is an inverse problem which requires the primary characteristics as inputs.

In such a context of multidisciplinary modeling with optimization, a unified modeling approach is required. A dimensional analysis and the Vaschy–Buckingham theorem (Theorem 1) are good candidates for it [29,30]. Indeed, they are extensively used in aerodynamics and fluid mechanics because they provide a more physical and unified framework with a reduced number of parameters. This section shows how they can be extended to other domains, such as the electrical motor and the housing of an EMA.

Theorem 1 (Vaschy–Buckingham theorem). *Any physical equation dealing with n physical variables depending on k fundamental units (mass, length, time, temperature, charge) can be formulated as an equivalent equation with $p = n - k$ dimensionless variables called "π-numbers" built from the initial variables.*

The development steps of an estimation model are presented in Figure 7. The starting point is the expression of one component characteristic y as an algebraic function f depending on the geometrical dimensions and material/physical properties p_i. L is a length and d_i the rest of dimensions.

$$y = f(\underbrace{L, d_1, d_2, ...}_{\text{dimensions}}, \underbrace{p_1, p_2, ...}_{\text{properties}}) \quad (2)$$

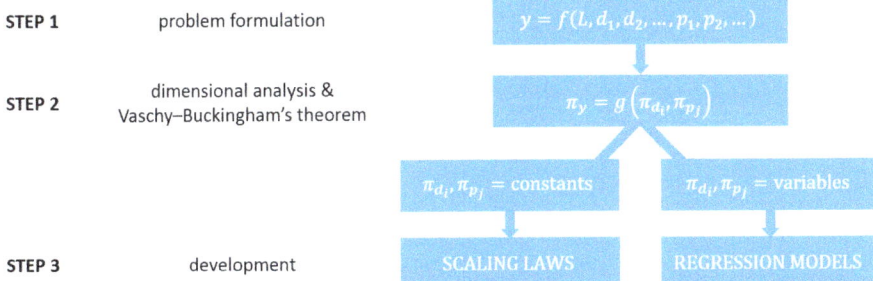

Figure 7. Estimation model development steps.

Applying the dimensional analysis and Theorem 1, the problem is rewritten into a reduced number of dimensionless parameters (Equation (3), [31]).

$$\pi_y = g(\underbrace{\pi_{d_1}, \pi_{d_2}, ..., \pi_{p_1}, \pi_{p_2}, ...}_{p \text{ variables}}) \text{ with } \begin{cases} \pi_y = y^{c_y} \cdot L^{c_L} \cdot \prod_i p_i^{c_{p_i}} \\ \pi_{d_i} = \frac{d_i}{L} \\ \pi_{p_i} = L^{c_{p_i,0}} \cdot \prod_j p_i^{c_{p_{i,j}}} \end{cases} \quad (3)$$

The next step is to develop the estimation models based on the π-groups.

Scaling law formulations are undertaken when the dimensionless numbers π_{d_i} and π_{p_i} remain constant around a given component product range. This means the geometrical and/or material similarities are satisfied. This is applicable for the electrical motor of the actuator as detailed in the following Section 5.

When the dimensionless numbers π_{d_i} and π_{p_i} are not considered as constant, the approximation of the function g can be achieved by performing data regressions [31,32]. The data may come from manufacturer product data, test measurements, or finite element simulation results based on design of Experiment (DoE) as presented in Section 6 for the housing vibratory sizing.

5. Scaling Laws

In this section, the scaling law theory is firstly developed and then applied to an electrical motor. The actuator sizing requires *estimation models* for its integration (motor dimensions) and its losses (copper and iron losses, inertia).

5.1. Fundamentals

In the literature, scaling laws are also called *similarity laws* or *allometric models* [33]. They estimate the component main characteristics requested for their selection without requiring a detailed design.

Scaling laws are based on three hypotheses:

- Geometric similarity: all the dimensions of the considered component to all the lengths of the component used for reference are constant. Thus, all corresponding aspect ratios are constant: $\pi_{d_i} = $ constant.
- Uniqueness of design drivers: only one main dominant physical phenomenon drives the evolution of the component secondary characteristic y. Thus, in most cases, there is not anymore dependency with any π_{p_i} (function g, Equation (3)).
- Material similarity: all material properties are assumed to be identical to those of the component used for reference. Thus, all corresponding scaling ratios are equal to 1: $\pi_{p_i} = 1$.

Once these assumptions are satisfied, π_y is stated to be constant because it depends on constant variables:

$$\pi_y = g(\pi_{d_1}, \pi_{d_2}, ..., \pi_{p_1}, \pi_{p_2}, ...) = \text{constant} \quad (4)$$

This gives the standard power-law shape of a scaling law:

$$\pi_y = y^{c_y} \cdot L^{c_L} \cdot \prod_i p_i^{c_{p_i}} = \text{constant} \implies y \propto L^c \quad (5)$$

with c a constant. Then, as proposed by [34], the "star" notation is introduced. It indicates the scaling ratio of a desired component characteristic x by the same characteristic x_{ref} of a component taken as a reference: $x^* = x/x_{ref}$. This component of reference is picked up into the supplier range of the considered product.

Thus, Equation (5) becomes:

$$y^* = L^{*c} \iff \frac{y}{y_{ref}} = \left(\frac{L}{L_{ref}}\right)^c \quad (6)$$

From a single component of reference and a reduced number of parameters (no detailed design required), the scaling laws quickly extrapolate the main characteristics y_{ref} of a known component toward the characteristic y of a possible component of the same technology:

$$y = y_{ref} \cdot \left(\frac{L}{L_{ref}}\right)^c \quad (7)$$

Consequently, the scaling laws level down the complexity of the inversion problem. All the useful relations are easily expressed as a function of a single *key design parameter* (also named *definition parameter*) that is associated with the component under design (Figure 6).

5.2. Electrical Motor Scaling Laws

The previously mentioned approach is applied to the electrical brushless motor. The motor mass-law formulation is detailed hereafter. Beforehand, some hypotheses need to be stated: the main design driver is the maximum continuous winding temperature; the natural convection is the dominant thermal phenomenon; the mean induction in the airgap is constant for a given magnet technology; the number of magnets is constant over the considered product range; and the geometric similarities are verified, and the material and boundary limits similarities are satisfied.

In Figure 8, following the approach mentioned in Section 5.1, step by step, the torque evolution is obtained.

As diameters and lengths are supposed to evolve similarly ($d^* = L^*$, Figure 9), the mass M of the motor is basically approximated by:

$$M = \int \rho_{eq} \, dV \implies M^* = L^{*3} \quad (8)$$

Using the torque expression (prerequisite of Figure 8), the motor mass M becomes:

$$M^* = T^{*3/3.5} \implies M = M_{ref} \cdot \left(\frac{T}{T_{ref}}\right)^{3/3.5} \quad (9)$$

Figure 10 compares the evolution of this law (Equation (9)) to real data from two manufactured motors: Parvex NK [35] and Kollmorgen RBE [36]. It is observed that a single reference of a motor allows to rebuild the evolution of the motor mass for a broad range of torque. This is possible with less than 10% of the mean relative error ϵ.

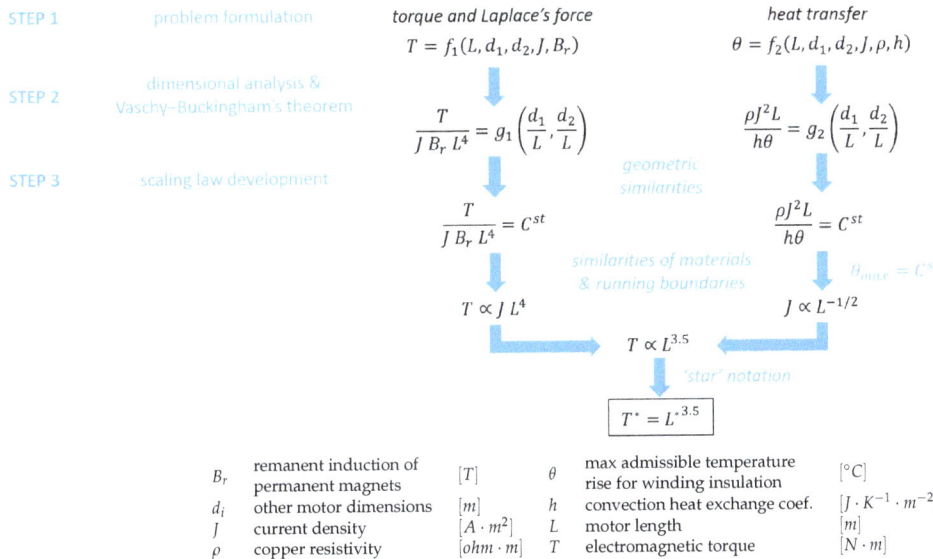

Figure 8. Electrical motor torque formulation.

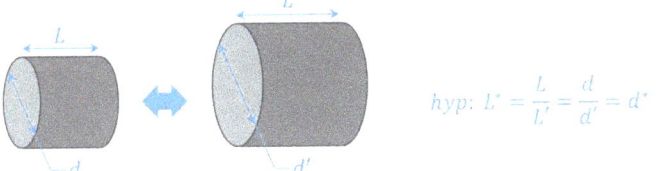

Figure 9. Electrical motor: homothety hypothesis.

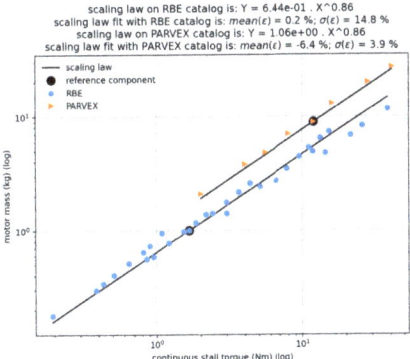

Figure 10. Motor mass: scaling law prediction compared to manufacturer catalogs (Parvex NX, Kollmorgen RBE).

In the same way, and following the same hypothesis, the other motor characteristic laws are formulated. Table 3 presents them with their prediction level compared to the Parvex NK catalog. The prediction levels compared to the RBE catalog are presented when data are missing in the Parvex catalog. Both catalogs validate the presented laws.

Table 3. Electrical motor scaling law sum up and prediction quality compared to manufacturer catalog data.

Component Characteristics	Unit	Scaling Law	ϵ	σ	
ELECTRICAL MOTOR (EM) (brushless, cylindrical)					
Note: error estimations performed using PARVEX NK and KOLLMORGEN RBE catalogue ranges ($T \in [0.5; 41]$ Nm)					
key design parameter					
continuous torque	Nm	$T^* = d^{*3.5} = K_m^{*3.5/5}$	-	-	
motor constant	$(N \cdot m)^2/W$	$K_m^* = d^{*5} = T^{*5/3.5}$	-	-	
integration parameter					
dimensions ($l^* = d^*$)	m	$d^* = T^{*1/3.5} = K_m^{*1/5}$	1.1%	7.5%	
mass	kg	$M^* = d^{*3} = T^{*3/3.5} = K_m^{*3/5}$	6.4%	3.9%	
simulation parameter					
inertia	kg·m²	$J^* = d^{*5} = T^{*5/3.5} = K_m^*$	0.1%	20%	
copper coef.	W/(Nm)²	$\alpha^* = d^{*-5} = T^{*-5/3.5} = K_m^{*-1}$	13%	16%	
Joules' losses	W	$P_J^* = d^{*2} = T^{*2/3.5} = K_m^{*2/5}$	13%	16%	
iron loss coef.	W/(rad/s)$^{1.5}$	$\beta^* = d^{*3} = T^{*3/3.5} = K_m^{*3/5}$	6.9%	19%	
resistance	Ω/(Nm/A)²	$\mathcal{R}^*/K_t^{*2} = d^{*-5} = T^{*-5/3.5} = K_m^{*-1}$	19%	14%	
inductance	H/(Nm/A)²	$\mathcal{L}^*/K_t^{*2} = d^{*-3} = T^{*-3/3.5} = K_m^{*-3/5}$	0.5%	17%	
number of pole pair	-	$p^* = 1$	0%	0%	
operational limit parameter					
peak torque	Nm	$T_p^* = d^{*3.5} = T^* = K_m^{*3.5/5}$	0%	2.3%	(1)
		$T_{p,mag}^* = d^{*3} = T^{*3/3.5} = K_m^{*3/5}$	-%	-%	(2)
		$T_{p,th}^* = d^{*4} = T^{*4/3.5} = K_m^{*4/5}$	13%	6.5%	(3)
max speed (4)	RPM	$\Omega_{max}^* = d^{*-1} = T^{*-1/3.5} = K_m^{*-1/5}$	1.6%	4.5%	

(1) Parvex specificity; (2) definition based on magnetic saturation criteria, missing data for validation; (3) definition based on thermal dissipation criteria; be careful that RBE catalog gathers different thermal integrations, T is dependent on them; (4) based on mechanical limitations; * the 'star' notation: $x^* = x/x_{ref}$.

In Table 3, the motor constant K_m $[(N \cdot m)^2/W]$ is introduced. It is the ratio of the squared torque provided per unit of heat generated. Moreover, it can be found in catalogs with another definition: $[(N \cdot m)/W^{0.5}]$. K_m is an interesting parameter because it directly links the application mechanical need (the required motor torque) with the motor losses without knowing the motor winding characteristics.

As far as the motor losses are concerned, [34,37] mention that the copper and iron losses bound the motor continuous operation domain. The operation domains of electrical motors can be found in [3,33]. Thus, at a steady state, the total heat generated by the electrical motor is the sum of Joules and iron losses, such as:

$$Q_{th} = Q_{Joules} + Q_{iron} = \alpha \cdot T^2 + \beta \cdot f_{elect}^b \tag{10}$$

where α and β are, respectively, copper and iron losses coefficients, and f_{elect} the electrical speed. b is a constant depending on the hysteresis and the Eddy's current contributions into iron losses, and [33,37] indicate a mean value of 1.5.

As far as the other components of the actuator are concerned, their scaling laws are found in the publications [20,33].

6. Regression Models

In a helicopter context, the resonance frequencies and stress under a vibratory environment is an unavoidable check to perform. This section presents a preliminary vibratory study of a simplified actuator housing. The approach goes through the synthesis of a regression model (also called the surrogate model) to implement into the actuator preliminary sizing loop.

6.1. Introduction

From a purely mechanical point of view, the design of the EMA housing has to focus on the elementary forces acting on the housing, which can be divided into two categories: the static stresses induced by the power transmission to the load, which have low frequencies, and the vibratory stresses induced by the vibratory environment, which have high frequencies ($f \in [5; 2000]$ hz, [7]).

The path of the various static or dynamic loads is represented for a generic actuator in Figure 11.

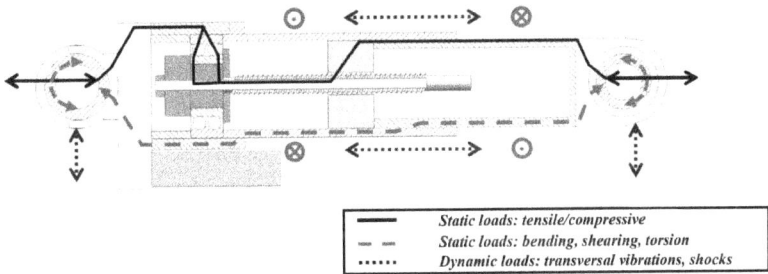

Figure 11. Load path in generic EMA [38].

The most significant static loadings are the tensile, compressive, buckling forces. They are transmitted through the rod to the nut, the screw, the bearings, and finally to the housing. The high number of cycles generally requires the fatigue limit of materials to be considered.

The dynamic stress is mainly generated by the transversal vibrations due to the vibratory environment which can generate important mechanical bending stresses. For a long actuator, as it can be for a direct-drive EMA, these stresses are prevailing.

6.2. Prior Considerations

A high-fidelity model of vibratory stress in the housing would be difficult to develop. Indeed, ball bearings and roller screw stiffnesses and plays are unknown and not supplied in datasheets. The contact at the linear bushing level is unclear.

Therefore, we propose a simple model based on some simplifications. The first one concerns the potential mechanical backlash in the actuator assembly and their effect with respect to excitation frequencies. The vibratory amplitude relies on the acceleration level, as illustrated in Table 4.

Table 4. Vibratory amplitudes for different acceleration and frequency.

number of g	-	6	6	20	20
acceleration a	m·s^{-1}	58.8	58.8	196	196
frequency f	hz	250	50	250	50
vibratory amplitude x	μm	24	596	79	1986

$$x = \frac{a}{w^2} \quad \begin{cases} x & \text{amplitude [m]} \\ a & \text{acceleration [m·s}^{-2}\text{]} \\ w = 2\pi f & \text{pulsation [rad·s}^{-1}\text{]} \end{cases} \quad (11)$$

The vibratory amplitudes are estimated in Table 4 with Equation (11) regarding commonly used accelerations found in the DO160 standard [7] (6 g for in-cabin equipment, 20 g for an under-swashplate location). The amplitudes can be lower and close to the typical plays. In this case, the vibratory phenomena becomes even more complex. Typical plays at the linear bushing level are roughly 100 μm. The linear bushing or sleeve bearing guides the output rod inside the housing.

To model the effect of the play, we propose a lumped-parameter simulation. It associates one or two mass–spring system(s) excited through an elastogap where the play is modeled as a non-linear stiffness (very low value into the play, and high value far from the play).

In Figure 12, the accelerations stated by the DO–160 standard [7] are plotted in terms of amplitude and frequency.

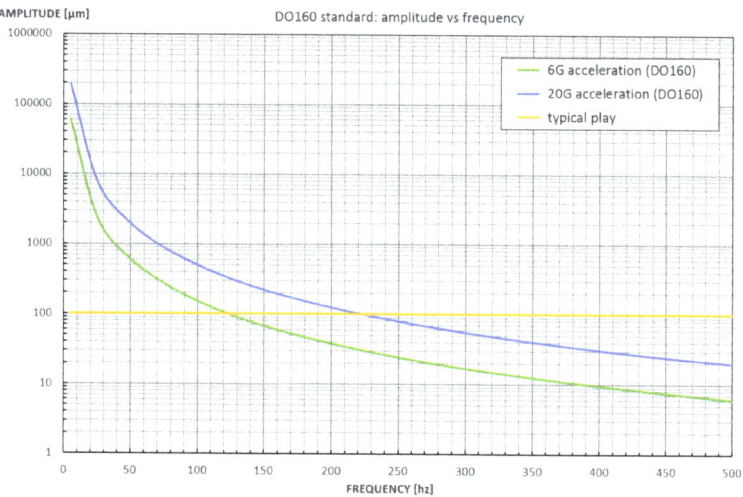

Figure 12. DO–160 accelerations: amplitude evolutions with respect to frequencies.

Two cases are studied and illustrated thereafter: *case 1* at 250 hz with $x \leq 100$ μm and *case 2* at 50 hz with $x \geq 100$ μm. Both cases consider an acceleration of 6 g.

First of all, the model considers only one mass–spring system.

Case 1 is simulated in Figure 13. The mass–spring resonance is set at 250 hz and the excitation is around this frequency. No resonance mode of the mass is observed, and the vibratory excitation is 'filtered' by the play.

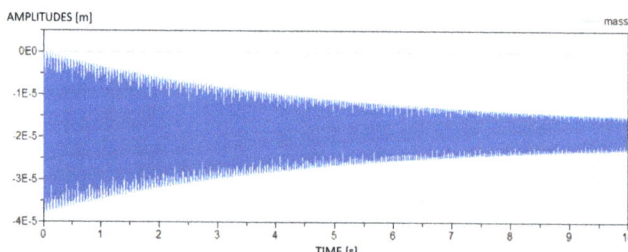

Figure 13. Mass amplitudes of the simple mass–spring coupled with play ($f_r = 250$ hz).

Case 2 is simulated in Figure 14. The mass–spring resonance is set at 50 hz and the excitation is around this frequency. The mass vibrations are more important, and the resonance is observed.

Now, for the plays inside the actuator, the model considers two mass–spring systems linked by an elastogap. Using this model, the displacement of each mass is plotted in Figure 15. For a resonance frequency around 50 hz, the amplitudes are such that the vibrating parts interfaced by the play can be considered as one and a single part.

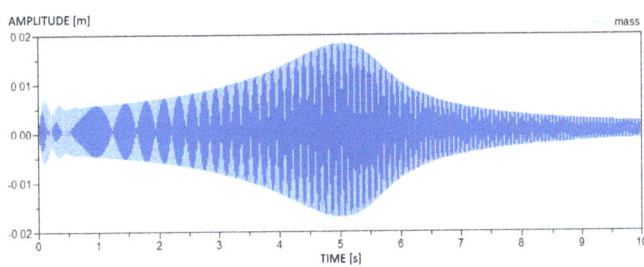

Figure 14. Mass amplitudes of the simple mass–spring coupled with play ($f_r = 50$ hz).

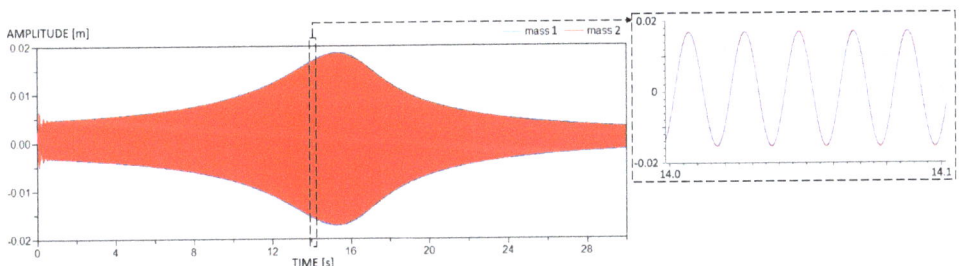

Figure 15. Mass amplitudes of the double mass–spring coupled with play ($f_r = 50$ hz).

For a resonance frequency around 250 hz, the vibratory amplitudes are much lower, and the two masses evolve within the play.

As a result, to keep the vibratory amplitudes smaller than the play between the parts, the resonance frequencies must be high (e.g., ≥ 200 hz; this limit depends on the amplitude of the acceleration). It is not the case for long and narrow actuators as direct-drive actuators can be. The resonance frequencies are low.

Thus, some simplifications can be introduced into the estimations of stresses of the actuator envelop under vibratory excitations. We propose to suppose, for housings with low resonance frequency, that the plays are negligible compared to the vibratory amplitudes. Consequently, the contact with linear bushing is modeled as an infinitely rigid contact.

6.3. Hypothesis

Now, an FEM model of reduced parameters is developed in order to represent it by a surrogate model. Some more hypotheses are formulated.

A simplified geometry is considered (Figure 16): two hollowed cylinders, one in the other. The connection between them is assumed to be perfect. The set (motor, brake, connectors, and bearings) is supposed to be a cylinder of 1/3rd of L_a with an equivalent density. This equivalent mass is modeled with Young's low modulus (1/10th of aluminium modulus). This choice is conservative; it is so as to not impact the stiffness of the housing.

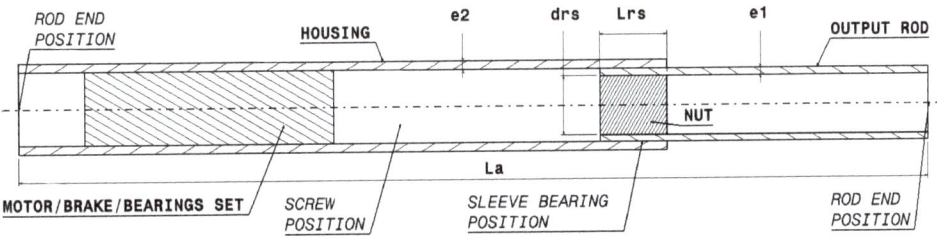

Figure 16. Actuator simplified geometry for modal analysis.

The housing and output rod are set with the same material properties (aluminium). Their lengths are, respectively, supposed to be 2/3rd and 1/3rd of L_a.

The nut is modeled as a full cylinder with 90% steel density (7000 kg·m^{-3}) to consider the air content between the rollers. The geometry of the nut evolves with the geometrical similarity assumption (scaling law). The cylinder representing the nut is modeled using a low Young's modulus (1/10th of steel modulus) so as to not influence the stiffness of the structure.

The rod ends allow rotation with no friction.

The antirotation key and the sealing leap at the interface output rod with housing are not modeled.

The three following sections develop a surrogate model using the surrogate modeling technic suggested in the paper [31].

6.4. Problem Formulation

Under vibration, the system can be associated to a basic damped mass–spring model. Figure 17 presents this model with U (m) the displacement of an equivalent mass M_{eq} (kg) evolving according to an excitation force F (N), a stiffness K_{eq} (N·m^{-1}), and a damping C_{eq} (N·m^{-1}·s).

The stress is linearly linked to the displacement:

$$\sigma = k_\sigma \cdot U \qquad\qquad k_\sigma = \frac{\sigma_0}{U_0} \qquad (12)$$

with σ_0 [Pa] and U_0 [m], the maximum stress and the maximum displacement at resonance frequency.

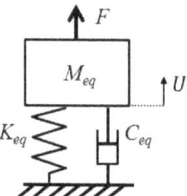

Figure 17. Mass–spring model.

Newton's second law applied to the moving body enables the Laplace function between the displacement $U(t)$ and the excitation load $F(t)$ to be estimated, such as:

$$\frac{U(p)}{F(p)} = \frac{1}{M_{eq} \cdot p^2 + C_{eq} \cdot p + K_{eq}} = \frac{1/K_{eq}}{p^2/\omega_r^2 + 2 \cdot \xi \cdot p/\omega_r + 1} \qquad (13)$$

Considering an excitation of the mass with a sinusoidal force $F(t) = F_0 \cdot sin(\omega \cdot t)$, the maximum displacement at the first resonance mode is:

$$U_0 = \frac{1}{2 \cdot \xi} \cdot \frac{F_0}{K_{eq}} = \frac{Q_m \cdot F_0}{K_{eq}} \qquad \xi = \frac{C_{eq}}{2 \cdot \sqrt{K_{eq} \cdot M_{eq}}} \approx \frac{1}{2 \cdot Q_m} \qquad (14)$$

where Q_m is the mechanical quality coefficient.

The article [38] reports that tests performed on industrial prototypes show a wide range of practical values for the equivalent mechanical quality coefficient Q_m. In addition, it reports that experiments give typical values for Q_m between 10 and 50, depending on the application and boundary conditions. For structural dynamic models, in the absence of better information, it is normally acceptable to assume a value of $Q_m = 30$ (according to [39]).

The equivalent force F_0 of the acceleration effect can be evaluated thanks to an equivalent work $\mathcal{W}_0$ [40,41]:

$$\mathcal{W}_0 = F_0 \cdot U_0 = \iiint_\mathcal{V} u_0 \cdot a \cdot \rho \cdot d\mathcal{V} \tag{15}$$

with $u_0(x)$ the deflection of the actuator envelop, a amplitude of the vibratory sinusoidal acceleration. A mass M_{acc} subjected to the acceleration can be defined:

$$F_0 = M_{acc} \cdot a \qquad M_{acc} = \frac{\iiint_\mathcal{V} u_0 \cdot \rho \cdot d\mathcal{V}}{U_0} \tag{16}$$

The mass subjected to the acceleration is not identical to the mass expressing the kinetic energy, M_{eq}, defined such as [40,41]:

$$\frac{1}{2} \cdot M_{eq} \cdot V_0^2 = \iiint_\mathcal{V} \frac{1}{2} \cdot \rho \cdot v_0^2 \cdot d\mathcal{V} \tag{17}$$

with, at the first mode resonance, V_0 the speed of M_{eq} and $v_0(x)$ the speed of each point of the actuator deflection. The speeds can be defined such as:

$$v = w_0 \cdot u \qquad V = w_0 \cdot U \tag{18}$$

Thus, we can easily define the equivalent mass such as:

$$M_{eq} = \frac{\iiint_\mathcal{V} \rho \cdot u_0^2 \cdot d\mathcal{V}}{U_0^2} \tag{19}$$

The following ratio is introduced:

$$k_{acc} = \frac{M_{acc}}{M_{eq}} = \begin{cases} U_0 \cdot \frac{\iiint_\mathcal{V} u_0 \cdot d\mathcal{V}}{\iiint_\mathcal{V} u_0^2 \cdot d\mathcal{V}} & \text{if } \rho \text{ is constant} \\ U_0 \cdot \frac{\iiint_\mathcal{V} u_0 \cdot \rho \cdot d\mathcal{V}}{\iiint_\mathcal{V} u_0^2 \cdot \rho \cdot d\mathcal{V}} & \text{if } \rho \text{ is not constant} \end{cases} \tag{20}$$

and the maximum displacement at the resonance can be approximated as it follows:

$$U_0 = \frac{Q_m \cdot k_{acc} \cdot M_{eq} \cdot a_0}{K_{eq}} = \frac{Q_m \cdot k_{acc} \cdot a_0}{w_0^2} \tag{21}$$

with $w_0^2 = K_{eq}/M_{eq}$ the resonance angular frequency.

6.5. Dimensional Analysis

As seen in Section 5, the use of a dimensional analysis and Buckingham's Theorem enable to reduce the number of variables expressing a physical problem. Here below, this approach is developed for the vibratory use case. By simplification and for a reduced number of parameters, a constant density ρ is assumed all along the actuator (Equation (20)).

The link between stress and displacement evolves according to the following variables:

$$\frac{\sigma}{U} = k_\sigma = f(E, d_{rs}, L_a, e_1, e_2, L_{rs}) \tag{22}$$

which can be rewritten with the following dimensionless numbers:

$$\pi_{k_\sigma} = \frac{\sigma \cdot d_{rs}}{U \cdot E} = f\left(\frac{L_a}{d_{rs}}, \frac{e_1}{d_{rs}}, \frac{e_2}{d_{rs}}, \frac{L_{rs}}{d_{rs}}\right) \tag{23}$$

The resonance angular frequency evolves according to:

$$w_0 = g(E, \rho, d_{rs}, L_a, e_1, e_2, L_{rs}) \tag{24}$$

which can be rewritten with the following dimensionless numbers:

$$\pi_{\omega_0} = \omega_0 \cdot \left(\frac{\rho}{E}\right)^{1/2} \cdot d_{rs} = g\left(\frac{L_a}{d_{rs}}, \frac{e_1}{d_{rs}}, \frac{e_2}{d_{rs}}, \frac{L_{rs}}{d_{rs}}\right) \tag{25}$$

The stress under a vibratory acceleration can be expressed as:

$$\sigma = k_\sigma \cdot U = k_\sigma \cdot \frac{Q_m \cdot k_{acc} \cdot a}{\omega_r^2} = \sigma_0 \cdot Q_m \cdot a \cdot \frac{\iiint_V u_0 \cdot dV}{\iiint_V u_0^2 \cdot dV} \tag{26}$$

The stress evolves according to:

$$\sigma = h(k_\sigma, \omega_0^2, a, Q_m) \tag{27}$$

which can be rewritten as:

$$\pi_0 = \frac{\sigma}{Q_m \cdot a \cdot d_{rs} \cdot \rho} = h\left(\frac{L_a}{d_{rs}}, \frac{e_1}{d_{rs}}, \frac{e_2}{d_{rs}}, \frac{L_{rs}}{d_{rs}}\right) \tag{28}$$

The expression of the stress is thus only function of four aspect ratios. One of these aspect ratios, L_{rs}/d_{rs}, can be assumed to be constant because of the geometrical similarity assumption used for roller screw component sizing (scaling law).

The final expression of the stress is a function dependent of three dimensionless quantities:

$$\pi_0 = g(\pi_1, \pi_2, \pi_3) \begin{cases} \pi_0 = \sigma/(Q_m \cdot a \cdot d_{rs} \cdot \rho) \\ \pi_1 = L_a/d_{rs} \\ \pi_2 = e_1/d_{rs} \\ \pi_3 = e_2/d_{rs} \end{cases} \tag{29}$$

It remains to determine this function g. The following section does the job.

6.6. FEM Software Model

Using a software for computation by finite elements, a model is parametrized according to the previous considerations and hypotheses. It enables to obtain:

- The resonance frequency f_r or the resonance angular frequency ω_r.
- The modal form characterized by a maximal displacement U_0.
- The corresponding maximum stress σ_0. The maximum stress is identified to be on the output rod tube (e_1 thickness).

The intersection of both cylinders has been taken care of by smooth and arced geometries to avoid numerical stress constraints. The boundaries are pinned at each extremity of the actuator model. The deflection is allowed within the plane of the section presented in Figure 18.

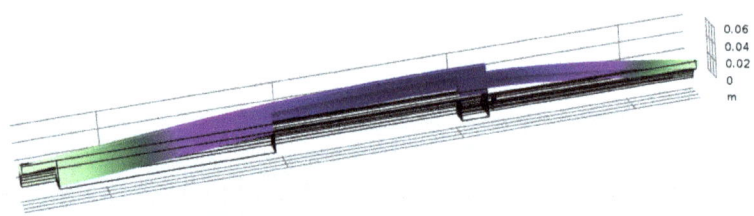

Figure 18. Modal analysis of the simplified actuator geometry.

The tandem topology of actuation is largely used in the aeronautic field because it complies with safety aviation rules. This topology involves two actuators stuck to each other at their basement, and their extension is in opposite directions (more details provided in Section 7.2). This topology increases the total length of the actuation unit. To consider this use case, another surrogate model needs to be developed.

The FEM model for the single actuator is reused. The boundary conditions are changed. A symmetry constraint is applied to the geometry at the actuator basement. The stress is picked up onto two critical points: at the interface between both actuators and at the interface between the output rod and the housing. Two surrogate models are expressed to determine the value of π_0 for each of these points.

6.7. DOE and Surrogate Synthesis

A design of experiments (DoE) is realized with e_1, e_2, d_{rs}, and L_a. The simulations (a modal analysis) generate the variable of interest π_0.

The dependent variable of the problem is approximated thanks to a linear regression (response surface model (RSM)) where the development takes into account a mean value, a first-order member (which represents the main effects of the problem), a combined member (representing the interactions), and a second-order member to consider further effects. The development takes the following form:

$$\pi_0 = \underbrace{a_0}_{\text{mean value}} + \underbrace{\sum a_i \pi_i}_{\text{main effect}} + \underbrace{\sum a_{ij} \pi_i \pi_j}_{\text{interactions}} + \underbrace{\sum a_{ii} \pi_i^2}_{\text{high order effect}} \tag{30}$$

The VPLM methodology (Variable Power-Law Metamodel) [31] is applied. A log transformation on variables is performed for the linearization which gives the form:

$$log(\pi_0) = a_0 + \sum a_i log(\pi_i) + \sum a_{ij} log(\pi_i) log(\pi_j) + \sum a_{ii} log(\pi_i)^2 \tag{31}$$

and can be rewritten as:

$$\pi_0 = 10^{a_0} \prod_{i=1}^{n} \pi_i^{a_i + a_{ii} log(\pi_i) + \sum_{j=i+1}^{n} a_{ij} log(\pi_j)} \tag{32}$$

This variable power-law form enables to deal with the large variation range of the dependent and independent variables.

The data set coming from the DoE is shared in two sets: one for the regression procedure so as to determine the coefficients a_i and a_{ij} (Equation (31)) and the other for the test of the final surrogate.

The regression gives the following surrogate model which determines the value of π_0 for the housing of a single actuator:

$$\begin{aligned}log_{10}(\pi_0) =& 68 \cdot log_{10}\left(\frac{L}{d}\right)^2 \cdot \frac{1}{293} + 64 \cdot log_{10}\left(\frac{L}{d}\right) \cdot log_{10}\left(\frac{e_1}{d}\right) \cdot \frac{1}{973} + log_{10}\left(\frac{L}{d}\right) \cdot log_{10}\left(\frac{e_2}{d}\right) \cdot \frac{1}{1000} \\ &+ 1099 \cdot log_{10}\left(\frac{L}{d}\right) \cdot \frac{1}{984} + 86 \cdot log_{10}\left(\frac{e_1}{d}\right)^2 \cdot \frac{1}{657} + 229 \cdot log_{10}\left(\frac{e_1}{d}\right) \cdot log_{10}\left(\frac{e_2}{d}\right) \cdot \frac{1}{884} \\ &- 101 \cdot log_{10}\left(\frac{e_1}{d}\right) \cdot \frac{1}{249} + 538 \cdot log_{10}\left(\frac{e_2}{d}\right)^2 \cdot \frac{1}{685} + 670 \cdot log_{10}\left(\frac{e_2}{d}\right) \cdot \frac{1}{359} + \frac{527}{551}\end{aligned} \tag{33}$$

with L the length L_a and d the diameter d_{rs}.

The two other surrogate models developed for the tandem actuator housing are shown in Appendix A.

6.8. Validation

In Figure 19, the prediction of the surrogate model (Equation (33)) is compared to the FEM simulation result data set. The prediction level is satisfying with $R^2 > 99\%$. The two other housing surrogate models show the same prediction level quality.

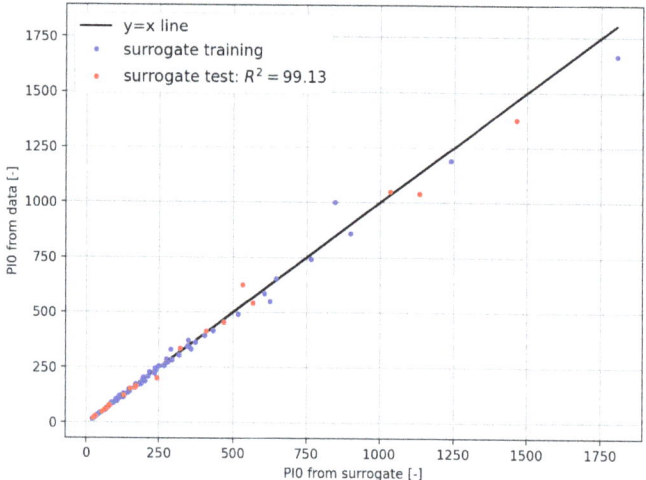

Figure 19. Surrogate model for single actuator housing: π_0 results (Equation (33)).

7. Results

In this section, some real application cases are discussed to illustrate the previously presented methodologies.

The methodology mentioned in Section 2 is implemented as a graphical user interface (GUI) web application based on a Jupyter Notebook calling functions through Python scripts. The GUI is developed with a dashboard named Voila. The tools developed in Section 4 are included with an optimization algorithm (differential evolution). Different specifications from real application cases are executed into this sizing code and the output results are presented. These specifications are arbitrary or linked to redundant topologies of actuation and to two different helicopter use cases: the main rotor and the tail rotor.

The considered actuator architecture is provided in Figure 20. In a housing (H), a frameless PMSM electrical motor (EM) is guided by two bearings (BB1 and BB2). BB1 is an angular contact double-row ball bearing; it withstands the entire axial load. BB2 is a radial contact single-row ball bearing; it ensures the motor rotor alignment. The electrical motor (EM) is linked to a screw mechanism (SM). In this paper, a planetary roller-screw (PRS) technology is considered. The screw spins and the nut moves linearly. The output rod (OR) is fixed onto the nut of the SM. A rod end (RE) (also named a spherical bearing) ensures the connexion of the output rod with the loading device. In this paper, the loading device is the helicopter swashplate. On the other side, another rod end (RE) ensures the connexion of the actuator housing (H) with a frame. Finally, the electromagnetic brake (EMB) satisfies a safety function in the case of a electricity supply cut-off.

For confidentiality matters, the inputs and results are presented as ratios of quantities. The observation of the sizing evolutions is the focus point of this section.

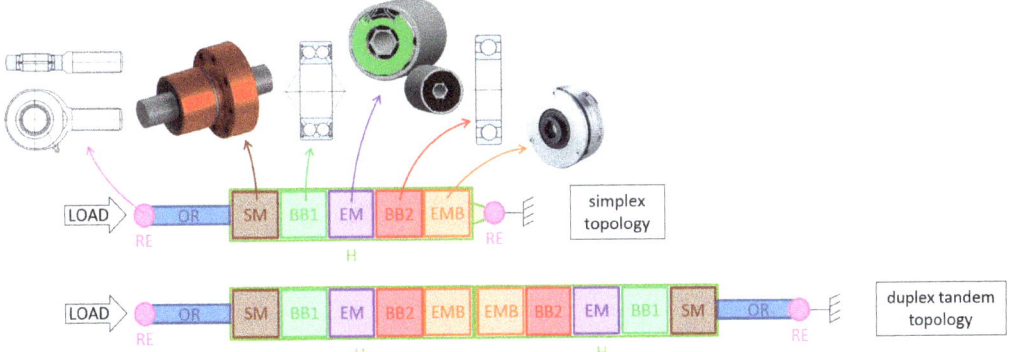

Figure 20. Actuator architecture considered in results.

7.1. Arbitrary Sizings

We suggest presenting some preliminary sizing results from an arbitrary specification. The arbitrary specification and the associated design hypothesis are given in Table 5. The actuator sizing results regarding the simplex topology (Figure 20) are available in Table 6, in the column named *sizing A*. To show the sensitivity of the values of the static load, stroke, and acceleration (RMS), the arbitrary specification is run again three times with, every time, a modification of one indicator value. *Sizing B* includes the arbitrary static load multiplied by 2. *Sizing C* includes the arbitrary stroke multiplied by 4. *Sizing D* includes the arbitrary RMS acceleration multiplied by 2.

The following lines comment the sizing results presented in Table 6. Although the mechanical components are not the focus point of this article, they must be mentioned to better understand the sizing choices performed by the optimization.

The arbitrary specification leads a design of the mechanical components mainly guided by the fatigue criteria (*sizing A*). This means that the static load criteria are satisfied with margins. Moreover, because of the low heat-transfer coefficient, the low emissivity, and the significant RMS acceleration level (see Table 5), the electrical motor design is driven by the heat dissipated through the actuator skin, by convection and radiation.

Table 5. Arbitrary specification and its design hypothesis.

Specification	Unit	Value	Design Hypothesis	Unit	Value
stroke s_{max}	mm	50	all safety coefficients	-	1
equivalent distance traveled L_{eq}	km	100	skin temperature max	°C	100
speed max v_{max}	m·s^{-1}	0.2	housing heat-transfer coef. (convection)	W·m^{-2}·K^{-1}	5
speed iron v_{iron}	m·s^{-1}	0.1	housing emissivity	-	0.4
acceleration max a_{max}	m·s^{-2}	5	housing & output rod density	kg·m^{-3}	7800
acceleration rms a_{rms}	m·s^{-1}	1	housing & output rod thickness min	mm	1
load max F_{max}	kN	3	housing & output rod fatigue stress	MPa	500
load rms F_{rms}	kN	1	quality coefficient Q_m	-	30
load rmc F_{rmc}	kN	1	vibratory acceleration	m·s^{-2}	98
load dynamic peak-to-peak $F_{pitting}$	kN	1	bus voltage max	Vdc	110
power rate mean $\overline{PR}$	W·s^{-1}	1	motor phase current max	Apeak-sine	10
pair (a_{PRmax}; F_{PRmax})	(m·s^{-2}; kN)	(3; 1.5)	time-to-stop speed (EMB)	s	0.05
equivalent load mass	kg	50	shaft density	kg·m^{-3}	7800
ambient temperature	°C	25			
load frequency	hz	20			
total lifespan t_{life}	hours	20,000			

Table 6. Sizing results from the arbitrary specification.

Component	Characteristic	Value of Characteristics			
		Sizing A	Sizing B (F_{max} X2)	Sizing C (s_{max} X4)	sizing D (a_{rms} X2)
MECHANICAL COMPONENTS					
screw mechanism (SM)	thread lead (mm/rev)	3.1	3.1	3.5	3.8
SM (nut, screw), BB1, BB2, and RE (x2)	total mass (kg)	0.72	0.72	0.91	0.65
ELECTRICAL COMPONENTS					
electrical motor (EM)	inertia (kg · m^2)	7.6×10^{-5}	1.1×10^{-4}	4.9×10^{-5}	1.1×10^{-4}
	external diameter (mm)	60	64	55	65
	motor constant K_m (10^{-2} (Nm)^2/W)	4.6	6.3	3.0	6.8
	torque constant K_t (Nm/Arms-sine)	0.23	0.26	0.24	0.26
	mass (kg)	1.1	1.4	0.87	1.4
electromagnetic brake (EMB)	inertia (kg · m^2)	1.0×10^{-5}	2.0×10^{-5}	1.0×10^{-5}	1.3×10^{-5}
	mass (kg)	0.72	1.1	0.73	0.83
ACTUATOR					
housing (H)	thickness (mm)	1.1	1.2	4.6	1.1
	mass (kg)	1.8	2.0	4.1	1.9
output rod (OR)	thickness (mm)	1.4	1.4	3.9	1.4
	mass (kg)	0.14	0.14	0.78	0.14
actuator (simplex)	total mass (kg)	4.6	5.5	7.5	5.1

In *sizing B*, the increase in the static load stands in the static load margin of the mechanical components; their sizing remains unchanged. However, the electromagnetic brake (EMB) must stop a higher torque. Indeed, the torque to stop in an emergency relies on two specified values: the static load converted by the indirect efficiency of the screw (SM) and the screw maximum speed to stop within the specified time. Doubling the static load, the EMB requires to develop a higher braking force. It is bigger then. This involves the increase in the EMB disk inertia that sums to the total rotating inertia. Thus, a higher electrical motor performance is required. Because of the increased length of the electrical components, the housing is heavier.

Increasing the specified stroke (*sizing C*) makes the actuator longer. The ratio diameter by length is decreased. The housing and the output rod must have their thickness increased to withstand the vibratory accelerations. The screw of the SM is longer, involving an additional mass. What is more, the increased actuator length offers a more extended outer surface for dissipating the heat generated by the motor. The electrical motor (EM) can have lower performances if the thread lead is increased. The EM has a reduced mass then.

The motor heat generation is based on a continuous torque. The RMS acceleration highly contributes to this torque. Doubling the specified RMS acceleration (*sizing D*) involves an important specified continuous torque. The motor size must be increased to satisfy this specification. Increasing the lead limits the motor size increase for a reduced mass. With a bigger motor, the rotating inertia is higher and the EMB is required to be bigger. Moreover, increasing the lead reduces the fatigue phenomenon applied to the mechanical components. The mechanical components are chosen with slightly smaller fatigue capabilities. They are slightly smaller then. Moreover, the sized electrical components make the actuator longer. Thus, the housing (H) is slightly heavier.

7.2. Sizing of Redundant Topologies of Actuation

There are mainly two redundant topologies of actuation [6,42,43]. The first one is the force summing where the failed or passive actuator shall be free in motion. In this case, the actuator must be equipped with a clutch or any breaking fuse system. The second one is the position summing where the failed or passive actuator shall be locked from motion. In this case, the actuators must be equipped with a power-off brake. The position summing topology can be seen with actuators either installed in tandem (as presented in Figure 21a) or in parallel (Figure 21b). The tandem configuration is the one most commonly found in aeronautics. This equips the PFCS of fighter aircrafts and helicopters and the SFCS of commercial aircrafts (except the spoiler) [6]. Meanwhile, the parallel configuration is much less used [6].

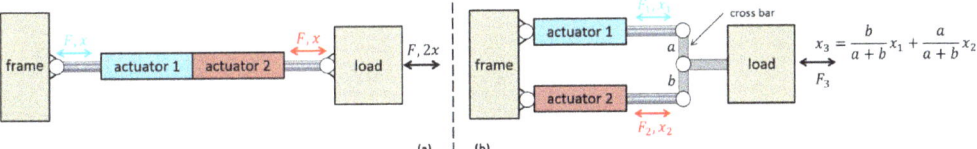

Figure 21. Redundant topolgies of actuator in position summing (passive/failed actuator shall be locked). (a) One tandem actuator, (b) two simplex actuators linked by a cross bar ($a = b$ considered, $F_1 = F_2 = F_3/2$).

This paper proposes to study the impact on the housing mass and output rod mass involved in the consideration of topology (b) compared to topology (a) (Figure 21). It is clear that topology (b) introduces potential additional plays and wear points compared to topology (a). However, it is important to estimate if this topology involves any mass gain that would compensate these drawbacks.

From tandem to parallel topologies, the force is halved; meanwhile, the stroke, speed, and acceleration are doubled. The work produced by the actuator remains the same.

The bar chart presented in Figure 22 presents the sizing results involved in both actuation topologies. The results are given as a ratio, with the sizing results obtained for a simplex actuator not equipped with any electromagnetic brake.

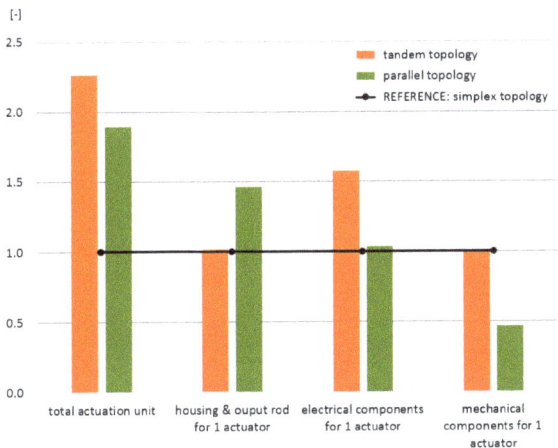

Figure 22. Mass evolution of tandem or parallel topologies compared to simplex topology.

As the parallel topology requires a doubled stroke and half load, the actuator is longer and smaller in diameter than in the tandem topology. The thicknesses of the housing and the output rod are nearly doubled. The mass of the set (housing and rod) results to be half heavier than the one of the tandem topology. The parallel topology shows electrical components and mechanical ones with reduced characteristics and reduced masses compared to the tandem topology. Therefore, the contribution of the set (housing and rod) mass on the total actuator mass is much higher for the parallel topology. Figure 23 confirms it with a contribution of a third of the actuator mass concerning the tandem topology against more than half the actuator mass concerning the parallel topology.

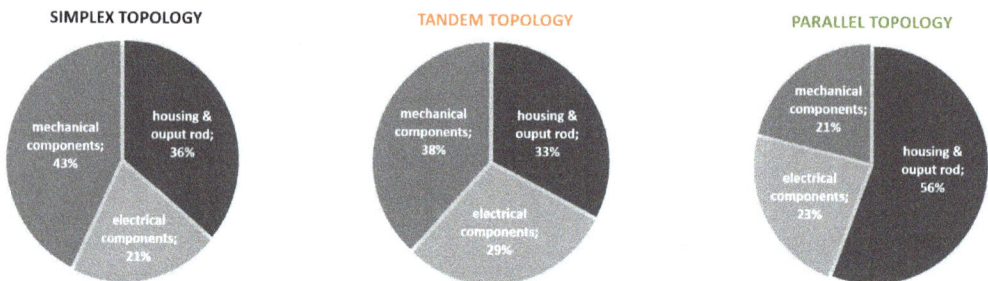

Figure 23. Component mass distribution for one actuator.

The parallel configuration seemed to be a relevant choice because of the load reduction. However, this finally penalizes the actuator because the load reduction involves a small actuator diameter. With the increase in the stroke, the diameter-to-length ratio is not interesting anymore. There is not any mass gain on the housing and output rod. The cross bar has to be considered in the mass statement too.

Finally, the total actuation mass gain involved by the parallel topology is not significant enough relatively to the potential drawbacks it introduces.

7.3. Sizing of Main and Tail Rotor Actuators

Section 3 showed that the TRA application was distinguished from the MRA in terms of dynamism, load, and especially in terms of induced motor losses. To illustrate this difference between the MRA and TRA applications (Table 1), their specifications are executed into the sizing code. The sized motor characteristics and an actuator mass distribution are displayed in Figures 24 and 25.

Figure 25 shows the important mass decrease among components induced by the TRA application. The TRA total mass shows to be 80% smaller than the actuator, satisfying both applications.

For confidentiality reasons, the results of the MRA and TRA sizing are normalized. The reference is taken as the actuator sizing which satisfies both applications at the same time.

Figure 24 globally shows that the selected motor for the TRA has reduced characteristics compared to the one selected for the MRA. The rotor inertia, which induces motor losses, is drastically lowered compared to the MRA-selected motor.

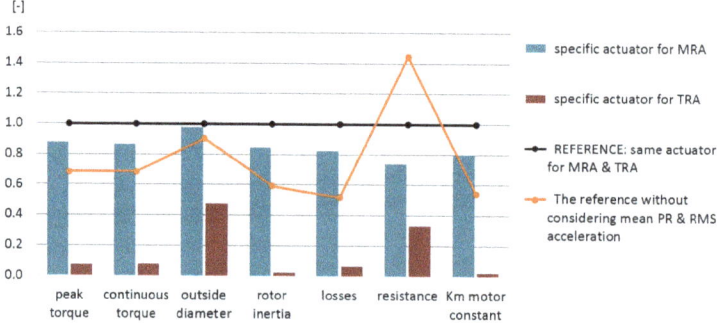

Figure 24. Motor characteristic evolution of specific actuators for the MRA and TRA applications compared to a unique actuator for both applications (reference).

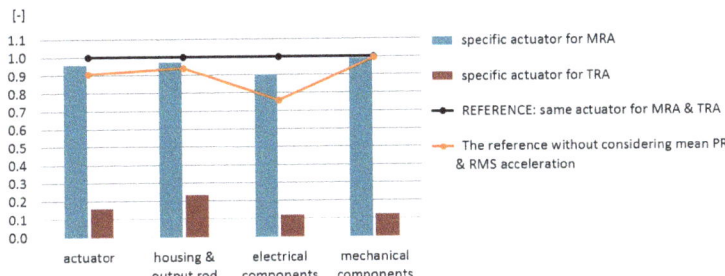

Figure 25. Mass distribution of specific actuators for the MRA and TRA applications compared to a unique actuator for both applications.

Section 3 suggests the mean power rate $\overline{PR}$ and the RMS acceleration a_{rms} as a way to consider the induced motor losses. In Figure 24, the orange line shows the sizing resulting from the reference actuator specification regardless of $\overline{PR}$ and a_{rms}. The losses result to be estimated at roughly 40% of what they are for the reference actuator, and the continuous torque is at roughly 70%. The performance difference is also seen through the actuator mass statement in Figure 25. It is clear that not considering the values $\overline{PR}$ and a_{rms} in the actuator specification involves a significant risk of undersizing the electrical motor of the actuator.

The MRA and TRA of the considered helicopter PFCS lead to two significantly different sizes. The MRA and TRA are two applications to be distinguished. Designing a specific actuator for each application will benefit the helicopter mass and its electrical network.

8. Discussion

This paper presents a design methodology supported by tools for the preliminary sizing of critical actuators. This methodology can be applied to any actuator architectures. This methodology finds an extension to the design of multirotor drones as presented in [44]. Moreover, it completes, at the component level, the methodology proposed by [45] at the vehicle level. The author of [45] developed a power system architecture sizing process for the preliminary design phase of civil aircraft.

The first tool presented is the one from [3]. It is a data-driven specification methodology which draws a parallel between measurement data on flight and EMA technologies using indicators to estimate over mission profiles.

Nowadays, the industrial context comes to a digital twin. Data content is globally exponentially increasing. As a result, there is a necessity to develop such data analysis methodologies based on synthetic values. What is more, the industrial trend is to reduce the number of helicopter flight tests to shrink development costs. Consequently, the methodology must extract as much added value as possible from any available data. The paper [3] presents a statistical approach getting interest in this area. From a reduced data set, the statistical laws are formulated. The laws are used then to express the load, speed, and acceleration domain limits of the application.

Furthermore, as seen in this paper, the specification methodology applied on main and tail rotor applications taught about the helicopter specificities. It showed a clear dynamism and load difference between the MRA and the TRA. The indicators quantified the importance of the motor losses on the TRA compared to the MRA. These losses are induced by the mean power rate $\overline{PR}$ and especially the equivalent continuous acceleration a_{rms}. The last part of the paper showed the risk of undersizing if a_{rms} and $\overline{PR}$ are not taken into account. In addition, it concludes that the helicopter mass could benefit from a specific sizing for the MRA and the TRA. The TRA shall be designed with a rotor inertia as low as possible.

The second tool presented is the scaling laws based on a dimensional analysis. The electrical motor is chosen as an applicative example. The scaling laws capture the main physical phenomena driving the component design and easily estimate the main component characteristics from a reduced number of parameters. As illustrated by the last part of this paper, the scaling laws are useful for the exploration study of a design domain: sizing with optimization, a scenario analysis, an integration study (mass and dimensions), and a technology or architecture comparison. In addition, scaling laws can be useful to replace component catalogs when they are unavailable or to complete missing contents. In the context of negotiation, scaling laws are easy enough to exchange with suppliers and to challenge them.

The third tool presented is a surrogate model or response surface model (RSM). It is set up using dimensionless numbers (or π-numbers). This has several advantages. First, it decreases the number of variables to be manipulated and therefore it drastically decreases the number of physical or numerical experiments to be carried out. Secondly, it increases the regression robustness [32], in particular if the RSM is built within the logarithmic space. Paper [17] mentions that the logarithmic shows good results in interpolation and in extrapolation because of the power-law form. This is the case for the VPLM methodology (Variable Power-Law Metamodel) [31] used in the surrogate model setup in this paper.

The complex vibratory problem of the actuator housing is suggested to be addressed by a surrogate model because it is adapted to preliminary studies. Indeed, it is a simple way of modeling with a reduced number of parameters. The suggested model easily provides design trends for architecture decision making. This is shown in the last part of this paper where the actuator topologies are compared. Obviously, the final actuator housing design will require laboratory experiments on a vibratory test bench.

Finally, as perspectives, the presented design methodology can be applied to any other actuator architectures. Thus, for a given application, it allows to study and select the best actuator architecture. For critical actuators, an analysis among different redundant topologies of actuation and a safety analysis with failure trees are important features to develop. In addition, a price assessment is a relevant feature to implement.

Author Contributions: Conceptualization, J.R., M.B. and L.R.; methodology, J.R., M.B. and L.R.; software, J.R. and M.B.; validation, J.R., M.B. and L.R.; formal analysis, J.R. and M.B.; investigation, J.R. and M.B.; resources, J.R. and M.B.; data curation, J.R. and M.B.; writing—original draft preparation, J.R.; writing—review and editing, J.R. and M.B.; visualization, J.R.; supervision, M.B. and L.R.; project administration, J.R.; funding acquisition, J.R. All authors have read and agreed to the published version of the manuscript.

Funding: This research was funded by Airbus Helicopters and the French Minister of Higher Education, Research and Innovation (MESRI) through the French National Association of Research and Technology (ANRT).

Institutional Review Board Statement: Not applicable.

Informed Consent Statement: Not applicable.

Data Availability Statement: Not applicable.

Acknowledgments: The authors would like to acknowledge Airbus and the ANRT who funded the PhD thesis that produced the work presented in this paper.

Conflicts of Interest: The authors declare no conflict of interest.

Nomenclature

Symbol	Unit	Name	Symbol	Unit	Name
		Section 3			
$a(t)$	$m \cdot s^{-2}$	acceleration on mission profiles	$F(t)$	N	load on mission profiles
s	m	stroke	$v(t)$	$m \cdot s^{-1}$	speed on mission profiles
PR	$N \cdot m \cdot s^{-2}$	power rate	$\overline{PR}$	$N \cdot m \cdot s^{-2}$	mean power rate
a_{PRmax}	$m \cdot s^{-2}$	acceleration at maximum power rate	F_{PRmax}	N	load at maximum power rate

Symbol	Unit	Name	Symbol	Unit	Name
		Section 5			
x^*	-	scaling ratio x/x_{ref}	π_i	-	π-number
x_{ref}	[x]	characteristic of component of reference used in scaling law			
d	m	dimension or diameter	p	[p]	material property
L	m	characteristic length	M	kg	mass
ρ	$kg \cdot m^{-3}$	density	V	m^3	volume
T	$m \cdot N$	torque	T_p	$m \cdot N$	peak torque
α	$m^{-1} \cdot N^{-1} \cdot W$	copper losses coefficient	β	$rad^{-1} \cdot s \cdot W$	iron losses coefficient
f_{elect}	$rad \cdot s^{-1}$	electrical speed	Ω	$rad \cdot s^{-1}$	angular speed
K_m	$m^2 \cdot N^2 \cdot W^{-1}$	motor constant	Q	W	heat
		Section 6			
x	m	vibratory displacement amplitude	f_r	hz	first resonance mode frequency
a	$m \cdot s^{-2}$	amplitude of sinusoidal vibratory acceleration			
ω	$rad \cdot s^{-1}$	angular frequency	L_a	m	actuator total length
e_1	m	housing thickness	e_2	m	output rod thickness
d_{rs}	m	roller-screw nut diameter	L_{rs}	m	roller-screw nut length
M_{eq}	kg	equivalent mass	K_{eq}	$N \cdot m^{-1}$	equivalent stiffness
C_{eq}	$N \cdot m^{-1} \cdot s$	equivalent damping	ζ	-	equivalent damping coefficient
x_0	[x]	value of x at first resonance mode	F	N	excitation load applied on M_{eq}
U	m	displacement of M_{eq}	u	m	actuator deflection
V	$m \cdot s^{-1}$	speed of M_{eq}	v	$m \cdot s^{-1}$	actuator deflection speed
σ	Pa	actuator housing stress	Q_m	-	actuator quality factor
E	Pa	Young's modulus of actuator material	ρ	$kg \cdot m^{-3}$	density of actuator material

Abbreviations

The following abbreviations are used in this manuscript:

BB1	Angular Contact Double-row Ball Bearing
BB2	Deep Groove Single-row Ball Bearing
DoE	Design of Experiment
DDV	Direct Drive Valve
EM	Electrical Motor (PMSM)
EMA	Electromechanical Actuator
EMB	Electromagnetic Brake
FCC	Flight Control Computer
FCL	Flight Control Links
H	Housing
IDF	Individual Disciplinary Feasible
KDD	Key Design Driver
MDO	Multidisciplinary Design Optimization
MRA	Main Rotor Actuator
OPV	Optionally Piloted Vehicle
OR	Output Rod
PFCS	Primary Flight Control System
RE	Rod End
RMC	Root Mean Cube
RMS	Root Mean Square
RSM	Response Surface Model
SFCS	Secondary Flight Control System
SM	Screw Mechanism
TRA	Tail Rotor Actuator
UAM	Urban Air Mobility
UAS	Unmanned Aerial System
UAV	Unmanned Aerial Vehicle
VPLM	Variable Power-Law Metamodel
VTOL	Vertical Take-off Landing

Appendix A. Surrogate Model: Housing of the Tandem Topology of Actuation

For the tandem topology of actuation and at the interface between output rod and housing, the regression gives the following surrogate model:

$$\begin{aligned}
log_{10}(\pi_0) = & 180 \cdot log_{10}\left(\frac{L}{d}\right)^2 \cdot \frac{1}{983} + 23 \cdot log_{10}\left(\frac{L}{d}\right) \cdot log_{10}\left(\frac{e_1}{d}\right) \cdot \frac{1}{243} + 43 \cdot log_{10}\left(\frac{L}{d}\right) \cdot log_{10}\left(\frac{e_2}{d}\right) \cdot \frac{1}{880} \\
& + 1187 \cdot log_{10}\left(\frac{L}{d}\right) \cdot \frac{1}{766} + 149 \cdot log_{10}\left(\frac{e_1}{d}\right)^2 \cdot \frac{1}{993} + 179 \cdot log_{10}\left(\frac{e_1}{d}\right) \cdot log_{10}\left(\frac{e_2}{d}\right) \cdot \frac{1}{712} \\
& - 175 \cdot log_{10}\left(\frac{e_1}{d}\right) \cdot \frac{1}{429} + 625 \cdot log_{10}\left(\frac{e_2}{d}\right)^2 \cdot \frac{1}{791} + 1229 \cdot log_{10}\left(\frac{e_2}{d}\right) \cdot \frac{1}{694} + \frac{922}{999}
\end{aligned} \quad (A1)$$

with L the length L_a and d the diameter d_{rs}.

For the tandem topology of actuation and at the interface between both actuators, the regression gives the following surrogate model:

$$\begin{aligned}
log_{10}(\pi_0) = & 96 \cdot log_{10}\left(\frac{L}{d}\right)^2 \cdot \frac{1}{899} + 25 \cdot log_{10}\left(\frac{L}{d}\right) \cdot log_{10}\left(\frac{e_1}{d}\right) \cdot \frac{1}{974} + 8 \cdot log_{10}\left(\frac{L}{d}\right) \cdot log_{10}\left(\frac{e_2}{d}\right) \cdot \frac{1}{223} \\
& + 1012 \cdot log_{10}\left(\frac{L}{d}\right) \cdot \frac{1}{585} - 8 \cdot log_{10}\left(\frac{e_1}{d}\right)^2 \cdot \frac{1}{819} - 43 \cdot log_{10}\left(\frac{e_1}{d}\right) \cdot log_{10}\left(\frac{e_2}{d}\right) \cdot \frac{1}{974} \\
& - 84 \cdot log_{10}\left(\frac{e_1}{d}\right) \cdot \frac{1}{811} + 2 \cdot log_{10}\left(\frac{e_2}{d}\right)^2 \cdot \frac{1}{15} - 259 \cdot log_{10}\left(\frac{e_2}{d}\right) \cdot \frac{1}{372} - \frac{175}{731}
\end{aligned} \quad (A2)$$

with L the length L_a and d the diameter d_{rs}.

References

1. Qiao, G.; Liu, G.; Shi, Z.; Wang, Y.; Ma, S.; Lim, T.C. A review of electromechanical actuators for More/All Electric aircraft systems. *Proc. Inst. Mech. Eng. Part J. Mech. Eng. Sci.* **2018**, *232*, 4128–4151. [CrossRef]
2. Cochoy, O.; Carl, U.B.; Thielecke, F. Integration and control of electromechanical and electrohydraulic actuators in a hybrid primary flight control architecture. In Proceedings of the International Conference on Recent Advances in Aerospace Actuation Systems and Components (R3ASC), Toulouse, France, 13–15 June 2007; INSA: Toulouse, France, 2007; pp. 1–8.
3. Roussel, J.; Budinger, M.; Ruet, L. Unmanned helicopter flight control actuator specification through mission profile analysis. In *Proceedings of the IOP Conference Series: Materials Science and Engineering*; IOP Publishing: Bristol, UK, 2022; Volume 1226; p. 012100. Available online: https://iopscience.iop.org/article/10.1088/1757-899X/1226/1/012100/meta (accessed on 18 August 2022).
4. Raletz, R. *Basic Theory of the Helicopter*; CEPADUES: Toulouse, France, 2010. ISBN 9782854289374.
5. FAA. *Helicopter Flying Handbook*; US Department of Transportation, 2019. Available online: https://www.faa.gov/sites/faa.gov/files/regulations_policies/handbooks_manuals/aviation/helicopter_flying_handbook/helicopter_flying_handbook.pdf (accessed on 30 July 2022).
6. Maré, J.C. *Aerospace Actuators 2: Signal-by-Wire and Power-by-Wire*; ISTE Group: London, UK, 2017; Volume 2.
7. RTCA. *Environmental Conditions and Test Procedures for Airborne Equipment, DO-160E, EUROCAE ED-14E*; RTCA: Washington, DC, USA, 2005.
8. Mazzoleni, M.; Di Rito, G.; Previdi, F. *Electro-Mechanical Actuators for the More Electric Aircraft*; Springer: Berlin/Heidelberg, Germany, 2021.
9. Liscouët, J.; Maré, J.C.; Budinger, M. An integrated methodology for the preliminary design of highly reliable electromechanical actuators: Search for architecture solutions. *Aerosp. Sci. Technol.* **2012**, *22*, 9–18. Available online: https://scholar.google.fr/citations?view_op=view_citation&hl=fr&user=nkGEGZgAAAAJ&citation_for_view=nkGEGZgAAAAJ:ufrVoPGSRksC (accessed on 30 July 2022). [CrossRef]
10. Andersson, J.; Krus, P.; Nilsson, K. Optimization as a support for selection and design of aircraft actuation systems. In Proceedings of the 7th AIAA/USAF/NASA/ISSMO Symposium on Multidisciplinary Analysis and Optimization, St. Louis, MO, USA, 2–4 September 1998; p. 4887.
11. Jiao, Z.; Yu, B.; Wu, S.; Shang, Y.; Huang, H.; Tang, Z.; Wei, R.; Li, C. An intelligent design method for actuation system architecture optimization for more electrical aircraft. *Aerosp. Sci. Technol.* **2019**, *93*, 105079. [CrossRef]
12. Roos, F. On Design Methods for Mechatronics: Servo Motor and Gearhead. Ph.D. Thesis, Royal Institute of Technology, Stockholm, Sweden, 2005.

13. Roos, F.; Johansson, H.; Wikander, J. Optimal selection of motor and gearhead in mechatronic applications. *Mechatronics* **2006**, *16*, 63–72. [CrossRef]
14. Pfennig, M.; Carl, U.B.; Thielecke, F. Recent advances towards an integrated and optimized design of high lift actuation systems. *SAE Int.l J. Aerosp.* **2010**, *3*, 55. [CrossRef]
15. Wu, S.; Yu, B.; Jiao, Z.; Shang, Y.; Luk, P. Preliminary design and multi-objective optimization of electro-hydrostatic actuator. *Proc. Inst. Mech. Eng. Part J. Aerosp. Eng.* **2017**, *231*, 1258–1268. [CrossRef]
16. Raymer, D. *Aircraft Design: A Conceptual Approach*; American Institute of Aeronautics and Astronautics, Inc.: Reston, VA, USA, 2012.
17. Budinger, M.; Reysset, A.; Ochotorena, A.; Delbecq, S. Scaling laws and similarity models for the preliminary design of multirotor drones. *Aerosp. Sci. Technol.* **2020**, *98*, 105658. [CrossRef]
18. Massachusetts, I.o.T. Thermodynamic and Propulsion—Aircraft Engine Performance— Performance of Propellers. Online MIT Resource. 2022. Available online: https://web.mit.edu/16.unified/www/FALL/thermodynamics/notes/node86.html (accessed on 30 July 2022).
19. McCormick, B.W.; Aljabri, A.S.; Jumper, S.J.; Martinovic, Z.N. The Analysis of Propellers Including Interaction Effects. *NASA Scientific and Technical Information Facility*. 1979. Available online: https://www.researchgate.net/publication/23913137_The_Analysis_of_Propellers_Including_Interaction_Effects:2022/07 (accessed on 30 July 2022).
20. Saerens, E.; Crispel, S.; Garcia, P.L.; Verstraten, T.; Ducastel, V.; Vanderborght, B.; Lefeber, D. Scaling laws for robotic transmissions. *Mech. Mach. Theory* **2019**, *140*, 601–621. [CrossRef]
21. Papalambros, P.Y.; Wilde, D.J. *Principles of Optimal Design: Modeling and Computation*, 2nd ed.; Cambridge University Press: Cambridge, UK; New York, NY, USA, 2000.
22. Martins, J.R.; Lambe, A.B. Multidisciplinary design optimization: A survey of architectures. *AIAA J.* **2013**, *51*, 2049–2075. [CrossRef]
23. Budinger, M.; Reysset, A.; Halabi, T.E.; Vasiliu, C.; Maré, J.C. Optimal preliminary design of electromechanical actuators. *Proc. Inst. Mech. Eng. Part G J. Aerosp. Eng.* **2014**, *228*, 1598–1616. [CrossRef]
24. Reysset, A.; Budinger, M.; Maré, J.C. Computer-aided definition of sizing procedures and optimization problems of mechatronic systems. *Concurr. Eng.* **2015**, *23*, 320–332. [CrossRef]
25. Delbecq, S.; Budinger, M.; Reysset, A. Benchmarking of monolithic MDO formulations and derivative computation techniques using OpenMDAO. *Struct. Multidiscip. Optim.* **2020**, *62*, 645–666. [CrossRef]
26. Jaafar, A. Analysis of Mission Profiles and Environment Variables for Integration into the Systemic Design Processus. Ph.D. Thesis, INP, Toulouse, France, 2011.
27. Reysset, A. Preliminary Design of Electromechanical Actuators—Tools to Support the Specification and the Generation of Sizing Procedures for the Optimization. Ph.D. Thesis, INSA, Toulouse, France, 2015.
28. Hehenberger, P.; Poltschak, F.; Zeman, K.; Amrhein, W. Hierarchical design models in the mechatronic product development process of synchronous machines. *Mechatronics* **2010**, *20*, 864–875. [CrossRef]
29. Van Groesen, E.; Molenaar, J. *Continuum Modeling in the Physical Sciences*; SIAM: Bangkok, Thailand, 2007. Volume 13; pp. 1–29.
30. Holmes, M.H. Dimensional Analysis (chapter). In *Introduction to the Foundations of Applied Mathematics*; Springer: Berlin/Heidelberg, Germany, 2019; pp. 1–47.
31. Sanchez, F.; Budinger, M.; Hazyuk, I. Dimensional analysis and surrogate models for the thermal modeling of Multiphysics systems. *Appl. Therm. Eng.* **2017**, *110*, 758–771. [CrossRef]
32. Lacey, D.; Steele, C. The use of dimensional analysis to augment design of experiments for optimization and robustification. *J. Eng. Des.* **2006**, *17*, 55–73. [CrossRef]
33. Budinger, M.; Liscouët, J.; Hospital, F.; Maré, J. Estimation models for the preliminary design of electromechanical actuators. *Proc. Inst. Mech. Eng. Part J. Aerosp. Eng.* **2012**, *226*, 243–259. [CrossRef]
34. Jufer, M. Design and losses-scaling law approach. In Proceedings of the Nordic Research Symposium Energy Efficient Electric Motors and Drives, Skagen, Denmark, January, 1996; pp. 21–25. Available online: https://www.researchgate.net/profile/Marcel-Jufer/publication/288811698_Design_and_Losses-Scaling_Law_Approach/links/6146d76e3c6cb310697a4154/Design-and-Losses-Scaling-Law-Approach.pdf (accessed on 18 August 2022).
35. PARKER. Frameless Low Cogging Servo Motors—NK Series. 2022. Available online: https://ph.parker.com/us/en/frameless-low-cogging-servo-motors-nk-series/nk630esrr1000 (accessed on 30 July 2022).
36. KOLLMORGEN. RBE Series Brushless Motors for Frameless DDR (Direct Drive Rotary) Motor Applications, 2003. Available online: http://www.kollmorgen.com/ (accessed on 30 July 2022).
37. Grellet, G. *TI Technics of the Engineer—d3450v1—Losses in the Spinning Machines*; Technics of the Engineer Editions: Saint-Denis, France, 1989.
38. Budinger, M.; Reysset, A.; Maré, J.C. Preliminary design of aerospace linear actuator housings. *Aircr. Eng. Aerosp. Technol. Int. J.* **2015**, *87*, 224–237.
39. EASA. *Certification Specifications and Acceptable Means of Compliance for Large Aeroplanes CS-25 Amendment 4*; EASA: Cologne, Germany, 2007.
40. Nicolas, M. Continuum and Discrete Mechanics. Available online: https://cel.archives-ouvertes.fr/cel-00612360v1/document. (accessed on 30 July 2022).

41. Spencer, A.J.M. *Continuum Mechanics*; Courier Corporation, North Chelmsford, MA, USA, 2004; ISBN 978-0486435947.
42. Naubert, A.; Bachmann, M.; Binz, H.; Christmann, M.; Perni, F.; Toro, S. Disconnect device design options for jam-tolerant EMAs. In Proceedings of the International Conference on Recent Advances in Aerospace Actuation Systems and Components (R3ASC), Toulouse, France, 16–17 March 2016; INSA Toulouse: Toulouse, France, 2016; pp. 187–192.
43. Seemann, S.; Christmann, M.; Jänker, P.J. Control and monitoring concept for a fault-tolerant electromechanical actuation system, EADS Innovation Work. In Proceedings of the International Conference on Recent Advances in Aerospace Actuation Systems and Components (R3ASC), Toulouse, France, 13–14 June 2012; INSA Toulouse: Toulouse, France, 2012; pp. 39–43.
44. Delbecq, S.; Budinger, M.; Ochotorena, A.; Reysset, A.; Defaÿ, F. Efficient sizing and optimization of multirotor drones based on scaling laws and similarity models. *Aerosp. Sci. Technol.* **2020**, *102*, 105873. [CrossRef]
45. Liscouet-Hanke, S. A Model-Based Methodology for Integrated Preliminary Sizing and Analysis of Aircraft Power System Architectures. Ph.D. Thesis, Institut National des Sciences Appliquées de Toulouse, Toulouse, France, 2008.

Article

Incremental Nonlinear Dynamic Inversion Attitude Control for Helicopter with Actuator Delay and Saturation

Shaojie Zhang *, Han Zhang and Kun Ji

College of Automation Engineering, Nanjing University of Aeronautics and Astronautics, Nanjing 210016, China
* Correspondence: zhangsj@nuaa.edu.cn

Abstract: In this paper, an incremental nonlinear dynamic inversion (INDI) control scheme is proposed for the attitude tracking of a helicopter with model uncertainties, and actuator delay and saturation constraints. A finite integral compensation based on model reduction is used to compensate the actuator delay, and the proposed scheme can guarantee the semi-globally uniformly ultimately bounded tracking. The overall attitude controller is separated into a rate, an attitude, and a collective pitch controller. The rate and collective pitch controllers combine the proposed method and INDI to enhance the robustness to actuator delay and model uncertainties. Considering the dynamic of physical actuators, pseudo-control hedging (PCH) is introduced both in the rate and attitude controller to improve tracking performance. By using the proposed controller, the helicopter shows good dynamics under the multiple restrictions of the actuators.

Keywords: incremental nonlinear dynamic inversion (INDI); actuator compensation; model reduction; pseudo-control hedging (PCH); helicopter attitude control

1. Introduction

The helicopter, aircraft with one or more power-driven horizontal propellers or rotors that enable it to take off and land vertically, to move in any direction, or to remain stationary in the air, has become very popular for a wide range of services, including air–sea rescue, firefighting, traffic control, oil platform resupply, and business transportation [1]. However, these tasks often bring heavy workloads to pilots, especially in situations of high crosswinds or low light. Furthermore, subject to a complicated dynamic response, multiple flight modes, system uncertainties, and rapidly varying flight conditions, the helicopter is a highly complex system. For the reasons above, a highly reliable and effective flight control system which allows the helicopter to execute multiple tasks in adverse flying conditions becomes more in demand [2].

In the past few years, various fight control methodologies are developed for the helicopter [3–6]. Feedback linearization, as the most widely used nonlinear control method in the aircraft control systems, is often combined with adaptive control to deal with the model uncertainties [7–10]. However, it is hard to guarantee that the control system can recover from a failure in adaptation [11]. Therefore, whether it can be applied in the systems with high security requirements is worthy of consideration. In order to overcome the shortcomings above, the INDI technique is adopted in this paper for helicopter flight control.

By producing the incremental form of the control command by calculating the error between the virtual control law and the acceleration of the system state, INDI is robust to model uncertainties [12–19]. In [17], the stability and robustness of the INDI technique has been proven. The INDI scheme was first used in the design of a six-degree-of-freedom helicopter controller in [18], and its robustness to model uncertainties was verified by simulation. Ref. [19] uses the INDI technique to redesign the existing Apache flight control and improve the handling quality.

However, the weakness of the INDI controller is the accuracy of the onboard measurement and actuator delay. The current measurement technology combined with data

processing algorithms (such as Kalman filtering) has been able to reduce the uncertainties brought by these sensors. However, there is no effective solution to the poor performance caused by the existence of actuator delay. According to [1], the delay time is approximately 100 ms in the helicopter. In a control system operating at 100 Hz, there will be a difference of about ten samples between the command signal and the actuator, which seriously deteriorates the tracking performance of the system and even puts the system stability in risk. Therefore, various control approaches are proposed for the systems subject to actuator delay [20–27]. Ref. [25] extends the Artstein model reduction method in [26] to nonlinear systems, and designs a compensation control law for known and unknown systems, respectively. Then, the Lyapunov–Krasovskii functional is adopted to prove the stability of them. Ref. [27] designs a feedback robust tracking controller with delay compensation for a class of systems with actuator delay and external disturbance, and achieved the desired effect. In [28,29], Rohollah Moghadam proposed an ADP-based solution to the optimal adaptive adjustment problem of systems with state delay and input delay, which can be applied to the optimal control problem of a class of nonlinear time-delay systems.

Besides the constraint of time delay, actuator saturation is very common and a more general problem in the helicopter due to the limitation of actuators in terms of the position and rate saturation. Therefore, many recent works have been carried out on actuator saturation nonlinearity as it causes the windup phenomenon [30–35]. Based on nonlinear partial differential inequalities, an optimal saturation compensator was developed in [32]. In [11], the pseudo-control hedging method is proposed to offset the virtual control input when the actuators are saturated. Ref. [33] further expanded the PCH theory and [34] applied the method to the development of the Boeing 747 flight control system.

In this paper, combining with model reduction and the PCH technique, a novel INDI-based controller is proposed for helicopter attitude control with actuator delay and saturation. The main contributions of this paper are the following: (1) a novel INDI-based actuator delay compensation control scheme with guaranteed stability is proposed, which can be applied to nonlinear systems with actuator delay in a certain range; and (2) aiming at helicopter characteristics, the proposed method and PCH are adopted to design the controller for the helicopter which is subjected to model uncertainty, actuator delay, and saturation.

The overall structure of this paper is as follows: The problem is formulated in Section 2. Section 3 presents an actuator delay compensation scheme for the INDI controller. The introducing of the helicopter model and the design of the anti-windup helicopter attitude controller by the proposed method is given in Section 4. Section 5 focuses on the display of the simulation results and related explanations. The conclusions are presented in Section 6.

2. Problem Formulation

Consider a class of the nonlinear system with input delay in the following form:

$$\begin{cases} \dot{x} = f(x) + g(x, u_\tau) \\ y = x \end{cases} \quad (1)$$

where x is the state vector in $\mathbb{R}^n$. $f(\cdot)$ and $g(\cdot)$ are nonsingular, bounded continuous smooth nonlinear functions in $\mathbb{R}^n$. The delayed input vector $u_\tau = u(t - \tau)$ and output vector y are both functions in $\mathbb{R}^n$, where $\tau \in \mathbb{R}^+$ is the time-delay constant. The following assumptions are used for further development.

Assumption 1. *The desired trajectory vector $x_d(t)$ is bounded and has bounded time derivatives up to i-th for $i = 1, 2, 3$; that is, there exists η_i such that $\left\|x_d^{(i)}(t)\right\| \leq \eta_i$ for all t, where $\|\cdot\|$ represents the standard Euclidean norm.*

Assumption 2. *Both state x and the delayed input u_τ are measurable. Furthermore, the first derivative of x in the last sampling time, denoted by $\dot{x}_0$, can also be measured by acceleration sensors.*

Assumption 3. *If $x(t)$, $\dot{x}(t)$, and $u(t)$ are bounded, then $f(x)$ and $g(x,u)$ are bounded. Moreover, the first partial derivatives of $f(x)$ and $g(x,u)$ with respect to their arguments x and u exist and are bounded. The infinity norm of the partial derivative of $g(x,u)$ to u around (x_0, u_0) and its inverse can be upper-bounded as*

$$\left\| \frac{\partial g(x,u)}{\partial u} \bigg|_{x_0, u_0} \right\|_\infty \leq \bar{\zeta}_1, \quad \left\| \left(\frac{\partial g(x,u)}{\partial u} \right)^{-1} \bigg|_{x_0, u_0} \right\|_\infty \leq \bar{\zeta}_2 \tag{2}$$

where $\bar{\zeta}_1, \bar{\zeta}_2 \in \mathbb{R}^+$ are known constants.

3. Control Law Design

Before the development of the control law, we define some variables for subsequent analysis. We denote the tracking error as

$$e = x_d(t) - x(t) \tag{3}$$

Then, we give the virtual control from the tracking error, denoted by

$$\nu = \dot{x}_d + k_v e \tag{4}$$

where $k_v \in \mathbb{R}^{n \times n}$ denotes a positive gain matrix.

To facilitate the subsequent analysis, an auxiliary tracking error which is inspired from the model reduction method is defined as

$$r = \dot{e} + k_r e + g(x, u(t-\tau)) - g(x, u(t)) \tag{5}$$

where $k_r \in \mathbb{R}^{n \times n}$ is a known positive gain matrix.

Submitting the expressions in (1) and (3), the transformed open-loop tracking error system can be represented in an input delay free form as

$$r = \dot{x}_d + k_r e - f(x) - g(x, u) \tag{6}$$

We rewrite (6) as its partial first-order Taylor series expansion around the current solution of the system, denoted by (x_0, u_0):

$$r = \dot{x}_d + k_r e - f(x_0) - g(x_0, u_0) - \frac{\partial}{\partial x}[f(x) + g(x,u)]\bigg|_{x_0, u_0}(x - x_0) \\ - \frac{\partial}{\partial u}[f(x) + g(x,u)]\bigg|_{x_0, u_0}(u - u_0) \tag{7}$$

where x_0 and u_0 denote the values in the last control step of x and u, respectively. Based on the assumption on INDI, the variation $x - x_0$ can be neglected for each small time increment. Then, (7) can be simplified by

$$r = \dot{x}_d + k_r e - f(x_0) - g(x_0, u_0) - G(x_0, u_0)(u - u_0) \tag{8}$$

where $G(x_0, u_0) = \frac{\partial g(x,u)}{\partial u}\bigg|_{x_0, u_0}$. Based on (8) and the INDI control law, the control law $u(t)$ can be given by

$$u = u_0 + G(x_0, u_0)^{-1}\left[(\nu - \dot{x}_0) + k_u \int_{t_0}^t [\dot{e} + k_r e + g(x, u(\theta - \tau)) - g(x, u(\theta))]d\theta\right] \tag{9}$$

where $k_u \in \mathbb{R}^{n \times n}$ is a positive control gain matrix. The control law in (9) can be thought of as a combination of an INDI-based term through an online state measurement and a predictor term which can stabilize the system and compensate the input delay. Note that the control law in (9) does not directly depend on $f(x)$ anymore, which means the controller is robust to the uncertainty of the model that only depends on the states of the system. However, the uncertainty in the control effectiveness matrix ΔG should meet $\|\Delta G\| \leq 0.5\|G\|$ [17]. Compared with the model-based feedforward control method in [27], the proposed controller achieves a good control effect even when the model is uncertain. After applying the Expression (4) and (9), (8) can be expressed as

$$r = (k_r - k_v)e + g(x_0, u_{\tau_0}) - g(x_0, u_0) - k_u \int_{t_0}^{t} [\dot{e} + k_r e + g(x, u(\theta - \tau)) - g(x, u(\theta))] d\theta \tag{10}$$

where $g(x_0, u_{\tau_0})$ is the control effectiveness function under the last state vector x_0 and the last delayed input vector u_{τ_0}. Then, the time derivative of (10) can be obtained by

$$\dot{r} = (k_r - k_v)\dot{e} - k_u r \tag{11}$$

In addition, we can also get the time derivative of (9) by using (8):

$$\dot{u} = G(x_0, u_0)^{-1}(\dot{v} + k_u r) \tag{12}$$

Recall the Assumption 3: one thing that can be determined is that the discrepancy between the $g(x, u)$ and $g(x, u_\tau)$ upper bound is

$$g(x, u) - g(x, u_\tau) \leq \rho(\|e_u\|)\|e_u\| \tag{13}$$

where $e_u \in \mathbb{R}^n$ is defined as

$$e_u = u - u_\tau = \int_{t-\tau}^{t} \dot{u}(\theta) d\theta \tag{14}$$

In addition, the bounding function $\rho(e_u) \in \mathbb{R}$ is a known positive globally invertible nondecreasing function.

Theorem 1. *The control law given in (9) can ensure the semi-globally uniformly ultimately bounded (SUUB) tracking in the case that*

$$0 \leq \tau \leq \frac{\chi}{2\bar{\zeta}_2^2} \tag{15}$$

where χ is subsequently defined control gains, provided the control gains k_v, k_r, k_u, ω are selected with the following sufficient conditions:

$$\begin{cases} k_r > \frac{\chi^2 + 2}{4} \\ k_v = k_r \\ \frac{1-\delta}{2\omega\tau\bar{\zeta}_2^2} < k_u < \frac{1+\delta}{2\omega\tau\bar{\zeta}_2^2} \\ \tau \leq \omega \leq \frac{\chi^2}{4\tau\bar{\zeta}_2^2} \end{cases} \tag{16}$$

where $\delta = \sqrt{1 - \frac{4}{\chi^2}\omega\tau\bar{\zeta}_2^2}$ and ω are subsequently defined constants.

Proof. Define $y_a \in \mathbb{R}^{2n+1}$ as

$$y_a = \begin{bmatrix} e^T & r^T & \sqrt{Q} \end{bmatrix}^T \tag{17}$$

where the positive definite LK functional Q is defined by

$$Q = \omega \int_{t-\tau}^{t} \left(\int_{s}^{t} \|\dot{u}(\gamma)\|^2 d\gamma \right) ds \tag{18}$$

where $\omega \in \mathbb{R}$ is a positive constant. Then, the positive definite Lyapunov functional V is defined as

$$V(y_a) = \frac{1}{2}e^T e + \frac{1}{2}r^T r + Q \tag{19}$$

satisfying the following inequalities

$$c_1\|y_a\|^2 \leq V \leq c_2\|y_a\|^2 \tag{20}$$

where $c_1, c_2 \in \mathbb{R}^+$ are known constants.

After submitting the Equations (5) and (11), we have the time derivative of (19) as

$$\begin{aligned}\dot{V} &= e^T r - k_r e^T e + e^T(g(x,u) - g(x,u_\tau)) \\ &+ (k_r k_v - k_r^2) r^T e + (k_r - k_v) r^T (g(x,u) - g(x,u_\tau)) \\ &- (k_v - k_r) r^T r - k_u r^T r + \omega\tau\|\dot{u}\|^2 - \omega \int_{t-\tau}^{t} \|\dot{u}(\gamma)\| d\gamma\end{aligned} \tag{21}$$

Combining (13) and canceling common terms, we can upper-bound (21) in the case of $k_r = k_v$ as

$$\dot{V} \leq \|e\|\|r\| - k_r\|e\|^2 - k_u\|r\|^2 + \omega\tau\|\dot{u}\|^2 + \rho(\|e_u\|)\|e\|\|e_u\| - \omega\int_{t-\tau}^{t}\|\dot{u}(\gamma)\|^2 d\gamma \tag{22}$$

According to Young's inequality, the following relation can be obtained

$$\|e\|\|r\| \leq \frac{\chi^2}{4}\|e\|^2 + \frac{1}{\chi^2}\|r\|^2 \tag{23}$$

$$\rho(\|e_u\|)\|e\|\|e_u\| \leq \frac{1}{2}\|e\|^2 + \frac{\rho^2(\|e_u\|)}{2}\|e_u\|^2 \tag{24}$$

where $\chi \in R^+$ is a known constant. Moreover, using the Cauchy–Schwarz inequality, the terms in (24) can be further upper-bounded as

$$\|e_u\| \leq \tau \int_{t-\tau}^{t}\|\dot{u}(\gamma)\|^2 d\gamma \tag{25}$$

By adding and subtracting $\tau\int_{t-\tau}^{t}\|\dot{u}(\gamma)\|^2 d\gamma$, inequality (22) can be upper-bounded as

$$\begin{aligned}\dot{V} &\leq -(k_r - \tfrac{\chi^2}{4} - \tfrac{1}{2})\|e\|^2 + \omega\tau\|\dot{u}\|^2 - \left(k_u - \tfrac{1}{\chi^2}\right)\|r\|^2 \\ &- \tau\int_{t-\tau}^{t}\|\dot{u}(\gamma)\|^2 d\gamma - \left(\tfrac{\omega}{\tau} - \tfrac{1}{2}\rho^2(\|e_u\|) - 1\right)\|e_u\|^2\end{aligned} \tag{26}$$

Recall the Equation (12) and Assumption 3: there exists a positive constant $\lambda \in \mathbb{R}$ such that

$$\|\dot{u}\|^2 \leq \lambda + \xi_2^2 k_u^2 \|r\|^2 \tag{27}$$

Then, Equation (26) can be rewritten as

$$\dot{V} \leq -\kappa\|y_a\|^2 - \left(\frac{\omega}{\tau} - \frac{1}{2}\rho^2(\|e_u\|) - 1\right)\|e_u\|^2 + \omega\tau\lambda \tag{28}$$

where the function $\kappa \in \mathbb{R}^+$ in (28) is defined as

$$\kappa = \min\left[\left(k_r - \frac{\chi^2}{4} - \frac{1}{2}\right), \left(k_u - \frac{1}{\chi^2} - \omega\tau\xi_2^2 k_u^2\right), \left(\frac{1}{\omega}\right)\right] \tag{29}$$

where the inequation

$$\begin{aligned}Q &\leq \omega\tau \sup_{s\in[t,t-\tau]}\left[\int_s^t\|\dot{u}(\gamma)\|^2 d\gamma\right] \\ &= \omega\tau\int_{t-\tau}^{t}\|\dot{u}(\gamma)\|^2 d\gamma\end{aligned} \tag{30}$$

is used. Substituting the bound given in (20), the inequality in (28) can be further upper-bounded as

$$\dot{V} \leq -\frac{\kappa}{c_2}V + \omega\tau\lambda \tag{31}$$

in the set S defined by

$$S = \left\{ e_u(t) | \|e_u\| < \rho^{-1}\left(\sqrt{\frac{2\omega}{\tau}} - 2\right) \right\} \quad (32)$$

Hence, Expression (31) can lead to the solution as

$$V \leq V(0)e^{-\frac{\kappa}{c_2}t} + \omega\tau\lambda\frac{c_2}{\varphi}(1 - e^{-\frac{\kappa}{c_2}t}) \quad (33)$$

In the case of $e_u(0) = 0 \in S$ and according to the definition of V in (19) and the solution in (19), it can be concluded that e, r are bounded. From Assumption 1 and the bounded r, we can infer that the variable $e_u = \int_{t-\tau}^{t} \dot{u}(\theta)d\theta = G(x_0, u_0)^{-1}\int_{t-\tau}^{t} (\dot{x}_d + k_v e - \ddot{x}_0 + k_u r)d\theta$ is bounded. Then, using the definition of r in (5) and combining the bounded tracking error e and Assumption 1, we can infer that both $x(t)$ and $\dot{x}(t)$ are also bounded. Finally, we can obtain from Equation (9) that u is bounded with the initial condition $u(0) = 0$. Therefore, all of variables in the closed loop is guaranteed to be bounded by the proposed control law. □

4. Attitude Controller Design for Helicopter
4.1. Helicopter Model

The attitude control of the Messerschmitt–Bölkow–Blohm (MBB) Bö-105 helicopter is considered in this paper. Here, we give the attitude model in the equations of motion from

$$\dot{\omega}_h = \begin{bmatrix} \dot{p} \\ \dot{q} \\ \dot{r} \end{bmatrix} = J^{-1}(m - \omega_h \times J\omega_h) \quad (34)$$

$$\dot{\theta} = \begin{bmatrix} \dot{\phi} \\ \dot{\theta} \\ \dot{\psi} \end{bmatrix} = \Omega\omega_h = \begin{bmatrix} 1 & \sin\phi\tan\theta & \cos\phi\tan\theta \\ 0 & \cos\phi & -\sin\phi \\ 0 & \sin\phi/\cos\theta & \cos\phi/\cos\theta \end{bmatrix}\omega_h \quad (35)$$

where $m = m_{fus}(x) + m_{ht}(x) + m_{vt}(x) + m_{mr}(x, u) + m_{tr}(x, u)$ represents the total moments with respect to the gravity center of the helicopter. It consists of the moments produced by the fuselage, horizontal tail, vertical tail, main rotor, and tail rotor, which are represented by the subscript fus, ht, vt, mr, and tr, respectively. $\omega_h = \begin{bmatrix} p & q & r \end{bmatrix}^T$ and $\theta = \begin{bmatrix} \phi & \theta & \psi \end{bmatrix}^T$ indicate the roll, pitch and yaw angular rate and attitude angle of helicopter, respectively. J is the inertia matrix of the helicopter which is given by

$$J = \begin{bmatrix} 1433 & 0 & -660 \\ 0 & 4973 & 0 \\ -660 & 0 & 4099 \end{bmatrix} \quad (36)$$

The controller proposed in this paper is based on the principle of time-scale separation, which assumes that the state variables in the inner loop are preforming fast while the related parameters in the outer loop change more slowly under the same actuator inputs. Table 1 can verify that this hypothesis is reasonable.

Table 1. Time-scale separation between attitude angles and rates.

Different Control Channel	Maximum Rates of Attitude Angles	Maximum Rates of Attitude Rates
Lateral cyclic pitch	$\frac{d\phi}{dt} = 5.7$ deg/s	$\frac{dp}{dt} = 145.8$ deg/s^2
Longitudinal cyclic pitch	$\frac{d\theta}{dt} = -2.5$ deg/s	$\frac{dq}{dt} = -60.1$ deg/s^2
Collective of the tail rotor	$\frac{d\psi}{dt} = -0.7$ deg/s	$\frac{dr}{dt} = -16.5$ deg/s^2

From Table 1 we can see that there exists a time-scale separation between the time derivative of attitude angles and rates. Therefore, we divide these six state variables into two loops for the controller design, namely, the rate loop and the attitude loop. This type of assumption is often carried out for flight dynamics and control applications. Between two loops, the parameters associated with the slow dynamics are treated as constants by the fast dynamics and its dynamic inversion is performed assuming that the states controlled in the inner loop achieve their commanded values instantaneously. The fast variables are thus used as control inputs to the slow dynamics.

However, what needs to be considered is the dynamic of the actuators. In fact, there exists a time delay between the controller delivering the signal to the actuators. Moreover, actuators are often limited by their moving rate, which is shown in Table 2. If these issues are not considered, the tracking performance of the control system will be severely degraded and even face stability problems. To overcome these problems, the proposed method and PCH are adopted in the next sections.

Table 2. Actuators' rate limits of Bö-105.

Actuator Name	Variable Name	Maximum Rate Limit
Main rotor	θ_0	16.0 deg/s
Longitudinal cyclic	θ_{1s}	28.8 deg/s
Lateral cyclic	θ_{1c}	16.0 deg/s
Tail rotor	θ_{0tr}	32.0 deg/s

4.2. Anti-Windup Design

To overcome the effect caused by the actuator saturation, the pseudo-control hedging (PCH) scheme is adopted. The PCH solves the effect of actuator saturation by modifying the virtual control input instead of directly influencing the actuator input. When the input error signal of the control system is too radical for actuators, it allows the production of a signal v_s which is opposite to the virtual control law to the first-order reference model, so that it can prevent the system from still trying to track the commanded references when actuator saturation occurs.

In order to achieve the control hedging, a first-order reference model is introduced as Figure 1. I_{com} represents the command signal and I_{sat} is the filtered signal to ensure the input is within the acceptable range of the system. v_s is defined, corresponding to the input under the actuator dynamics. The maximum rate change allowed for this helicopter follows the ADS-33E-PRF standard, that is, the limits of 40 degrees per second on the roll and pitch rates and 80 degrees per second on the yaw rates. By using the reference model, we can obtain the time derivative of I_{com} such that the virtual control v in (4) can be made easily when no saturation occurs.

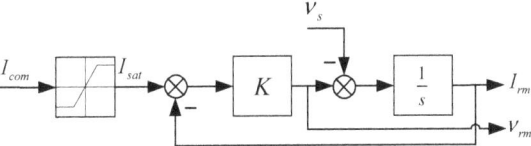

Figure 1. First-order reference model with saturation filter.

The PCH signal v_s in the affine nonlinear systems can be calculated by

$$\begin{aligned} v_s &= [\dot{x}_0 + G(x_0, u_0)(u_{des} - u_0)] - [\dot{x}_0 + G(x_0, u_0)(u - u_0)] \\ &= G(x_0, u_0)(u_{des} - u) \end{aligned} \quad (37)$$

where u_{des} represents the desired actuator input, which can be produced by the rate controller.

4.3. Rate Loop

For the rate loop, it is expected that the helicopter can track the input angular rate signal in real time, which requires an error e_{rat} to be defined between the reference signal and the system output, yielding

$$e_{rat} = \omega_{des} - y_{rat} \tag{38}$$

where ω_{des} is the reference command which can be produced by the attitude controller. The rate of angular change between the body frame to the North–East–Down (NED) co-ordinate system $y_{rat} = x = \omega_h = \begin{bmatrix} p & q & r \end{bmatrix}^T$ represents the angular rate output of helicopter.

Parallel to the INDI design procedure, we differentiate the output expression to obtain its explicit dependence on the actuator inputs u. This yields

$$\dot{y}_{rat} = \dot{\omega}_h = J^{-1}(m - \omega_h \times J\omega_h) \tag{39}$$

(39) can be rewritten as

$$\dot{y}_{rat} = \dot{\omega}_h = f_{rat}(x) + g_{rat}(x, u) \tag{40}$$

where

$$f_{rat}(x) = J^{-1}[m_{fus}(x) + m_{ht}(x) + m_{vt}(x) - \omega \times J\omega] \tag{41}$$

$$g_{rat}(x, u) = J^{-1}[m_{mr}(x, u) + m_{tr}(x, u)] \tag{42}$$

Note that the number of actuators of the helicopter is four, which is not equal to the output state number in the rate loop. This means a control allocation scheme must be used to deal with this overdetermined system. However, because the change of the collective pitch of the main rotor, denoted by θ_0, is always accompanied by the alteration of total lift, the value of θ_0 can be determined by the velocity on the z-axis under the NED reference frame. Therefore, we separately give the control law of the main rotor in the subsequent design. Now, we define the other three actuators as $u' = \begin{bmatrix} \theta_{1s} & \theta_{1c} & \theta_{0tr} \end{bmatrix}^T$.

Combining the analysis of the helicopter model before, we can obtain the rotational dynamics under actuator delay:

$$\dot{y}_{rat} = \dot{\omega}_h = f_{rat}(x) + g_{rat}(x, u_\tau) \tag{43}$$

Choosing the virtual control $v_{rat} = \dot{\omega}_{des} + k_{vr}e_{ret}$ with the control gain matrix $k_{vr} \in R^{3 \times 3}$, the controller u' can be given by

$$u' = u'_0 + \Delta u'_n + \Delta u'_c \tag{44}$$

$$\Delta u'_n = G_{rat}(x_0, u_0)^{-1}(v - \dot{\omega}_{h_0}) \tag{45}$$

$$\Delta u'_c = k_{ur}G_{rat}(x_0, u_0)^{-1} \int_{t_0}^{t} [\dot{e}_{rat} + k_{rr}e_{rat} + g_{rat}(x, u_\tau) - g_{rat}(x, u)]d\theta \tag{46}$$

where $G_{rat}(x_0, u_0) = J^{-1}\frac{\partial}{\partial u'}[m_{mr}(x, u) + m_{tr}(x, u)]\Big|_{x_0, u_0}$ and $k_{ur}, k_{rr} \in R^{3 \times 3}$ are also diagonal matrices. Note that there is a very complicated relationship between control input u and the moment generated by the main rotor and tail rotor because of the aerodynamics of the rotors. Hence, we adopt the central finite differences to calculate the control effectiveness matrix denoted by D, yielding

$$D = \frac{\partial}{\partial u'}[m_{mr}(x,u) + m_{tr}(x,u)]\bigg|_{x_0,u_0} = \begin{bmatrix} \dfrac{m_{mr}^T\left(x_0,u'_0+\begin{bmatrix}\delta_{\theta_{1s}} & 0 & 0\end{bmatrix}^T\right) - m_{mr}^T\left(x_0,u'_0-\begin{bmatrix}\delta_{\theta_{1s}} & 0 & 0\end{bmatrix}^T\right)}{2\delta_{\theta_{1s}}} \\ \dfrac{m_{mr}^T\left(x_0,u'_0+\begin{bmatrix}0 & \delta_{\theta_{1c}} & 0\end{bmatrix}^T\right) - m_{mr}^T\left(x_0,u'_0-\begin{bmatrix}0 & \delta_{\theta_{1c}} & 0\end{bmatrix}^T\right)}{2\delta_{\theta_{1c}}} \\ \dfrac{m_{tr}^T\left(x_0,u'_0+\begin{bmatrix}0 & 0 & \delta_{\theta_{0tr}}\end{bmatrix}^T\right) - m_{tr}^T\left(x_0,u'_0-\begin{bmatrix}0 & 0 & \delta_{\theta_{0tr}}\end{bmatrix}^T\right)}{2\delta_{\theta_{0tr}}} \end{bmatrix} \quad (47)$$

where $\delta_{\theta_{1s}}$, $\delta_{\theta_{1c}}$, and $\delta_{\theta_{1c}}$ are a small percent of each actuator input value.

For actuator dynamics, a pseudo-control hedge command is generated to provide the control system from trying to track the reference command when saturation occurs. According to (37), the pseudo-control hedge command v_r for the rate loop can be obtained by

$$\begin{aligned} v_r &= (\dot{x}_0 + J^{-1}D(u'_{des} - u'_0)) - (\dot{x}_0 + J^{-1}D(u' - u'_0)) \\ &= J^{-1}D(u'_{des} - u') \end{aligned} \quad (48)$$

where u'_{des} represents the desired input vector produced by the rate controller.

After completing this, the whole rate-loop control system is accomplished. However, the helicopter still faces stability issues, for the reason that the Euler angle in the attitude loop is not closed-loop stable.

4.4. Attitude Loop

Then, for the attitude loop, we can use the NDI control on account of no model uncertainty existing here. In this loop, the system can be given by

$$\dot{\theta} = \begin{bmatrix} 1 & \sin\phi\tan\theta & \cos\phi\tan\theta \\ 0 & \cos\phi & -\sin\phi \\ 0 & \sin\phi/\cos\theta & \cos\phi/\cos\theta \end{bmatrix}\omega_h \quad (49)$$

$$y_{att} = x = \theta = \begin{bmatrix}\phi & \theta & \psi\end{bmatrix}^T \quad (50)$$

where y_{att} represents the attitude angle output of the helicopter.

Unlike the rate loop, there is no model uncertainty or time delay in the attitude loop. Therefore, the attitude controller only needs to convert the attitude angle tracking error into the desired angular rate command as the inner loop input signal through the NDI method. Considering the state Equation (49), the reference input signal of rate loop ω_{ref} can be obtained by

$$\omega_{ref} = \begin{bmatrix} 1 & 0 & -\sin\theta \\ 0 & \cos\phi & \sin\phi\cos\theta \\ 0 & -\sin\phi & \cos\phi\cos\theta \end{bmatrix} v_{att} \quad (51)$$

where v_{att} is the virtual control for the attitude loop and it can be given by the attitude tracking error e_{att} as

$$v_{att} = \dot{\theta}_{des} + k_{va}e_{att} \quad (52)$$

in which $k_{va} \in \mathbb{R}^{3\times 3}$ is the control gain matrix and θ_{des} is the attitude reference command for the helicopter. Note that the inverse of the transformation matrix always exists for $\det\Omega = 1/\cos\theta \neq 0$.

In the attitude loop, the pseudo-control hedge command v_a is

$$v_a = \Omega(\omega_{ref} - \omega_h) \quad (53)$$

4.5. Control Law for Collective Pitch of Main Rotor

As mentioned in the rate loop, the operation of the collective pitch of the main rotor θ_0 will change the lift of the helicopter directly. Hence, the actuator θ_0 should be related to the vertical velocity of the helicopter. In the z-axis direction, the following equation is made

$$\dot{V}_z - g = \begin{bmatrix} -\sin\theta & \cos\theta\sin\phi & \cos\theta\cos\phi \end{bmatrix} \frac{F}{m} \tag{54}$$

where g is the gravitational acceleration and $F = \begin{bmatrix} F_x & F_y & F_z \end{bmatrix}^T$ represents the total force in the three axes. It contains the contributions of all the main parts of the helicopter, yielding

$$F = F_{fus} + F_{ht} + F_{vt} + F_{mr} + F_{tr} \tag{55}$$

Once again, F consists of the force produced by the fuselage, horizontal tail, vertical tail, main rotor, and tail rotor, respectively.

Note that, although the total force F contains many parts, the main part is the force generated by the main rotor since it resists gravity while allowing the helicopter to move flexibly. Based on the assumption above, we can obtain the direct dependence about $\dot{V}_z$ to the delayed input θ_{0_τ} as

$$\dot{V}_z = g + \frac{1}{m} \begin{bmatrix} -\sin\theta \\ \cos\theta\sin\phi \\ \cos\theta\cos\phi \end{bmatrix}^T F_{mr}(x, u_\tau) \tag{56}$$

Choosing the virtual control $v_{vz} = \dot{V}_{zdes} + k_{vz} e_{vz}$ with the control gain constant $k_{vz} \in R^+$, the controller for θ_0 can be given by

$$\theta_0 = \theta_{0,0} + \Delta\theta_{0,n} + \Delta\theta_{0,c} \tag{57}$$

$$\Delta\theta_{0,n} = m \left(\begin{bmatrix} -\sin\theta \\ \cos\theta\sin\phi \\ \cos\theta\cos\phi \end{bmatrix}^T H \right)^{-1} (v_z - \dot{V}_{z0}) \tag{58}$$

$$\Delta\theta_{0,c} = k_{uz} m \left(\begin{bmatrix} -\sin\theta \\ \cos\theta\sin\phi \\ \cos\theta\cos\phi \end{bmatrix}^T H \right)^{-1} \int_{t_0}^{t} [\dot{e}_z + k_{rz} e_z + F_{mr}(x, u_\tau) - F_{mr}(x, u)] d\theta \tag{59}$$

where $\theta_{0,0}$ also represents the previous sampling value of θ_0; $k_{uz}, k_{rz} \in R$ are positive control gains; and H is the control effectiveness matrix, which can be expressed as

$$H = \left. \frac{\partial F_{mr}(x, u)}{\partial \theta_0} \right|_{x_0, u_0} \tag{60}$$

It can also be calculated by using the central finite difference as

$$H = \left[\frac{F_{mr}^T(x_0, \theta_{0,0} + \delta_{\theta_0}) - F_{mr}^T(x_0, \theta_{0,0} - \delta_{\theta_0})}{2\delta_{\theta_0}} \right]^T \tag{61}$$

where δ_{θ_0} is a small percent of θ_0.

Now, the helicopter attitude control system under the multiple constraints of the actuators has been designed, which is shown in Figure 2.

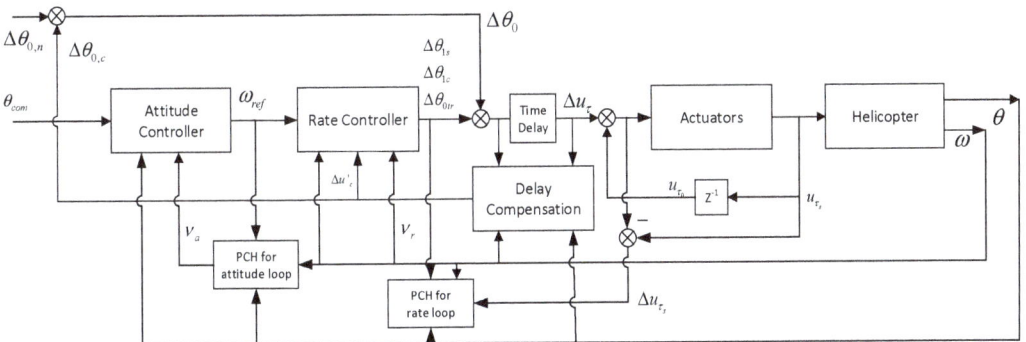

Figure 2. Schematic of the attitude control system based on the proposed method.

5. Simulation Results

In this section, several simulations will be given to verify the advantages of the proposed control law. We will simulate the attitude control of the helicopter in a hovering state. The model uncertainties are given as $\Delta C_{L_\alpha} = -0.1 C_{L_\alpha}$, $\Delta C_{L_{\alpha,tr}} = -0.1 C_{L_{\alpha,tr}}$, $\Delta C_{L_{\alpha,ht}} = 0.5 C_{L_{\alpha,ht}}$, and $\Delta C_{L_{\alpha,vt}} = 0.5 C_{L_{\alpha,vt}}$, where C_{L_α}, $C_{L_{\alpha,tr}}$, $C_{L_{\alpha,ht}}$, and $C_{L_{\alpha,vt}}$ are the lift curve slope of the blade of the main rotor, the blade of the tail rotor, the horizontal tail, and the vertical tail, respectively. The delay time is $\tau = 100 ms$, initial input $u_0 = \begin{bmatrix} 0.2484 & 0.0275 & -0.0135 & 0.1289 \end{bmatrix}^T$, diagonal element of control gain matrix k_{ur} is $\begin{bmatrix} 1 & 1 & 1.5 \end{bmatrix}^T$, and k_{rr} is $\begin{bmatrix} 2 & 2 & 3 \end{bmatrix}^T$.

The delay of the actuators will degrade the tracking performance and increase the control effort provided by the actuators, which always leads to state overshoot and actuator saturation, and even causes the system to become unstable when the delay time gradually increases. This phenomenon can be observed in Figure 3.

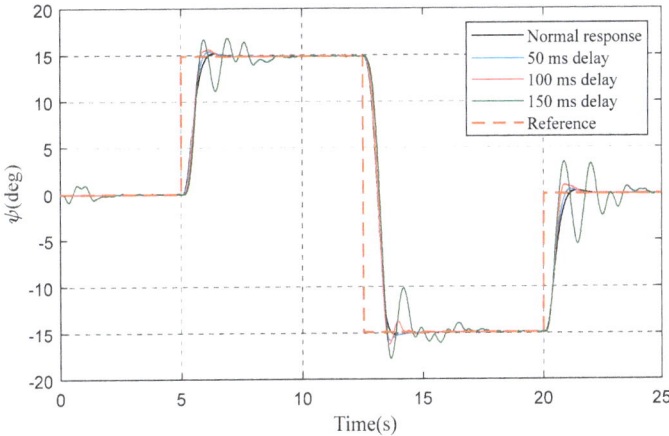

Figure 3. The response of the INDI-based control system under different actuators delay.

In this simulation, the operating frequency of the control system is set at 100 Hz. When the actuator delay is 50 ms, the INDI-based control system can barely maintain its tracking performance. A small range of oscillation appears in response when the delay time is 100 ms. However, actuators need to change frequency to maintain this steady state in the INDI scheme, which is shown in Figure 4. This can be understood as, when time delay exists, the control input does not correspond to the current input error, and it needs to be adjusted constantly within the whole time delay. In the case of the delay of 150 ms, the system response has already experienced a relatively large oscillation, and its dynamic characteristics are seriously degraded.

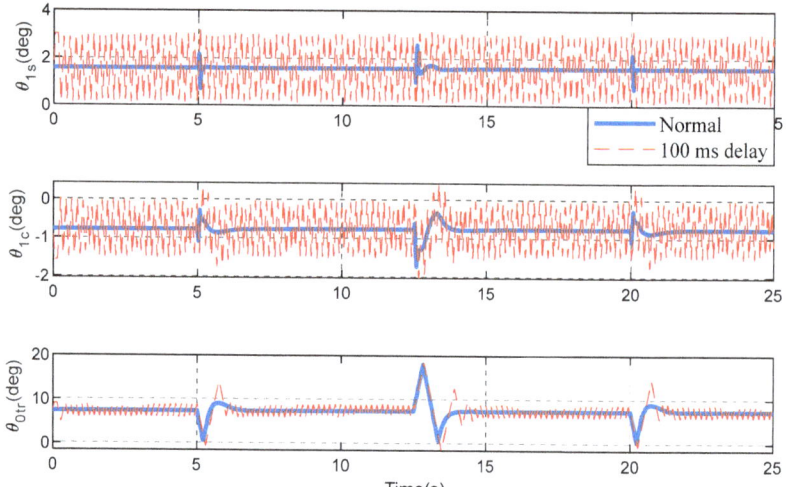

Figure 4. Actuator input of helicopter.

In the next simulation, the delay time between the controller and actuators is set at 100 ms. In Figure 5, it can be seen that, when the delay compensation is applied, the state overshoot is significantly reduced and the system's rapidity is also improved compared to the original response. Figure 6 shows that, in addition to the need for a larger control effect, the original phenomenon of rapid changes has disappeared.

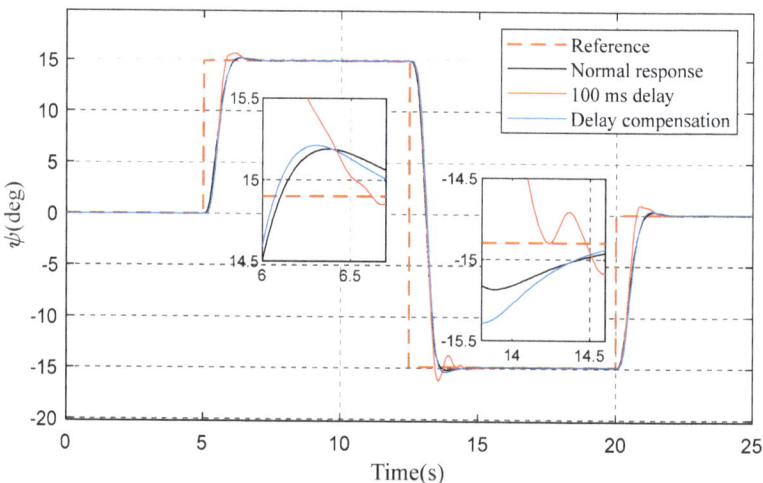

Figure 5. The response after delay compensation.

On the basis of delay compensation, we carry out the PCH design for the system. The advantages of PCH can be seen in Figure 7, which not only eliminates the overshoot, but also reduces the 0.7 s settling time of the system. Figure 8 is also given here to compare the changes of the actuators between the three cases.

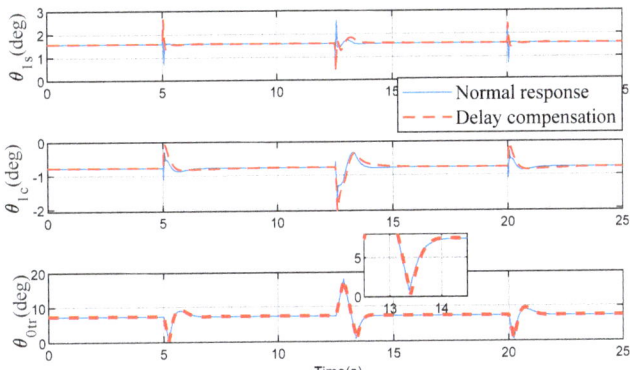

Figure 6. Actuator input after delay compensation.

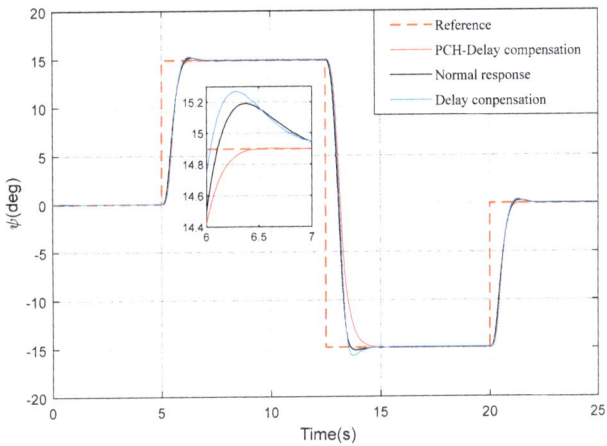

Figure 7. The response after the introduction of PCH.

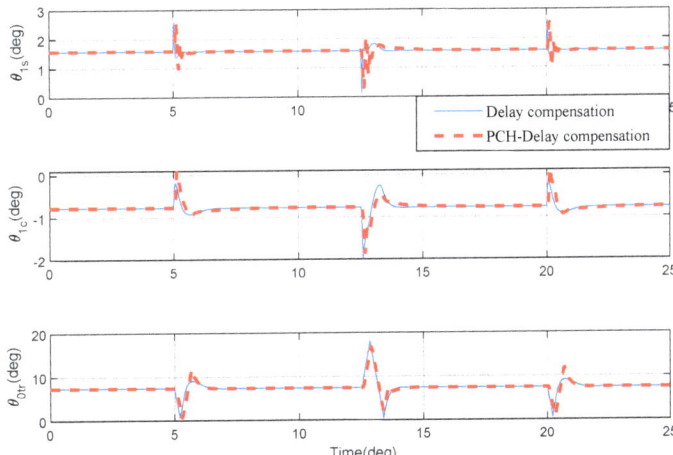

Figure 8. Actuator input after PCH.

Finally, the response of the three channels of the attitude angle under the proposed control scheme is given in Figure 9. It can be seen that the three Euler angles can be decoupled and show good dynamic characteristics. In the Figure 10, the angular rates also change regularly corresponding to the tracking of the three Euler angular rates. Furthermore, the system can also track the command signal well when the actuator input is saturated and delayed, which can be observed in Figure 10.

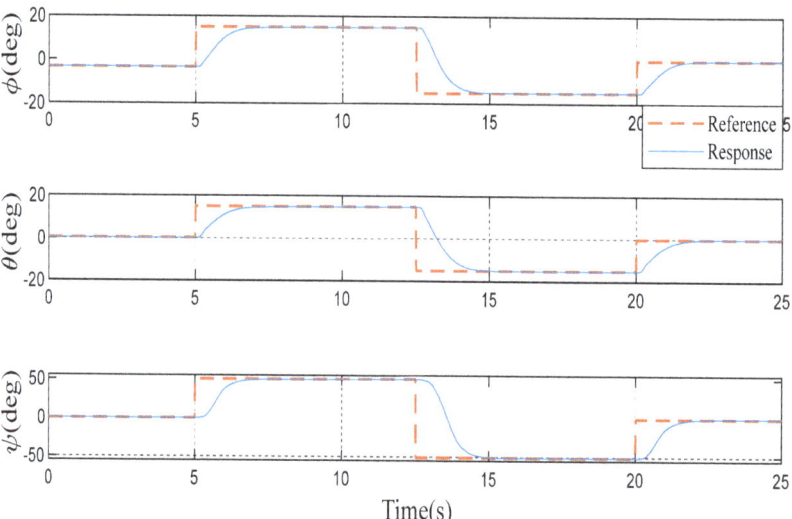

Figure 9. Three Euler angle responses based on the proposed method.

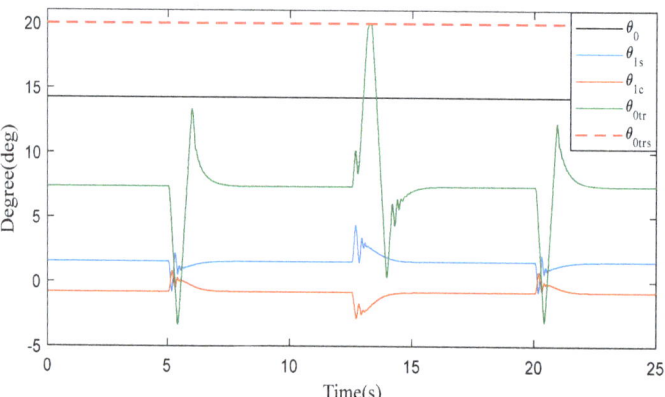

Figure 10. Actuator input corresponding to the responses in Figure 8.

6. Conclusions

In this paper, an INDI-based actuator compensation attitude controller is developed for the helicopter subject to the time delay, position, and rate saturations in actuators. The controller is composed of a rate controller which ensures the rate performance of the helicopter, an attitude controller which guarantees the attitude tracking performance, and a collective pitch controller which meets the needs of the vertical changes in the z-axis direction. The model reduction method is used to design an INDI-based controller for the rate loop and collective pitch of the main rotor, which improves the robustness to the time-varying actuator delay. Meanwhile, the PCH technique is introduced for both the rate loop and attitude loop to provide a filter such that the following commands hold within the capability of the controllers. Finally, simulations demonstrate the effectiveness and robustness of the proposed controller. In the future, the outer loop controller for the helicopter will be designed and a maneuver flight simulation analysis that meets the requirements of ADS-33E-PRF will be performed.

Author Contributions: Conceptualization, S.Z., H.Z. and K.J.; methodology, S.Z. and H.Z.; formal analysis, H.Z.; investigation, H.Z. and K.J.; resources, H.Z. and K.J.; data curation, H.Z.; writing—original draft preparation, H.Z.; writing—review and editing, H.Z. and K.J.; supervision, S.Z.; project administration, S.Z. All authors have read and agreed to the published version of the manuscript.

Funding: This research has received funding from the Natural Science Foundation of Jiangsu Province under Grant BK20201291, Aeronautical Science Foundation of China under Grant 201957052002, and Research and Practice Innovation Program of Nanjing University of Aeronautics and Astronautics under Grant xcxjh20210317.

Data Availability Statement: Not applicable.

Conflicts of Interest: The authors declare no conflict of interest.

References

1. Padfield, G.D. *Helicopter Flight Dynamics: The Theory and Application of Flying Qualities and Simulation Modelling*, 2nd ed.; American Institute of Aeronautics and Astronautics: Reston, VA, USA, 2007; Volume 113.
2. Prouty, R.W.; Curtiss, H.C. Helicopter Control Systems: A History. *J. Guid. Control. Dyn.* **2003**, *26*, 12–18. [CrossRef]
3. Hu, J.; Gu, H. Survey on Flight Control Technology for Large-Scale Helicopter. *Int. J. Aerosp. Eng.* **2017**, *2017*, 5309403. [CrossRef]
4. Kim, N.; Calise, A.J.; Hovakimyan, N.; Prasad, J.V.R.; Corban, E. Adaptive Output Feedback for High-Bandwidth Flight Control. *J. Guid. Control. Dyn.* **2002**, *25*, 993–1002. [CrossRef]
5. Lee, S.; Ha, C.; Kim, B. Adaptive nonlinear control system design for helicopter robust command augmentation. *Aerosp. Sci. Technol.* **2005**, *9*, 241–251. [CrossRef]
6. Amaral, T.G.; Crisóstomo, M.M.; Pires, V.F. Helicopter motion control using adaptive neuro-fuzzy inference controller. In Proceedings of the IEEE 2002 28th Annual Conference of the Industrial Electronics Society, Seville, Spain, 5–8 November 2002; Volume 3, pp. 2090–2095.
7. Leitner, J.; Calise, A.; Prasad, J.V.R. Analysis of Adaptive Neural Networks for Helicopter Flight Control. *J. Guid. Control Dyn.* **1997**, *20*, 972–979. [CrossRef]
8. Moelans, P. Adaptive Helicopter Control Using Feedback Linearization and Neural Networks. Master's Thesis, Delft University of Technology, Delft, The Netherlands, 2008.
9. Kutay, A.T.; Calise, A.J.; Idan, M.; Hovakimyan, N. Experimental results on adaptive output feedback control using a laboratory model helicopter. *IEEE Trans. Control Syst. Technol.* **2005**, *13*, 196–202. [CrossRef]
10. Johnson, E.N.; Kannan, S.K. Adaptive Trajectory Control for Autonomous Helicopters. *J. Guid. Control. Dyn.* **2005**, *28*, 524–538. [CrossRef]
11. Johnson, E.N.; Calise, A.J. Pseudo-control hedging: A new method for adaptive control. In Proceedings of the Advances in Navigation Guidance and Control Technology Workshop, Alabama, AL, USA, 1–2 November 2000.
12. Acquatella, B.P.; Chu, Q.P. Agile Spacecraft Attitude Control: An Incremental Nonlinear Dynamic Inversion Approach. *IFAC-PapersOnLine* **2020**, *53*, 5709–5716. [CrossRef]
13. Wang, X.; van Kampen, E.-J.; Chu, Q.P.; De Breuker, R. Flexible Aircraft Gust Load Alleviation with Incremental Nonlinear Dynamic Inversion. *J. Guid. Control. Dyn.* **2019**, *42*, 1519–1536. [CrossRef]
14. Smeur, E.; de Croon, G.; Chu, Q. Cascaded incremental nonlinear dynamic inversion for MAV disturbance rejection. *Control Eng. Pr.* **2018**, *73*, 79–90. [CrossRef]
15. Chen, G.; Liu, A.; Hu, J.; Feng, J.; Ma, Z. Attitude and Altitude Control of Unmanned Aerial-Underwater Vehicle Based on Incremental Nonlinear Dynamic Inversion. *IEEE Access* **2020**, *8*, 156129–156138. [CrossRef]
16. Cervantes, T.J.L.; Choi, S.H.; Kim, B.S. Flight Control Design using Incremental Nonlinear Dynamic Inversion with Fixed-lag Smoothing Estimation. *Int. J. Aeronaut. Space Sci.* **2020**, *21*, 1047–1058. [CrossRef]
17. Wang, X.; van Kampen, E.-J.; Chu, Q.; Lu, P. Stability Analysis for Incremental Nonlinear Dynamic Inversion Control. *J. Guid. Control Dyn.* **2019**, *42*, 1116–1129. [CrossRef]
18. Simplício, P.; Pavel, M.; van Kampen, E.; Chu, Q. An acceleration measurements-based approach for helicopter nonlinear flight control using Incremental Nonlinear Dynamic Inversion. *Control. Eng. Pr.* **2013**, *21*, 1065–1077. [CrossRef]
19. Pavel, M.D.; Perumal, S.; Olaf, S.; Mike, W.; Chu, Q.P.; Harm, C. Incremental Nonlinear Dynamic Inversion for the Apache AH-64 Helicopter Control. *J. Am. Helicopter Soc.* **2020**, *65*, 1–16. [CrossRef]
20. Richard, J.-P. Time-delay systems: An overview of some recent advances and open problems. *Automatica* **2003**, *39*, 1667–1694. [CrossRef]

21. Diagne, M.; Bekiaris-Liberis, N.; Krstic, M. Compensation of input delay that depends on delayed input. *Automatica* **2017**, *85*, 362–373. [CrossRef]
22. Krstic, M. Input Delay Compensation for Forward Complete and Strict-Feedforward Nonlinear Systems. *IEEE Trans. Autom. Control* **2009**, *55*, 287–303. [CrossRef]
23. Krstic, M. Lyapunov Stability of Linear Predictor Feedback for Time-Varying Input Delay. *IEEE Trans. Autom. Control* **2010**, *55*, 554–559. [CrossRef]
24. Yue, D.; Han, Q.-L. Delayed feedback control of uncertain systems with time-varying input delay. *Automatica* **2005**, *41*, 233–240. [CrossRef]
25. Sharma, N.; Bhasin, S.; Wang, Q.; Dixon, W.E. Predictor-based control for an uncertain Euler–Lagrange system with input delay. *Automatica* **2011**, *47*, 2332–2342. [CrossRef]
26. Artstein, Z. Linear systems with delayed controls: A reduction. *IEEE Trans. Autom. Control* **1982**, *27*, 869–879. [CrossRef]
27. Deng, W.; Yao, J.; Ma, D. Time-varying input delay compensation for nonlinear systems with additive disturbance: An output feedback approach. *Int. J. Robust Nonlinear Control* **2017**, *28*, 31–52. [CrossRef]
28. Rohollah, M.; Sarangapani, J. Optimal Adaptive Control of Uncertain Nonlinear Continuous-Time Systems with Input and State Delays. *IEEE Trans. Neural Netw. Learn. Syst.* **2021**, *2021*, 1–10.
29. Rohollah, M.; Sarangapani, J. Optimal control of linear continuous-time systems in the presence of state and input delays with application to a chemical reactor. In Proceedings of the 2020 American Control Conference (ACC), Denver, CO, USA, 1–3 July 2020; pp. 999–1004.
30. Tarbouriech, S.; Turner, M. Anti-windup design: An overview of some recent advances and open problems. *IET Control Theory Appl.* **2009**, *3*, 1–19. [CrossRef]
31. Ma, J.; Ge, S.S.; Zheng, Z.; Hu, D. Adaptive NN Control of a Class of Nonlinear Systems with Asymmetric Saturation Actuators. *IEEE Trans. Neural Networks Learn. Syst.* **2014**, *26*, 1532–1538. [CrossRef] [PubMed]
32. Zhang, S.; Meng, Q. An anti-windup INDI fault-tolerant control scheme for flying wing aircraft with actuator faults. *ISA Trans.* **2019**, *93*, 172–179. [CrossRef] [PubMed]
33. Quang, L.; Richard, H.; William, S.; Brett, R. Investigation and Preliminary Development of a Modified Pseudo Control Hedging for Missile Performance Enhancement. In Proceedings of the AIAA Guidance, Navigation, and Control Conference and Exhibit, San Francisco, CA, USA, 15–18 August 2005. [CrossRef]
34. Edwards, C.; Lombaerts, T.; Smaili, H. *Fault Tolerant Flight Control: A Benchmark Challenge*; Springer: Berlin/Heidelberg, Germany, 2010; pp. 1–560.
35. Hartjes, S. An Optimal Control Approach to Helicopter Noise and Emissions Abatement Terminal Procedures. Master's Thesis, Delft University of Technology, Delft, The Netherlands, 2015.

Disclaimer/Publisher's Note: The statements, opinions and data contained in all publications are solely those of the individual author(s) and contributor(s) and not of MDPI and/or the editor(s). MDPI and/or the editor(s) disclaim responsibility for any injury to people or property resulting from any ideas, methods, instructions or products referred to in the content.

Article

An Improved Fault Identification Method for Electromechanical Actuators

Gaetano Quattrocchi, Pier C. Berri, Matteo D. L. Dalla Vedova *, and Paolo Maggiore

Department of Aerospace and Mechanical Engineering (DIMEAS), Politecnico di Torino, 10129 Turin, Italy; gaetano.quattrocchi@polito.it (G.Q.); pier.berri@polito.it (P.C.B.); paolo.maggiore@polito.it (P.M.)
* Correspondence: matteo.dallavedova@polito.it

Abstract: Adoption of electromechanical actuation systems in aerospace is increasing, and so reliable diagnostic and prognostics schemes are required to ensure safe operations, especially in key, safety-critical systems such as primary flight controls. Furthermore, the use of prognostics methods can increase the system availability during the life cycle and thus reduce costs if implemented in a predictive maintenance framework. In this work, an improvement of an already presented algorithm will be introduced, whose scope is to predict the actual degradation state of a motor in an electromechanical actuator, also providing a temperature estimation. This objective is achieved by using a properly processed back-electromotive force signal and a simple feed-forward neural network. Good prediction of the motor health status is achieved with a small degree of inaccuracy.

Keywords: prognostics; electromechanical actuators; neural network; temperature

1. Introduction

Electromechanical actuators (EMA) are widely used in industry, and in recent decades are seeing increasing adoption in the aviation sector, especially when the more electric or all-electric [1] design philosophies are adopted, which seeks to reduce secondary power type usage, using mostly or exclusively electrical power for systems actuation [2].

An interesting application of the more electric philosophy is in the Airbus A380 where two electrical lines are used as backup sources of power in case of loss of the main hydraulic lines. However, electrohydrostatic actuators (EHAs) are used rather than EMAs, since one of EMAs' most common failure modes is jamming [3]. Jamming is characterized by a sudden actuator stop in a certain position, thus locking the flight control surface position, creating a dangerous moments imbalance and thus possibly uncontrollable yaw, pitch or roll. Furthermore, the estimation of jamming probability is harder than for EHAs, where the knowledge regarding current hydraulic actuators can be used [4].

For this reason, as of today the use of EMAs is still mainly limited to non-safety-critical systems, such as secondary flight systems, high-lift devices or airbrakes. However, EMAs have some merits that make them preferable over other architecture, such as complexity, weight and maintenance requirements, especially in low-power applications [5]. In fact, the simplest method for providing mechanical power using an electrical supply is an electromechanical actuator, which in its most basic form is an electrical motor with a mechanism converting rotary motion to linear (using for example a ball screw). Usually, in aviation, given the high actuation torques needed and the volume constraints, some form of a reducer is incorporated between the motor and the rotation-linear converter to multiply the motor torque. A schematic view of an EMA is shown in Figure 1.

Another important aspect to consider is the lack of extensive failure datasets in operating conditions, which further discourages the adoption of EMAs given the severity of some failure modes [6]. Even though EMAs, as previously stated, are already used in secondary flight control actuation, data obtained for this use case are not directly transferable for primary flight control due to the very different operating nature of the two applications.

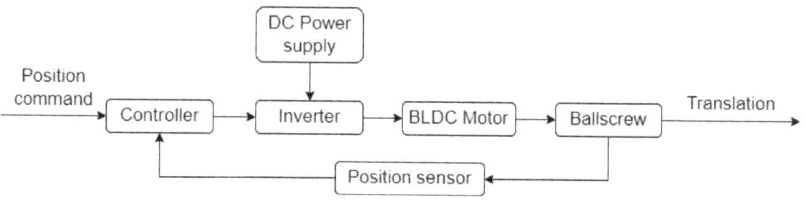

Figure 1. Block schematics of an EMA.

To increase safety and reliability, Prognostics and Health Management (PHM) could prove beneficial. The traditional definition of PHM can be found in [7], where it is defined as the ability to evaluate the current health status and predict the future behavior of a system, using knowledge of the current state; additionally, maintenance can be planned accordingly to properly maintain the system.

Several approaches to PHM exist and can be classified into three different categories: model-based, hybrid or data-driven, depending on the degree of knowledge of the system studied and the availability of data.

Model-based PHM uses a representation of the system (or component) of interest, using a set of differential Equations [8–10]. If properly modeled, the system status can be assessed with good accuracy and fault propagation is generally embedded in the fault model [11,12]. The main drawback of model-based approaches is the difficulty in creating accurate and representative models for complex systems; in particular in aeronautics, it is hard to model, or present in a simplified way, complex interactions such as aerodynamics or mechanical backlash.

The opposite approach is data-driven, where the system is treated as a black box, i.e., without detailed knowledge. In this case, machine learning methods are widely used to detect variations in the system's health status. One of the most used tools is Neural Networks of various kinds, such as in [13–16]. Some other tools used are combinations of filters and neural networks, as in [17,18], Bayesian inference [19] and Markov chains [20]. The main weakness of data-driven approaches is the need for a large dataset, which is generally not readily available, especially if run-to-failure of complex and expensive systems are needed.

Hybrid approaches use both model-based and data-driven techniques. The idea is to create a physical model of the system, extract relevant health indicators using physical quantities, and then use machine learning techniques to obtain an estimation of the actual system health status. Examples can be found in [21–24].

In this work, a hybrid prognostic method to evaluate the damage of a BLDC is presented, using the back-EMF signal as a prognostic indicator. In particular, the paper will describe an evolution of the algorithm presented in [25], which has been improved and can now also give a temperature estimate of the motor phases, besides the damage to individual stator coils and the static eccentricity of the rotor.

2. Materials and Methods

2.1. Scope of the Work and Novelties Introduced

The scope of this work is to present an evolution of the prognostic algorithm presented in [25]. The use of the back-EMF coefficient is supported by the fact that it is sensitive to the faults of interest (partial inter-turn shorts, static eccentricity), while not being affected by the command or load imposed on the actuator, as demonstrated in [26].

As previously stated, in this work the effect of temperature on phase resistance is also considered. The inclusion of temperature dependency requires a new approach to the evaluation of system status.

The algorithm can be described as follows:

1. Faults vectors are generated, and the system is simulated using these values, obtaining a simulations dataset;
2. Relevant physical quantities are logged for each simulation (e.g., voltages, currents, motor angular position and speed);
3. In each simulation, for each phase, an estimation of the back-EMF coefficient is calculated;
4. Estimation error is minimized by obtaining the real values of phase resistance and back-EMF coefficient;
5. These values are used in a neural network to predict the health status of the system.

The first two steps are self-explanatory; the faults are modeled as in [25], using a fault vector $f = [N_a, N_b, N_c, \xi_x, \xi_y, \Delta T]$, with 6 different components: N_a, N_b, N_c which are the fraction of turns shorted for each phase; ξ_x, ξ_y which are the components of static eccentricity in cartesian coordinates (in [25] polar coordinates are used) and finally ΔT which is the temperature deviation from reference conditions $T_0 = 20\ °C$. The variation ranges for the variables are the same as those in [25]. For stator windings temperature, the range $\Delta T = [-50; +70]\ °C$ is chosen, since it is representative of conditions encountered in aeronautical applications. However, the range chosen might be too restrictive to include sudden transitory temperature spikes and might need adjustments in future developments. The faults are injected into the model prior to each simulation.

In the third step, the estimation of the back-EMF coefficient is obtained using the following equation:

$$V - k'\omega = V - (k + k_e)\omega = R_0 i \qquad (1)$$

where k' is the estimated back-EMF coefficient and k_e is the estimation error. In this case, values of V, ω, i are those that can be measured from the simulation (and by extension, from a real system), while R_0 is the nominal resistance of a phase. There is an error in the estimation of the back-EMF coefficient since the nominal resistance is used, thus not considering the effect of both a temperature variation and partial phase to phase short, which changes this value.

The actual (or true) system condition is described by:

$$V - k\omega = R \cdot i = (R_0 + \Delta R)i \qquad (2)$$

where k is the true back-EMF coefficient, $R = (R_0 + \Delta R)$ is the true resistance, i.e., the effective resistance of the coil in the instantaneous conditions of temperature, fault, etc. and ΔR is the deviation of actual resistance from nominal value.

Now, subtracting Equation (2) from Equation (1) and rearranging, one can obtain:

$$k_e = \Delta R \frac{i}{\omega} \qquad (3)$$

Assuming that ΔR is constant (i.e., R varies slowly), which is reasonable in the framework presented, since each measurement is very short (in the order of one second), derivating Equation (3) we have:

$$\frac{\partial k_e}{\partial (i/\omega)} = \Delta R \qquad (4)$$

Furthermore, assuming that k is constant, which implies that the fault does not change during the simulation, we can obtain:

$$\frac{\partial (k + k_e)}{\partial (i/\omega)} = \frac{\partial k'}{\partial (i/\omega)} = \Delta R \qquad (5)$$

So, we have demonstrated that $k' = (k + k_e)$ is linearly dependent to i/ω with a slope equal to ΔR. Now, to obtain the real values of back-EMF coefficient and resistance, we have to iteratively reconstruct the value of k (Equation (2)), using a temporary R^* variable to optimize to make Equation (5) equal to zero. This stems from the definition of back-EMF:

$$|BEMF| = \frac{\partial \Phi}{\partial t} = \frac{\partial \Phi}{\partial \theta}\frac{\partial \theta}{\partial t} = \frac{\partial \Phi}{\partial \theta}\omega \quad \rightarrow \quad k_j(n_j, \theta, T) = \frac{\partial \Phi(n_j, \theta, T)}{\partial \theta} \tag{6}$$

where Φ is the concatenated magnetic flux, n is the number of shorted turns and T is the temperature. In other words, the concatenated magnetic flux is a function of angle, number of turns and temperature and so is the back-EMF coefficient k_j. It does depend only on motor geometry and on the magnetic properties of the magnets, and thus the temperature dependency. In this preliminary work, the temperature-induced variation of magnetic properties is not considered and will be added in further developments.

The following problem must then be solved:

$$\begin{cases} V - k^*\omega = R^* \cdot i \\ \dfrac{\partial(k^*)}{\partial(i/\omega)} = 0 \end{cases} \tag{7}$$

At convergence, we obtain $R^* = R$ which implies $k_e = 0$ (from Equation (2)): we are calculating true resistance and true back-EMF coefficient. In this work, a simple bisection method is used to perform the optimization.

Up to this point, we have considered $k_j(\theta, i/\omega)$, as visible in Figure 2, where the subscript j indicates one of the three phases; in this case, the number of variables to optimize will be $3 \cdot n$, where n is the number of subdivisions along the θ axis. The total number of variables will thus be $6 \cdot n$, i.e., $3n$ values of resistance and $3n$ values of back-EMF coefficient.

However, in order to simplify the computation, the dependence on θ has been dropped, thus collapsing the 3D graph into a 2D plot of $k_j(i/\omega)$, i.e., Figure 3. Now, a 'global' (or *generalized*) k_j approximation can be calculated, using least square fit. Values close to zero have been discarded, since they provide no additional information, and an absolute value on k_j has been applied. The final result is a reduction of the number of variables from $6 \cdot n$ to 6, i.e., 3 generalized resistances (one for each phase) and 3 generalized back-EMF coefficients.

These 6 values are used in a simple feed-forward neural network to perform an estimation of the fault vector f.

Figures 2 and 3 have been obtained using a parabolic position command (i.e., a speed ramp) with a constant angular acceleration of 0.3 rad/s^2; initial conditions are zero angular position and zero angular velocity ($\theta = \dot{\theta} = 0$). The following fault vector was seeded: $f = [0.0375, 0.0504, 0.0507, 0.0059, 2.8 \times 10^{-5}, 4]$.

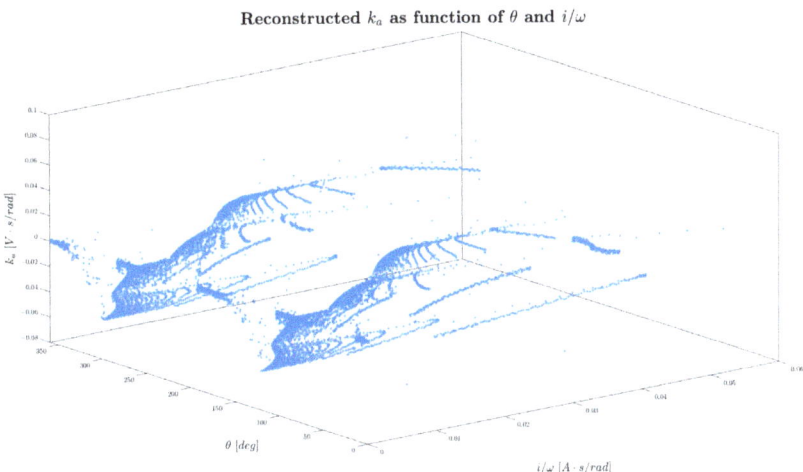

Figure 2. Reconstructed k_a as function of both θ and i/ω.

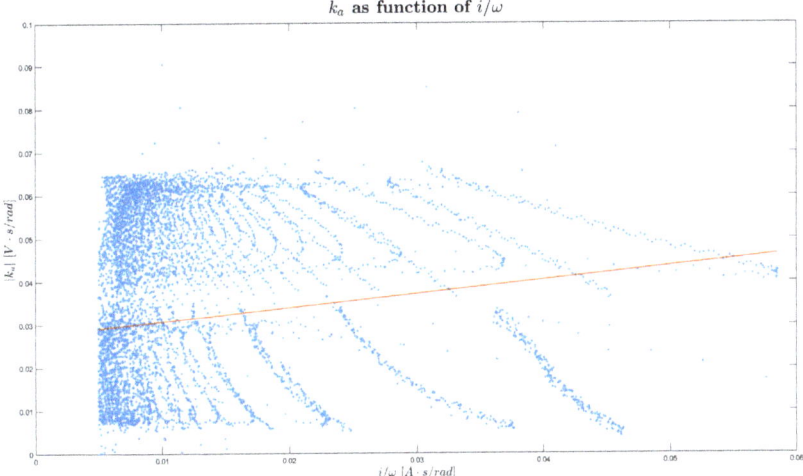

Figure 3. Reconstructed k_a as function of i/ω (non optimized).

2.2. Brief System Overview

As previously stated, the system used to test the algorithm is the same as that used in [25,26], with minor modifications (Figure 4); it represents an electromechanical actuator acting on a flight surface.

The model is a high fidelity representation of a trapezoidal EMA acting on secondary flight control. It is a detailed, component-level model with very limited assumptions, i.e., lumped parameters. Many non-linear phenomena are modeled, including but not limited to electronic noise, dry friction and current and speed saturations. The model has been validated on literature data as reported in [27,28]. For further details on the model, please refer to [26].

With respect to previous iterations, the main enhancement is the implementation of a temperature dependency on phase resistance (each of the three RL branches in Figure 5), using the classical equation:

$$R = R_{ref}(1 + \alpha \Delta T) \tag{8}$$

where $\alpha = 0.00404\ °C^{-1}$ is the resistance temperature coefficient of copper (the materials of which the coils are made), R_{ref} is the reference temperature resistance (in this case $T_{ref} = 20\ °C$) and ΔT is the temperature difference from T_{ref}.

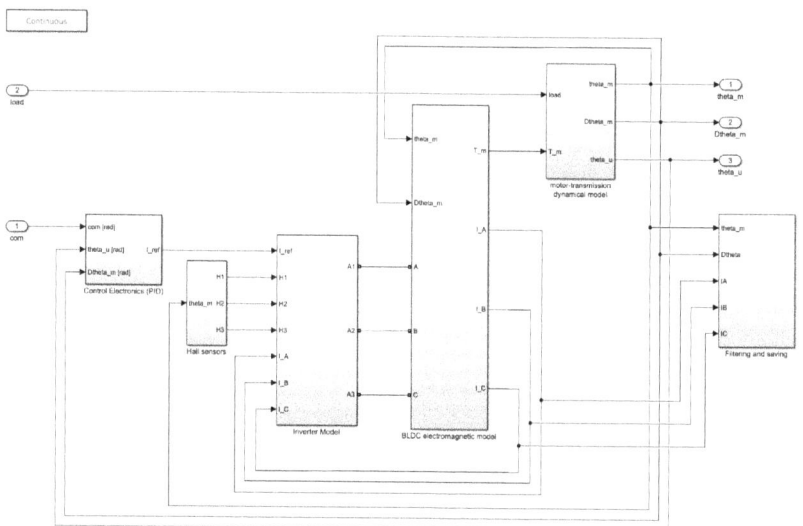

Figure 4. Top level view of the actuator model.

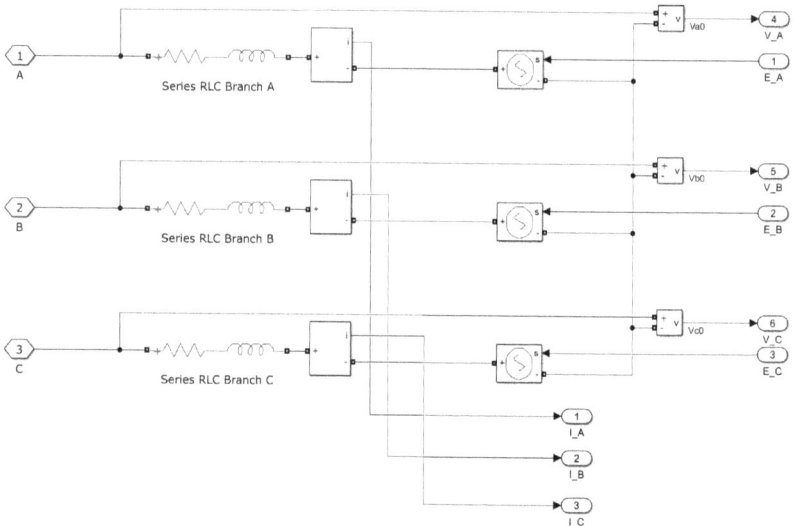

Figure 5. Phases model detail (a subsystem of 'BLDC electromagnetic model').

Furthermore, the external load is set to zero, supposing an actuation during ground operations, while the actuation command is now a parabolic position command (i.e., a velocity ramp, or constant acceleration), since using this type of command better covers the (i/ω) space and thus embed more information in the same simulation time as opposed to more classical commands such as position ramps or steps.

Some examples of the data obtained after simulations can be found in Figures 6 and 7. It can be noted how the first ~40 ms present strong fluctuations in the current and voltages values (but also angular velocity): this is due to the strong non-linear effects modeled in the system, e.g., dry friction. For this reason, the first ∼50 ms of each simulation are discarded before applying the algorithm presented.

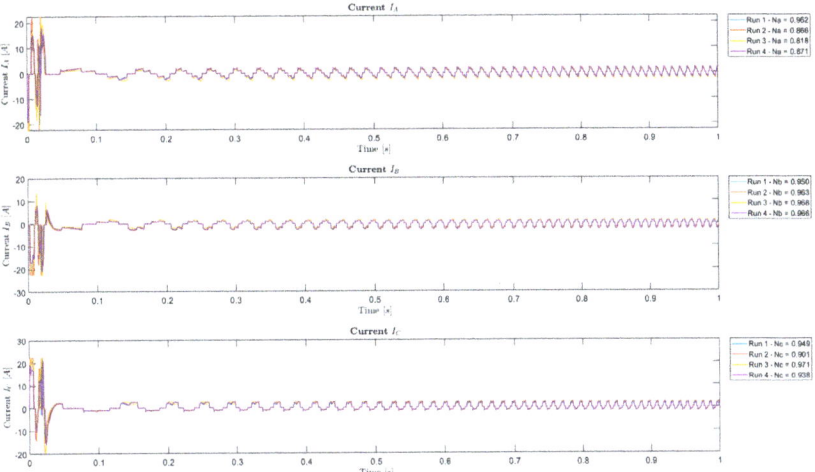

Figure 6. Graphs of phase current for different fault conditions.

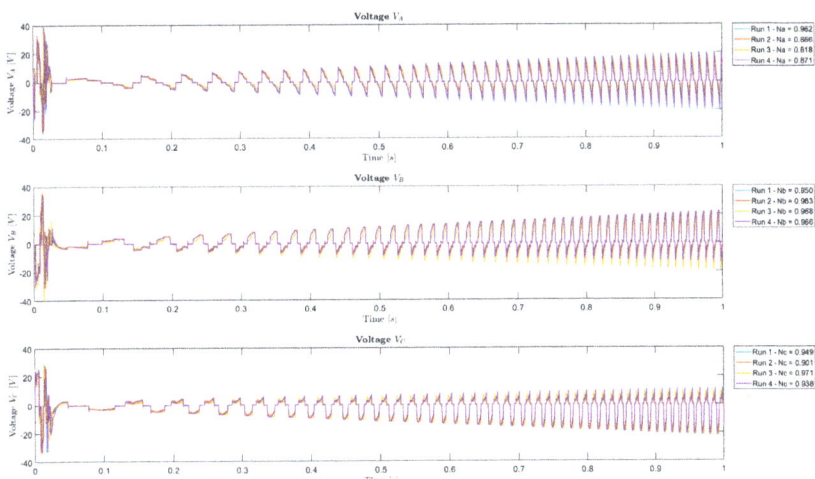

Figure 7. Graphs of phase voltage for different fault conditions.

2.3. Dataset Used

The number of different fault conditions simulated is 600; the dataset has been randomly divided into 70%, 15%, 15% subsets for training, validation and testing, respectively.

Dataset size has been empirically chosen to be a good representation of the 6-dimensional fault space; the dataset, regarding the first 5 variables of the fault vector (N_a, N_b, N_c, ξ_x, ξ_y) is the same presented in [25], with the two eccentricity components now transformed into cartesian coordinates from polar. For each fault vector, a temperature difference value (ΔT), randomly sampled from the interval considered, has been appended.

Using an AMD 5600X, each simulation takes about 50 s to run; parallel pooling has been used to reduce the total dataset generation time, combined with Simulink Accelerator mode. For this study, a full software workflow has been used. In other words, the full system is simulated and relevant data logged. The algorithm presented is then applied and neural network outputs are compared to the fault vectors injected in the model before simulation, which are considered ground truth.

In the future, we hope to implement the faults considered in this study on a real test bench, even though it is a complex task, especially for the partial phase shorts.

3. Results

In this section, the network used for fault regression will be described. Several tests on different hyperparameter combinations have been carried out, and the best performing set has been used for error distribution calculation. The network architecture is very similar to that presented in [25], with minor modifications to the values of the hyperparameters.

The architecture chosen is a feed-forward neural network with a single hidden layer of size 12 (Figure 8); the training function uses the Levenberg–Marquardt algorithm; the activation function for each neuron is the hyperbolic tangent sigmoid. The maximum size of failed validation checks is set to 10.

As expected, the network inputs vector is of size 6—including 3 generalized back-EMF coefficients and 3 generalized resistances, while the output is again of size 6 and is the fault vector used to generate the simulation.

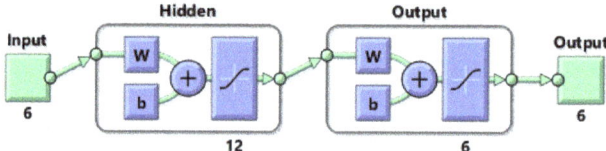

Figure 8. Neural network topology.

In Figure 9, the mean absolute error box plot for each variable is shown. For visualization purposes, a new subset of 10 rows is used and called an external validation set. This dataset has never been used during training of the neural network so it is a good representation of the predictive capabilities of the network with new data during operations. As visible, the mean absolute error is very small, in the order of 0.02 on normalized data. The error distribution is however uneven between variables and this might be caused by the relatively small dataset used in training.

In absolute terms, the mean error for the ΔT estimation, on the external verification dataset, is about 1.8 °C, which is a good result.

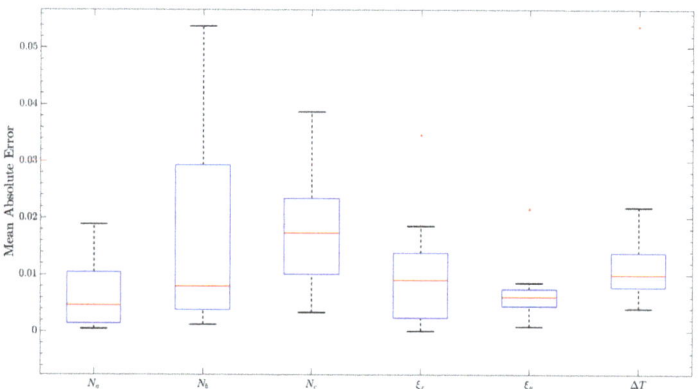

Figure 9. Mean absolute error for external validation set.

4. Discussion

As described in the previous section, the prediction accuracy of the system is promising. Even though the problem complexity has increased with respect to the previous algorithm [25], the neural network is capable of predicting the system status with adequate accuracy.

The network regression task is probably aided by the extensive pre-processing applied to raw data before being used as an input dataset, including filtering. In a real system, properly calibrated filtering will be needed to smooth out high-frequency noise in the signals of interest, before the application of the algorithm is presented.

Furthermore, the data compression techniques described in Section 2.1 are helpful in reducing the dimensions of the regression problem, even though the estimation obtained is a generalization of the health status rather than an estimation for each possible angular position.

5. Conclusions

In this work, an improved prognostic algorithm for EMA has been presented. The strong points of this paper include the ability to estimate the current health status of the motor in terms of fault variables including partial phase shorts, static eccentricity and temperature deviation from ambient conditions. Furthermore, no additional sensors besides those needed for normal operations are required. Motor damage estimation is carried out using a feed-forward neural network after the application of the algorithm to raw data.

As with any other work that includes neural networks in the pipeline, a parametric optimization of the network could yield additional benefits in the form of higher accuracy or a simpler network if the algorithm is to be implemented on embedded hardware.

Furthermore, several assumptions have been made to simplify the algorithm, mainly the temperature independence of magnetic properties. Even though it is a reasonable assumption for small variations of temperature (magnetic flux variation of ca. -4% for 100 °C for SaCo magnets), a generalized algorithm should include and simulate such variations.

An increase in the simulation dataset size, after considering what the best combination of actuation command and load is, could prove beneficial in further increasing the network accuracy and robustness.

Finally, an empirical validation on properly calibrated equipment is mandatory to test that the assumptions made are reasonable and can effectively represent the system status.

Author Contributions: Conceptualization, P.C.B.; methodology, G.Q. and P.C.B.; software, G.Q.; validation, G.Q.; formal analysis, G.Q. and P.C.B.; investigation, G.Q.; resources, G.Q.; data curation, G.Q.; writing—original draft preparation, G.Q.; writing—review and editing, G.Q. and M.D.L.D.V. visualization, G.Q.; supervision, M.D.L.D.V. and P.M.; project administration, M.D.L.D.V. and P.M.; funding acquisition, M.D.L.D.V. and P.M. All authors have read and agreed to the published version of the manuscript.

Funding: This research received no external funding.

Informed Consent Statement: Not applicable.

Data Availability Statement: The datasets needed to replicate what was presented in this study are openly available in FigShare at doi:10.6084/m9.figshare.19635378, accessed on 24 June 2022.

Conflicts of Interest: The authors declare no conflict of interest.

Abbreviations

The following abbreviations are used in this manuscript:

BEMF	Back Electro-Motive Force;
BLDC	BrushLess Direct Current;
EHA	Electrohydrostatic Actuator;
EMA	Electro-Mechanical Actuator;
FDI	Fault Detection and Identification;
PHM	Prognostics and Health Management.

References

1. Quigley, R.E.J. More Electric Aircraft. In Proceedings of the Eighth Annual Applied Power Electronics Conference and Exposition, Long Beach, CA, USA, 17–21 March 2013; pp. 906–911. [CrossRef]
2. Wheeler, P.; Bozhko, S. The more electric aircraft: Technology and challenges. *IEEE Electrif. Mag.* **2014**, *2*, 6–12. [CrossRef]
3. Balaban, E.; Bansal, P.; Stoelting, P.; Saxena, A.; Goebel, K.F.; Curran, S. A diagnostic approach for electro-mechanical actuators in aerospace systems. In Proceedings of the 2009 IEEE Aerospace Conference, Big Sky, MT, USA, 7–14 March 2009; pp. 1–13.
4. Van Den Bossche, D. The A380 flight control electrohydrostatic actuators, achievements and lessons learnt. In Proceedings of the 25th International Congress of the Aeronautical Sciences, Hamburg, Germany, 3–8 September 2006; International Council of Aeronautical Sciences (ICAS): Hamburg, Germany, 2006; pp. 1–8.
5. Botten, S.L.; Whitley, C.R.; King, A.D. Flight control actuation technology for next-generation all-electric aircraft. *Technol. Rev. J.* **2000**, *8*, 55–68.
6. Hussain, Y.M.; Burrow, S.; Henson, L.; Keogh, P. A review of techniques to mitigate jamming in electromechanical actuators for safety critical applications. *Int. J. Progn. Health Manag.* **2018**, *9*, 1–11. [CrossRef]
7. Vachtsevanos, G.J.; Vachtsevanos, G.J. *Intelligent Fault Diagnosis and Prognosis for Engineering Systems*; Wiley: Hoboken, NJ, USA, 2006; Volume 456.
8. Tinga, T.; Loendersloot, R. Physical model-based prognostics and health monitoring to enable predictive maintenance. In *Predictive Maintenance in Dynamic Systems*; Springer: Berlin/Heidelberg, Germany, 2019; pp. 313–353.
9. Yiwei, W.; Christian, G.; Binaud, N.; Christian, B.; Jian, F. A model-based prognostics method for fatigue crack growth in fuselage panels. *Chin. J. Aeronaut.* **2019**, *32*, 396–408.
10. Di Rito, G.; Luciano, B.; Borgarelli, N.; Nardeschi, M. Model-Based Condition-Monitoring and Jamming-Tolerant Control of an Electro-Mechanical Flight Actuator with Differential Ball Screws. *Actuators* **2021**, *10*, 230. [CrossRef]
11. Ray, A.; Tangirala, S. A nonlinear stochastic model of fatigue crack dynamics. *Probabilistic Eng. Mech.* **1997**, *12*, 33–40. [CrossRef]
12. Swindeman, R.; Swindeman, M. A comparison of creep models for nickel base alloys for advanced energy systems. *Int. J. Press. Vessel. Pip.* **2008**, *85*, 72–79. [CrossRef]
13. Li, X.; Ding, Q.; Sun, J.Q. Remaining useful life estimation in prognostics using deep convolution neural networks. *Reliab. Eng. Syst. Saf.* **2018**, *172*, 1–11. [CrossRef]
14. Guo, L.; Li, N.; Jia, F.; Lei, Y.; Lin, J. A recurrent neural network based health indicator for remaining useful life prediction of bearings. *Neurocomputing* **2017**, *240*, 98–109. [CrossRef]
15. Xia, M.; Zheng, X.; Imran, M.; Shoaib, M. Data-driven prognosis method using hybrid deep recurrent neural network. *Appl. Soft Comput.* **2020**, *93*, 106351. [CrossRef]
16. Al-Dulaimi, A.; Zabihi, S.; Asif, A.; Mohammadi, A. A multimodal and hybrid deep neural network model for Remaining Useful Life estimation. *Comput. Ind.* **2019**, *108*, 186–196. [CrossRef]
17. Liu, J.; Zhang, M.; Zuo, H.; Xie, J. Remaining useful life prognostics for aeroengine based on superstatistics and information fusion. *Chin. J. Aeronaut.* **2014**, *27*, 1086–1096. [CrossRef]
18. Baptista, M.; Henriques, E.M.; de Medeiros, I.P.; Malere, J.P.; Nascimento, C.L., Jr.; Prendinger, H. Remaining useful life estimation in aeronautics: Combining data-driven and Kalman filtering. *Reliab. Eng. Syst. Saf.* **2019**, *184*, 228–239. [CrossRef]
19. Jianzhong, S.; Fangyuan, W.; Shungang, N. Aircraft air conditioning system health state estimation and prediction for predictive maintenance. *Chin. J. Aeronaut.* **2020**, *33*, 947–955.
20. Li, Q.; Gao, Z.; Tang, D.; Li, B. Remaining useful life estimation for deteriorating systems with time-varying operational conditions and condition-specific failure zones. *Chin. J. Aeronaut.* **2016**, *29*, 662–674. [CrossRef]
21. Liu, J.; Wang, W.; Ma, F.; Yang, Y.; Yang, C. A data-model-fusion prognostic framework for dynamic system state forecasting. *Eng. Appl. Artif. Intell.* **2012**, *25*, 814–823. [CrossRef]
22. Wang, B.; Lei, Y.; Li, N.; Li, N. A hybrid prognostics approach for estimating remaining useful life of rolling element bearings. *IEEE Trans. Reliab.* **2018**, *69*, 401–412. [CrossRef]
23. Wang, P.; Long, Z.; Wang, G. A hybrid prognostics approach for estimating remaining useful life of wind turbine bearings. *Energy Rep.* **2020**, *6*, 173–182. [CrossRef]
24. Quattrocchi, G.; Iacono, A.; Berri, P.C.; Dalla Vedova, M.D.; Maggiore, P. A New Method for Friction Estimation in EMA Transmissions. *Actuators* **2021**, *10*, 194. [CrossRef]

25. Quattrocchi, G.; Berri, P.C.; Dalla Vedova, M.D.L.; Maggiore, P. Innovative Actuator Fault Identification Based on Back Electromotive Force Reconstruction. *Actuators* **2020**, *9*, 50. [CrossRef]
26. Berri, P.C.; Dalla Vedova, M.; Maggiore, P. A Lumped Parameter High Fidelity EMA Model for Model-Based Prognostics. In Proceedings of the 29th ESREL, Hannover, Germany, 22–26 September 2019; pp. 22–26.
27. Belmonte, D.; Dalla Vedova, M.; Maggiore, P. Electromechanical servomechanisms affected by motor static eccentricity: Proposal of fault evaluation algorithm based on spectral analysis techniques. In *Safety and Reliability of Complex Engineered Systems, Proceedings of the 25th European Safety and Reliability Conference, ESREL 2015, Zürich, Switzerland, 7–10 September 2015*; CRC Press: London, UK, 2015; pp. 2365–2372.
28. Belmonte, D.; Dalla Vedova, M.; Maggiore, P. Prognostics of Onboard Electromechanical Actuators: A New Approach Based on Spectral Analysis Techniques. *Int. Rev. Aerosp. Eng.* **2018**, *11*, 96–103. [CrossRef]

Article

Reliability-Oriented Configuration Optimization of More Electrical Control Systems

Zirui Liao [1,2,3], Shaoping Wang [1,2], Jian Shi [1,2,*], Dong Liu [1,2] and Rentong Chen [1,2]

1. School of Automation Science and Electrical Engineering, Beihang University, Beijing 100191, China; by2003110@buaa.edu.cn (Z.L.); shaopingwang@buaa.edu.cn (S.W.); ld_buaa@buaa.edu.cn (D.L.); rentongchen@buaa.edu.cn (R.C.)
2. Beijing Advanced Innovation Center for Big-Data Based Precision Medicine, Beihang University, Beijing 100191, China
3. Shenyuan Honors College, Beihang University, Beijing 100191, China
* Correspondence: shijian@buaa.edu.cn

Abstract: More electrical vehicles adopt dissimilar redundant control systems with dissimilar power supplies and dissimilar actuators to achieve high reliability and safety, but this introduces more intricacy into the configuration design. Currently, it is difficult to identify the optimum configuration via the conventional trial-and-error approach within an acceptable timeframe. Hence, it is imperative to discover novel methods for the configuration design of more electrical vehicles. This paper introduced the design specification of more electric vehicles and investigated the contribution of different kinds of actuators, presenting a new multi-objective configuration optimization approach on the foundation of system reliability, weight, power, and cost. By adopting the non-dominated sorting genetic algorithm-II (NSGA-II), the Pareto optimization design set was obtained. Then, the analytic hierarchy process (AHP) was introduced to make a comprehensive decision on the schemes in the Pareto set and determine the optimal system configuration. Eventually, numerical results indicated that the reliability of our designed configuration increased by 5.89% and 55.34%, respectively, compared with dual redundancies and single redundancy configurations, which verified the effectiveness and practicability of the proposed method.

Keywords: more electric vehicles; dissimilar redundant actuation system; NSGA-II algorithm; optimization design

1. Introduction

Safety critical systems in aircraft [1], carrier rockets [2], ships [3] and large machines [4] usually adopt redundant actuation systems to guarantee high reliability and safety [5]. In such designs, the rest systems can complete the work when one or more actuators fail. To reduce the number of common-cause faults and common-mode faults, dissimilar redundancy systems are adopted consequentially [6], in which different actuators with the same performances are used. With the rapid development of more electrical technologies, and more electric vehicles based on different power supply systems, actuation systems and computers have become increasingly common. In terms of actuation systems, hydraulic actuators (HAs), electro-hydrostatic actuators (EHAs) and electromechanical actuators (EMAs) [7] are leveraged in more electric vehicles. For instance, more electrical aircrafts are adopting heterogeneous actuation systems to move control surfaces based on HAs and EHAs [8]. Since HAs utilize hydraulic power supply systems while EHAs and EMAs adopt electrical power supply systems, the configuration design of more electric vehicles (MEVs) faces substantial challenges. The first one is how to design more electric control systems to maintain high mission reliability without permissive cost and weight. The second one is how to strike a balance between power supply systems and actuator distribution. The last one is how to determine the types of actuators for more electric control systems, which include HAs, EHAs, and EMAs. Given the aforementioned factors, the design of safety-critical system configurations becomes a daunting task. Therefore, an appropriate and effective approach should be adopted to solve the combination-explosion problem of system configurations, thus achieving the optimization design of a more electric control system (MECS) [9,10].

The optimization design of an electric control system needs to realize multi-objective optimization and comprehensively consider various contradictory indicators. Traditional methods include the weighting method [11,12], the constraint method [13,14], and goal programming [15–17]. However, these methods require the designer to master corresponding background knowledge to determine weights, energy dissipation, and other indicators, and the explosive configuration is another challenge that arises from those methods. In addition, the program needs to be run independently many times, which may lead to inconsistent results each time and render it more difficult for designers to make eventual decisions. In recent decades, various multi-objective intelligent optimization algorithms have emerged and vigorously developed, for example, the genetic algorithm (GA) [18], differential evolution algorithm (DE) [19], particle swarm optimization algorithm (PSO) [20,21], and non-dominated sorting genetic algorithm (NSGA) [22]. Among them, NSGA is famous for its good performance due to the adoption of non-dominated sorting and fitness-sharing strategy. The former increases the possibility of superior characteristics inherited by the next generation, while the latter sustains population diversity, overcoming the overmuch reproduction of super individuals and preventing prematurity convergence [23]. Nevertheless, there are also disadvantages such as high computational complexity in NSGA. Thus, in 2002, Deb et al. proposed NSGA-II [24]. In contrast to NSGA, the rapid non-dominated sorting method, elitism preservation strategy and congestion comparison were introduced in NSGA-II, which greatly reduced the computational complexity of the algorithm. Moreover, NSGA-II also expanded the distribution space of the solution set in the Pareto frontier, thus maintaining population diversity [25]. At present, NSGA-II has been widely applied to tackle multi-objective optimization problems in various fields. Xia et al. [26] investigated the multi-objective optimization problem for AUV conceptual design, where NSGA-II was applied to find the optimal Pareto frontier, with a comprehensive consideration of cost, effectiveness, and risks. The result verified the effectiveness of the algorithm. Alam et al. [27] studied the problem of AUV design and construction and employed NSGA-II to determine the optimum design of a torpedo-like AUV with a total length of 1.3 m. In addition, a kind of heavier-than-water underwater vehicle (HUV) was regarded as a research object by Liu et al. [28], and NSGA-II was adopted to establish a global approximation model, thus assisting the eventual optimal design of HUV. Nevertheless, these methods ignored the optimization designs of the global architectures of dissimilar control systems. As mentioned above, the optimization design of MECS is to solve the multi-objective optimization issue, where some contradictory indicators require all-round considerations. Therefore, the NSGA-II, combined with the AHP [29–32] method, was adopted in this paper to solve this problem, in which the binary encoding mode was employed and the decision variables were confined in discrete space. Moreover, the mutation process was modified to better match the characteristics of the problem. Eventually, the case study verified the validity of our approach.

The other sections of the paper are arranged as follows: Section 2 demonstrates the system descriptions. Section 3 presents the optimization approach on the foundation of NSGA-II and AHP method. Section 4 displays the case study, while the conclusions are presented in Section 5.

2. Mathematical Modeling of More Electric Control System (MECS)

According to the design specification of safety critical systems, more electric control systems require high reliability and safety. For example, the possibility of mission failure per flight owing to certain material damages in the flight control system should not surpass the upper limit. Normally, the failure rate of a flight control system is $10^{-9}/h$ to $10^{-10}/h$ for a commercial aircraft. Hence, it is necessary to utilize a dissimilar flight control system to maintain high reliability.

2.1. Single Control System Structure

A schematic diagram of the basic structure of a more electric control system, in which the controller, actuator, and sensor are main components, is illustrated in Figure 1. Since the electric power supply and hydraulic power supply are provided simultaneously, the actuator can be selected from HAs, EHAs, or EMAs, as shown in Figure 1. A HA is powered by a hydraulic supply system while an EHA or EMA is powered by an electric supply system. Since a centralized hydraulic power supply system has the characteristics of high-power density and fast response, HAs are widely used in control systems. However, the control system is heavy and easily exposed to oil contamination when HAs are adopted as they need pipeline transmission from the centralized power supply to the actuator. An EHA consists of the motor, pump, and cylinder that replaces the pipeline transmission by wire transmission. Hence, the weight of an EHA is light, whereas its heat dissipation is poor. An EMA consists of the motor and the ball screws. The command is transmitted from the wire controlling the motor and ball screws to drive the load. Therefore, an EMA is simple and light weight, but it gets stuck easy. Hence, to determine which type of actuator should be used, one has to comprehensively consider the weight, performance, and reliability.

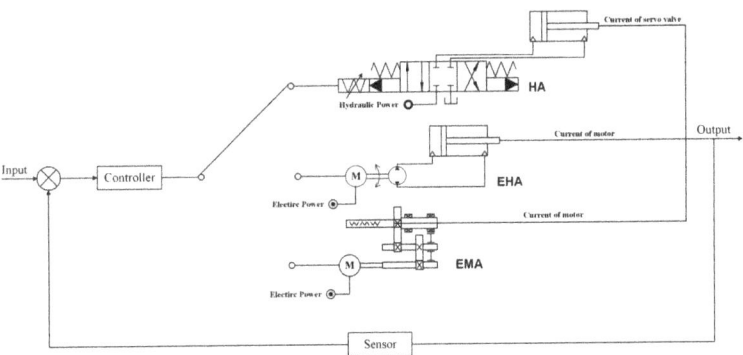

Figure 1. Structure of typical actuation system.

2.2. Redundant Configuration of Actuation System

The more electric control of safety critical systems not only requires the high-precision performance but also requires extremely high reliability and safety. Thus, redundant designs are often adopted, such as dual-redundant actuators and triple-redundant actuators. In order to integrate the advantages of hydraulic power and electric power, the typical isomorphic actuation system and heterogeneous actuation system shown in Figures 2–4 are used in real-world applications. In an aircraft, the HA/HA, HA/EHA, and HA/EMA are classic dual-redundant actuation systems, as shown in Figures 2–4.

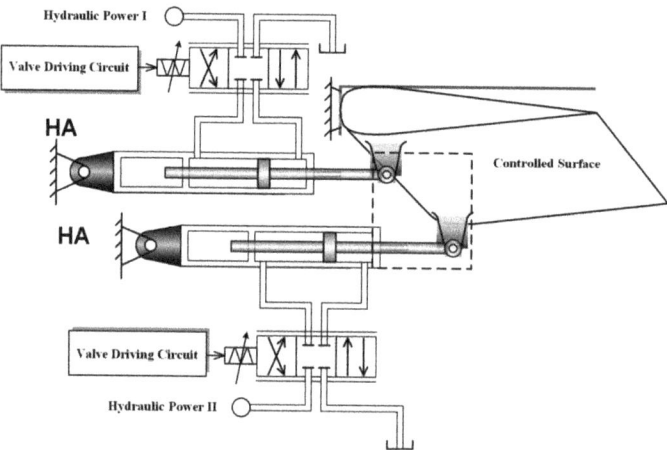

Figure 2. Configuration of actuation system with two HAs.

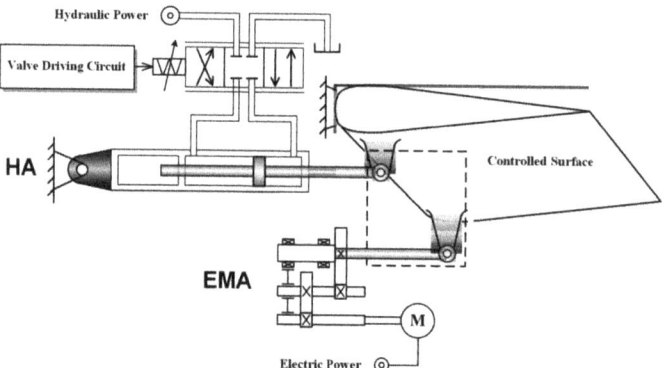

Figure 3. Configuration of actuation system with one HA and one EMA.

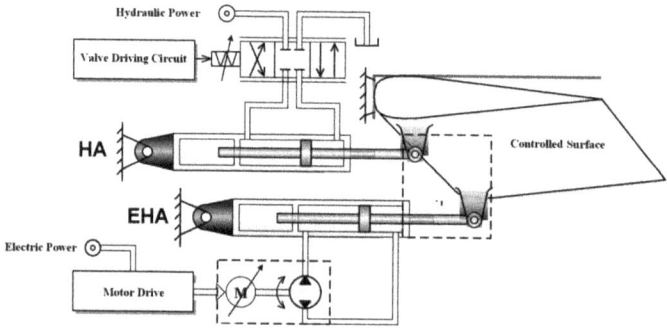

Figure 4. Configuration of actuation system with one HA and one EHA.

Generally, the regulation of a dissimilar redundant actuator is to design the different actuator powered by a different power supply for one surface. In such design, when any actuator or power supply system fails, the other can fulfill the task through fault switching. Table 1 presents the typical configuration of a dissimilar redundant actuation system.

Table 1. Typical configuration of dissimilar redundant actuation system.

Redundancy	Actuator Type	Power Supply
Dual redundancies	HA, HA	Hydraulic power
	HA, EMA	Hydraulic and electric power
	HA, EHA	Hydraulic and electric power
Triple redundancies	HA, EHA, and EMA	Hydraulic and electric power

Remark 1. *Triple redundancy was considered in this research because when the number of redundancies increases, the performance indicators of an MECS will increase accordingly, which is not beneficial for system operation. Besides, an actuation system with more than three redundancies is rarely used in practice.*

2.3. Redundant Configuration of More Electric Control System

As mentioned above, dissimilar redundancy technology is widely used in safety critical system design in order to improve system reliability. In a commercial aircraft, various controllers, actuators, and power supplies are applied in the aircraft actuation system. A typical redundant actuation system configuration based on high reliability for a commercial aircraft is illustrated in Figure 5. Although multiple control computers, actuators, and power supply systems can achieve high reliability, weights and costs will increase correspondingly, with maintainability decreased as well. Therefore, it is imperative to optimize the quantity of redundant control computers, power supply systems, and actuators. At same time, designers have to solve the common faults of dissimilar power supplies and dissimilar actuators as shown in Figures 2–4.

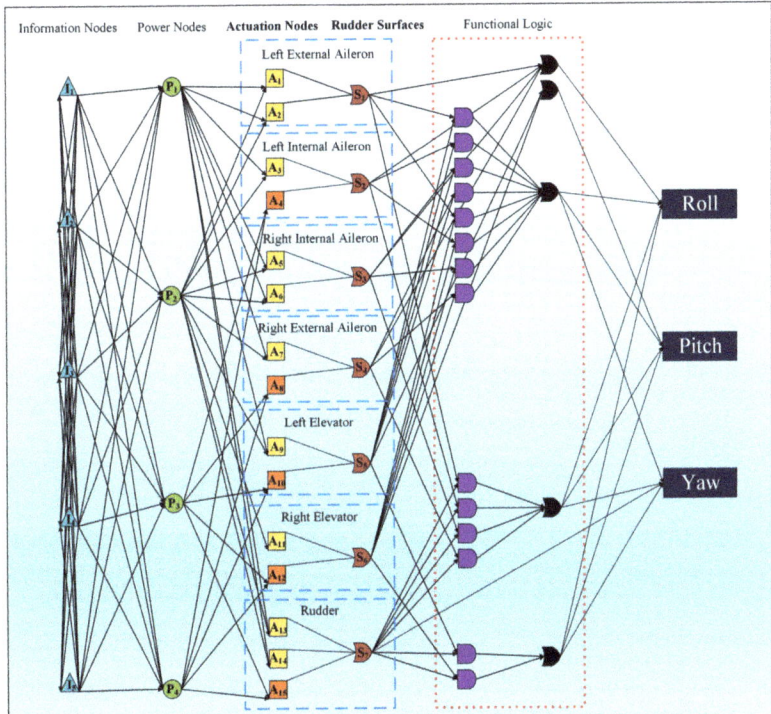

Figure 5. Configuration of commercial aircraft redundant actuation system.

In Figure 5, I expresses the information nodes of the control computer; P describes the power supply nodes; A is the actuator nodes; and S shows the surface of the aircraft. Herein, the actuator includes an HA, EHA, and EMA, as shown in Figures 2–4. Both the fonts of actuation nodes and rudder surfaces are bold since they are our main considerations.

Definition 1. *Power supply module set is* $\mathbf{P} = \{P_i\}(i = 1, 2, \ldots, n)$. *For aircraft shown in Figure 5, $P_1 = H$ means the power supply; P_1 is the hydraulic power; and $P_2 = E$ means that the power supply P_2 is electric power.*

Definition 2. *Actuator module set is* $\mathbf{A} = \{A_j\}(j = 1, 2, \ldots, m)$. *For aircraft shown in Figure 5, $A_1 = HA$ means that the actuator A_1 is a hydraulic actuator. $A_2 = EHA$ means that the actuator A_2 is an electro-hydrostatic actuator, and $A_3 = EMA$ means that the actuator A_3 is a mechatronic actuator.*

Remark 2. *Here, the commercial aircraft redundant actuation system is introduced as a typical MESC merely. In fact, our method is extensible and can be applied to more kinds of systems. In Section 4, we abstract a typical MECS with 3 hydraulic power supplies and 2 electric power supplies as the research object, which can be also extended to more combinations.*

3. Multi-Objective Optimization of MECS Based on NSGA-II and AHP

Though an MECS can improve system reliability and ensure safety, when the system structure is more sophisticated, the corresponding structural indicators such as weight and cost will increase enormously. The design of an MECS may also increase the system volume and faults derived from the heavy weight. Therefore, we need to optimize factors such as the weight, power dissipation, and reliability of an MECS to ensure a compromise between these indicators.

3.1. Multi-Objective Optimization Modeling of MECS

The primary task for optimization design is to transfer the optimization problem into mathematical descriptions. Generally, an optimization problem is composed of three elements: objective functions, decision variables, and constraint conditions. The optimization problem of an MECS can be expressed as

$$
\begin{aligned}
& J = \max R_s(t) \\
& R_s(t) = f(H_i, E_i, HA_j, EHA_j, EMA_j) \\
& s.t.\ 0 \leq m(H_i, E_i, HA_j, EHA_j, EMA_j) \leq M \\
& \quad 0 \leq \psi(H_i, E_i, HA_j, EHA_j, EMA_j) \leq \Psi \\
& \quad 0 \leq c(H_i, E_i, HA_j, EHA_j, EMA_j) \leq C \\
& \quad \vdots \\
& \quad 0 \leq g(H_i, E_i, HA_j, EHA_j, EMA_j) \leq G \\
& \quad i \in = \mathcal{L} = \{1, \ldots, m\} \subset \mathbb{N} \\
& \quad j \in = \mathcal{F} = \{1, \ldots, n\} \subset \mathbb{N}
\end{aligned}
\tag{1}
$$

where function J is the maximum reliability; $m(H_i, E_i, HA_j, EHA_j, EMA_j)$ is the actual evaluation mass of the MECS; M is the superior limit of mass; $\psi(H_i, E_i, HA_j, EHA_j, EMA_j)$ expresses the actual evaluation power dissipation of the MECS; Ψ is the superior limit of power dissipation; $c(H_i, E_i, HA_j, EHA_j, EMA_j)$ describes the actual evaluation cost of the MECS; c represents the superior limit of cost; $g(H_i, E_i, HA_j, EHA_j, EMA_j)$ represents the other actual evaluation indicators of MECS; and G represents the superior limit of other indicators. The specific evaluation methods of objective functions and constraints are given below.

3.1.1. Objective Function

According to Equation (1), reliability is the most essential indicator and the objective function. System reliability refers to the ability of the system to meet the specified functions within the specified time and under the specified conditions. Only when specific functions of an MECS are controlled effectively at the same time can the overall function of the system be guaranteed. Therefore, the overall functional reliability of an MECS is defined as the ability to effectively control the specific functions simultaneously within the specified time and under the specified conditions.

Based on the definition above and by considering an MECS with eight actuation functions, the functional reliability of an MECS can be expressed as

$$R_S = \Pr\{F_1 \cap F_2 \cap \cdots \cap F_8\} = \Pr\{S\} \qquad (2)$$

where $F_1, \ldots, F_8$ represent the pivotal functions that MECS should accomplish.

By considering the multiple control surface combinations involved in the realization of system functions, Equation (2) can be rewritten as

$$R_S = \Pr\left\{ \left(\bigcup_{x=1}^{p} F_1\right) \cap \left(\bigcup_{y=1}^{q} F_2\right) \cdots \cap \left(\bigcup_{z=1}^{m} F_8\right) \right\} \qquad (3)$$

where $\left(\bigcup_{x=1}^{p} F_1\right) \cap \left(\bigcup_{y=1}^{q} F_2\right) \cdots \cap \left(\bigcup_{z=1}^{m} F_8\right)$ represents a certain combination that realizes the pivotal functions.

To simplify the calculation, the program first calculated the overall reliability of electric power supplies E_1, E_2 and then calculated the overall reliability of hydraulic power supplies H_1, H_2. Next, we calculated the minimum path and the program performing disjoint operation. Through decoupling analysis, nine minimal paths were obtained. Consequently, the functional reliability of an MECS can be calculated by

$$R_S = \Pr\left\{\bigcup_{i=1}^{9} S_i\right\} = q_{E1}q_{E2}q_{E5}q_{E6}p_{EB}p_{ED}p_{EH}p_{EJ}p_{EN}p_{EO}p_{EP}p_{EQ}p_{ER}p_{ES}p_{ET}p_{EU} \\
p_{EV}p_{EW}p_{EX}p_{H1}p_{H3}p_{P1}p_{P2}p_{K1}p_{K2}p_{DD}p_{YY}p_{C1}p_{C2}p_{C3} + q_{E1}p_{E2}q_{E5}q_{E6}p_{EB}p_{ED}p_{EH} \\
p_{EJ}p_{EN}p_{EO}p_{EP}p_{EQ}p_{ER}p_{ES}p_{ET}p_{EU}p_{EV}p_{EW}p_{EX}p_{H3}p_{P1}p_{P2}p_{K1}p_{K2}p_{DD}p_{YY}p_{C1}p_{C2} \\
p_{C3} + \cdots + q_{E1}p_{E2}p_{E5}p_{EB}p_{ED}p_{EH}p_{EJ}p_{EN}p_{EO}p_{EP}p_{EQ}p_{ER}p_{ES}p_{ET}p_{EU}p_{EV}p_{EW}p_{EX} \\
p_{H1}p_{H3}p_{P1}p_{P2}p_{K1}p_{K2}p_{DD}p_{C1}p_{C2}p_{C3} + p_{E1}p_{E5}p_{EB}p_{ED}p_{EH}p_{EJ}p_{EN}p_{EO}p_{EP}p_{EQ}p_{ER} \\
p_{ES}p_{ET}p_{EU}p_{EV}p_{EW}p_{EX}p_{H1}p_{H3}p_{P1}p_{P2}p_{K1}p_{K2}p_{DD}p_{C1}p_{C2}p_{C3} \qquad (4)$$

Remark 3. *Equation (4) gives the relationship between system reliability and component system. In fact, system reliability is determined by each component reliability, that is, the actuator reliability, while actuator reliability is associated with weight, power consumption, and cost. These indicators will eventually have an influence on system reliability.*

3.1.2. Constraint Conditions

The aim of optimization design is to identify the optimal solution from the practical solutions. The optimal solution can meet the goal of design as far as possible. The elevation of weight will restrict the MECS installation power; hence, weight is a vital index for system property. Power dissipation is another vital property index. In the optimization design of an MECS, particularly for a highly-efficient MECS, how to decrease the power dissipation with the existent technology and apparatus is a challenge which need to be tackled by designers. Meanwhile, as an MECS has to harbor high performance and reliability, so the reliability of an MECS should be considered. For practical application, the cost is an indispensable indicator that must be evaluated. Eventually, weight, power dissipation, cost, and reliability are determined as constraint conditions.

- Weight

Weight is a vital evaluation index and decisive factor of an MECS. When we evaluate weight, the difficulties usually include various components, many of which exhibit a nonlinear growth relationship with design requirements, and serious coupling with other systems with different types of actuators. For an EHA, since it consists of the integration block, cylinder, pump, and motor, its weight evaluation can be presented as

$$M_{EHA} = M_{block} + M_{cylinder} + M_{pump} + M_{motor} \tag{5}$$

where M_{block}, $M_{cylinder}$, M_{pump} refer to the weight of the integration block, cylinder, and pump, respectively. M_{motor} denotes the weight of the brushless direct-current motor (BLDC). All of them have the same unit, kilograms.

A permanent magnet brushless direct-current motor (BLDC) is utilized in the EHA due to its satisfactory control property and great power-to-mass ratio. The weight forecast of the BLDC is expressed by

$$M_{motor} = 0.628 T^{3/3.5} + 0.783 \tag{6}$$

in which T represents the torque of BLDC.

The pump weight is proportionate to pump output. Thus, M_{pump} can be expressed as [33]

$$M_{pump} = 0.339 D + 2.038 \tag{7}$$

The integration block is the EHA frame and involves the indispensable parts such as the nonreturn valve, filter, and accumulator. The overall weight is speculated by the EHA power, which is expressed as [33]

$$M_{block} = 0.105 P_{EHA} + 2 \tag{8}$$

where P_{EHA} denotes the maximal power of EHA.

The fluid cylinder weight includes the four parts stated below:

$$M_{cylinder} = M_{cover} + M_{shell} + M_{piston} + M_{rod} \tag{9}$$

where M_{cover}, M_{shell}, M_{piston}, M_{rod} are the weights of the cylinder cover, shell, piston, and plunger rod, respectively. All of them can be calculated by

$$\begin{cases} M_{rod} = \frac{\pi}{4} \times d_{rod}^2 \times L_{rod} \times \rho_{steel} \\ M_{piston} = A \times t_{piston} \times \rho_{copper} \\ M_{shell} = \frac{\pi}{4} \times \left(d_{shell}^2 - \frac{4A}{\pi}\right) \times L_{shell} \times \rho_{steel} \\ M_{cover} = 2 \times \frac{\pi}{4} \times d_{shell}^2 \times t_{cover} \times \rho_{steel} \end{cases} \tag{10}$$

Thus, the total weight can be evaluated as

$$M = \sum_{i=1}^{n_{HA}} M_{HA} + \sum_{i=1}^{n_{EHA}} M_{EHA} + \sum_{i=1}^{n_{EMA}} M_{EMA} + M_{elec} + M_{pipe} + M_{wire} \tag{11}$$

where M_{HA}, M_{EHA}, M_{EMA} represent the weights of HA, EHA, and EMA, respectively, and M_{elec}, M_{pipe}, M_{wire} represent the weights of electric equipment, pipe and wire, respectively.

- Power efficiency

For an EHA, the motor converts electric energy to hydraulic power, pushing the cylinder to realize the movement of the control surface. For an EHA system, when its control surface load is certain, it saves more energy, resulting in lower power dissipation. The control surface torque T_S, control surface velocity ω_S, and pressure P_S are known design specifications. The aim of decreasing the power dissipation of an EHA is to decrease the power output of the motor P, which is written as

$$P = T \times W \qquad (12)$$

The motor is directly connected to the pump, so the torque and speed of the motor and pump are the same. The torque of the pump is expressed as

$$T = J_{pm} \times \frac{d\omega}{dt} + K_{fric} \times \omega + (p_f + 2p_{pipe}) \times \frac{D}{2\pi} \qquad (13)$$

where J_{pm} is the pump rotational inertia; K_{fric} denotes the viscosity factor, $J_{pm} \times d\omega/dt = 0$; the pump rotation velocity is constant; p_f denotes the differential pressure between the 2 cylinder cavities; and p_{pipe} denotes the pressure consumption within the pipe. Hence, Equation (13) is written as

$$T = K_{fric} \times \omega + (p_f + 2p_{pipe}) \times \frac{D}{2\pi} \qquad (14)$$

in which $p_f = F/A$, F is the loading force of an EHA and A denotes the piston effective area, which can be calculated by

$$A = \frac{kF}{P_S} \qquad (15)$$

in which $k > 1$ denotes a logical excess margin, and F can be described by

$$F = \frac{T_S \times \sin(\theta + 30°)}{L} \qquad (16)$$

For an EMA, the power dissipation is determined by the ball screw. Therefore, Equation (17) gives the calculation method for the drive torque of the ball screw:

$$T_a = \frac{F_a \times I}{2\pi \times n_1} \qquad (17)$$

where T_a represents the drive torque; I is the lead screw; n_1 is the positive efficiency of the feed screw; and F_a refers to the axial load, which can be expressed by

$$F_a = F + \mu m g \qquad (18)$$

where F means the axial cutting force of the lead screw, μ is the comprehensive friction coefficient, and m refers the weight of the worktable and workpiece.

After the calculation of the drive torque T_a, the motor power can be determined accordingly; thus, the corresponding power dissipation of the motor can be obtained.

For the power dissipation evaluation of a HA, the volumetric efficiency of the pump η_0 is the main affecting factor, which can be described by

$$\eta_0 = \frac{Q}{Q_0} \times 100\% \qquad (19)$$

where Q_0 means the theoretical flow of the pump, and Q is the actual pump flow, which is written as

$$Q = Q_{pump} + Q_{pipe} + Q_{cylinder} \qquad (20)$$

where Q_{pump} denotes the pump leak, Q_{pipe} denotes the loss of flow within the hydraulic tube, and $Q_{cylinder}$ denotes the cylinder flow.

$$\begin{cases} Q_{pump} = \xi \times (p_1 - p_2) = \xi \times F/A \\ Q_{cylinder} = A \times v \\ Q_{pipe} = 2 \times \xi \times p_{pipe} \end{cases} \qquad (21)$$

where ξ is the pump leakance coefficient, p_{pipe} denotes the pipe pressure drop, and v denotes the cylinder speed.

- Cost

For an MECS, cost is another index worthy of consideration. In this paper, the total costs are divided into two parts:

$$C = C_{manu} + C_{oper} \tag{22}$$

where C represents the total cost of an MECS, C_{manu} refers to the manufacturing cost, and C_{oper} represents the operation cost.

However, when we evaluated the total cost, the component cost exhibited a nonlinear growth relationship with the requirement. To simplify this process, this paper utilized different configurations to obtain the costs of all components. In the evaluation of manufacturing costs, the main components considered included the HA, EHA, EMA, oil boxes, pipelines, cables, engine-driven pump (EDP), and electric motor driven pump (EMP). The evaluation was conducted mainly based on the similarity principle. Thus, the manufacturing costs is expressed as

$$C_{manu} = \sum_{i=1}^{n_{edp}} C_{edp} + \sum_{i=1}^{n_{emp}} C_{emp} + \sum_{i=1}^{n_{tank}} C_{tank} + \sum_{i=1}^{n_{motor}} C_{motor} + \sum_{i=1}^{n_{act}} C_{act} + \sum_{i=1}^{n_{pipe}} C_{pipe} + \sum_{i=1}^{n_{wire}} C_{wire} \tag{23}$$

where C_{edp} represents the cost of the EDP; C_{emp} refers to the cost of the EMP; C_{tank} is the cost of the tank; C_{motor} is the cost of the motor; C_{act} describes the cost of the actuator; C_{pipe} expresses the cost of the pipe; and C_{wire} denotes the wire cost.

The operation expense primarily denotes the Direct Operating Costs (DOCs), which are changeable costs directly derived from operating the aircraft [34]. The fuel expense and maintenance expenditure of DOCs are changeable. Hence, the DOCs are predominantly related to fuel expense and maintenance expenditure herein. The operation expense speculation is written as

$$C_{oper} = C_{oil} + C_{main} \tag{24}$$

where C_{oil} and C_{main} represent the fuel cost and maintenance cost, respectively.

Based on the statistical proportion of main flight control maintenance cost in the total maintenance expenditure, the fuel expense speculation primarily considers the fuel consumption during the life of the commercial aircraft. The function of fuel cost with respect to time is expressed as

$$C_{oil}(t) = 50 \times C_0 \times Weight \times (1.02^t - 1) \tag{25}$$

where C_0 means the average annual fuel cost of the aircraft and $Weight$ represents the fuel weight.

The maintenance cost of the aircraft in the whole life cycle mainly has two parts [35]. One is called the cyclic cost C_C, which is related to the take-off and landing of the aircraft, such as the maintenance of its braking device, flap system, and landing gear. The other cost is related to the flight time of the aircraft, such as the regular loss and replacement of parts caused by each hour of flight. This part of the cost is called hourly cost C_H. The evaluation of the maintenance cost is usually accomplished through the statistic of man-hour cost. Equation (26) provides the method for man-hour cost calculation

$$C_{main} = C_H \times t + C_C \times t + M \times (1/\alpha) \times R \times t \tag{26}$$

3.1.3. Design Variables

The expression of a goal suggests that there are substantial that which ought to be identified in configuration designs. Nevertheless, considering all those variables during optimization makes convergence difficult and is quite time-consuming. Hence, merely vital variables having a remarkable influence on MECS property ought to be optimized. Overall, the selective principles of the design variates are stated below:

1. The quantity of design variates ought to be decreased to the utmost extent. Overall, the quantity of design variates in mechanical optimization design should not surpass 5.
2. The variables ought to exert a remarkable impact on the goal function. Indexes affecting the constraint and property directly ought to be chosen as design variates.
3. The chosen variates ought to be independent.
4. The variates ought to be chosen as per the optimization goal.

According to the aforesaid principles, the optimization variates are chosen via:

$$x = \left\{ \begin{array}{c} x_1 \\ x_2 \\ x_3 \end{array} \right\} = \left\{ \begin{array}{c} n_{HA} \\ n_{EHA} \\ n_{EMA} \end{array} \right\} \tag{27}$$

3.2. Multi-Objective Optimization Based on NSGA-II

The optimization issue herein aimed at the configuration design of an MECS. NSGA-II is proper for optimizing multi-objective issues as it exhibits strong distributed capability and rapid convergence. The flow chart of NSGA-II is shown in Figure 6. Subsequently, the encoding/decoding mode and fast non-dominated sorting process are elaborated.

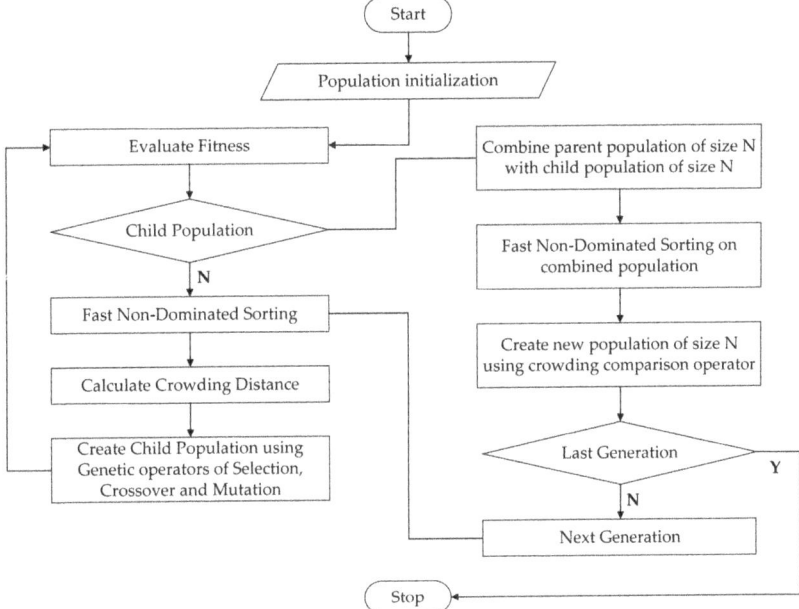

Figure 6. Flow chart of NSGA-II.

3.2.1. Encoding and Decoding

In our research, the binary encoding mode was adopted in NSGA-II, which was simple, efficient, and easy to design and use for implementing crossover and mutation operations. Two types of energy information were encoded through this mode, while the configuration of each actuation system was also expressed in binary. Figure 7 illustrates this coding mode.

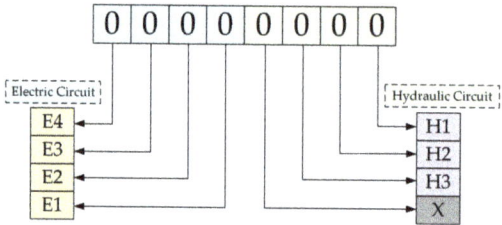

Figure 7. Binary encoding mode adopted in NSGA-I.

3.2.2. Fast Non-Dominated Sorting

Compared with non-dominated sorting approach utilized in NSGA arithmetic, the rapid non-dominated sorting utilized in NSGA-II algorithm requires lower computational complexity, which is only about $O(MN^2)$. The overall algorithm flow was presented as follow:

Algorithm 1: Fast non-dominated sorting

1: Fast-non-dominated-sort (P)
2: for each $p \in P$
3:   for each $q \in P$
4:     if $p \leq_n q$, then   #   if p is dominated by q, then add q to S_p
5:       $S_p = S_p \cup q$
6:     else if $q \leq_n p$, then
7:       $n_p = n_p + 1$
8:   if $n_p = 0$, then
9:     $p_{rank} = 1, F_1 = F_1 \cup P$ #   when n_p of the individual is 0, then this individual is the first level of Pareto
10: The comparison of dominating relationships between individuals, S_p and n_p, are introduced for storage and records, respectively; $\leq_n$ represents the comparison of dominating relationships. The solution of $n_p = 0$ is stored in the records of level 1, and the solution of level 1 has higher priority than that of level 2.
11: $i = 1$.
12: while $F_i = \varnothing$ do
13:   H=$\varnothing$
14:   for each $p \in F_i$
15:     for each $q \in S_p$ #   Sort all the individuals in S_p
16:       $n_q = n_q - 1$
17:     if $n_q = 0$, then   #   when n_q of the individual is 0, it is a non-dominated individual
18:       $q_{rank} = i + 2, H = H \cup q$ #   The Pareto level of this individual is the current highest level plus 1. At this moment, the initial value of i was 0, so we added 2.
19: end while
20: $i = i + 1, F_i = H$
21: Loop the program to obtain level 2, level 3 ... The computational complexity is $O(MN^2)$

3.2.3. Crowding Degree Calculation

Crowing degree I_d refers to the local crowding distance between any two adjacent points whose level are the same in the target space. The purpose of introducing crowding degree is to improve the distribution uniformity of the Pareto solution set. Furthermore, it could not only boost the diversity of population but also enhance system robustness. The crowding degree I_d could be expressed as the length of the maximum rectangle of individual i on both sides, where the rectangle only includes i itself.

The main procedure to determine the degree of individual congestion involves three steps:

- Define the crowding degree of every individual in population i as $I_d = 0$;
- Define the crowding degree 0_d and i_d of boundary individuals as ∞ according to each evaluation indicator;

- Define the crowding degree of marginal individuals as a larger number $L(1)_d = L(l)_d = M$ to prioritize individuals on the sorting edge; thus, the crowding degree of any other individual I_d can be expressed as

$$I_d = \sum_{j=1}^{m} \left(\left| f_j^{i+1} - f_j^{i-1} \right| \right) \tag{28}$$

where j refers to each evaluation indicator; f_j^{i+1} represents the jth evaluation indicator value of individual $i + 1$; and f_j^{i+1} represents the jth evaluation indicator value of individual $i - 1$.

Through the aforesaid steps and calculation, every individual was endowed with two attributes, i.e., the crowding distance and the rank. This laid the foundation for the follow-up processes.

3.2.4. Optimal Selection

The optimal selection avoids the loss of effective factors and ensures the survival rate of high-performance individuals, so it can ensure that the Pareto optimal solution is continuously optimized. The optimal selection can not only improve the efficiency and convergence of the optimization but also ensure the uniformity of the optimization process. The selection process is completed by comparing the results of fast non-dominated sorting and crowding calculation, and better individuals are selected after comparison. Two steps are involved during this process.

- The first step is the rank comparison. Select two individuals a and b randomly and make comparison between A_{rank} (the non-dominated rank of individual a) and B_{rank} (the non-dominated rank of individual b). When $A_{rank} < B_{rank}$, a is better than b and vice versa. Moreover, the crowding degree requires to be compared when $A_{rank} = B_{rank}$;
- The second step is the crowding degree comparison. When condition $A_d > B_d$ is satisfied, it indicates that individual a is better; otherwise, individual b is better. Then, the better individual is selected to continue the following optimal processes.

3.2.5. Crossover

The main function of crossover is to simulate gene recombination in the process of heredity and evolution. There are various approaches to implement it, such as uniform crossover, multi-point crossover, and binary crossover. In our research, the simulated binary crossover is adopted for crossover operation, which is expressed as

$$\begin{aligned} y_{1j} &= 0.5 \times \left[(1 + \gamma_j) x_{1j} + (1 - \gamma_j) x_{2j} \right] \\ y_{2j} &= 0.5 \times \left[(1 - \gamma_j) x_{1j} + (1 + \gamma_j) x_{2j} \right] \\ &\begin{cases} (2u_j)^{\frac{1}{\eta+1}} \text{ if } u_j \leq 0.5 \\ \left(\frac{1}{2(1-u_j)} \right)^{\frac{1}{\eta+1}} \text{ else} \end{cases} \end{aligned} \tag{29}$$

where $u_j \in (0,1)$; x_{1j} and x_{2j} are parent individuals; y_{1j} and y_{2j} are offspring individuals; and $\eta > 0$ refers to the cross-distribution index. Generally, $\eta = 20$ is the best value of the cross-distribution index by default.

3.2.6. Mutation

Mutation is widely applied to simulate variation links in the process of biological heredity and evolution, which refers to the replacement between genes, that is, substitute the other gene values with the alleles on the locus to create new individuals. Through mutation, not only could the local searchability be improved but also the population diversity is guaranteed.

This paper introduced the polynomial mutation method into the research on multi-objective optimization problems. The form of mutation operator is expressed as

$$V'_k = V_k + \delta(u_k - L_k)$$
$$\delta = \begin{cases} \left[2u + (1-2u)(1\mid -\delta_1)^{\eta_m+1}\right]^{\frac{1}{\eta_m+1}} - 1 \; if \; u \leq 0.5 \\ 1 - \left[2(1-u) + 2(u-0.5)(1-\delta_2)^{\eta_m+1}\right]^{\frac{1}{\eta_m+1}} else \; if \; u > 0.5 \end{cases} \quad (30)$$

where $\delta_1 = \frac{(V_k - L_k)}{(u_k - L_k)}$, $\delta_2 = \frac{(u_k - V_k)}{(u_k - L_k)}$, u denotes a stochastic number in interval [0,1], η_m denotes the distribution index, and V_k is a parent individual.

3.3. Comprehensive Evaluation of System Configuration Based on AHP

The analytic hierarchy process (AHP) is a multi-objective decision analysis approach that combines qualitation and quantitation analyses, where elements associated with task decisions are divided into the object level, criterion level and scheme level. The AHP mathematizes the decision-making via a few quantitation data based on deep analyses, such as the influential factors and inner association of the decision-making problems. AHP can offer an easy decision approach for intricate decision issues with several standards and no evident structure features. Questions to be evaluated in MECS optimization design ought to be methodized and layered, and the structure model of hierarchy analysis ought to be constructed. A three-level hierarchy structure of AHP was presented by Figure 8. Level 1 is the general objective of decision-making, which denotes the optimum design of MECS. Level 2 denotes the criterion layer, where the weight, power dissipation, expense, and dependability are utilized as evaluation standards of system multi-objective optimization design. Level 3 denotes the scheme layer, where the solutions are determined in the Pareto frontier acquired by NSGA-II. Besides, the process of AHP mainly includes the judgment matrix structure, weight coefficient calculation, consistency of judgment matrix verification, and weight coefficients of goal level calculation. The flow chart of AHP was shown in Figure 9.

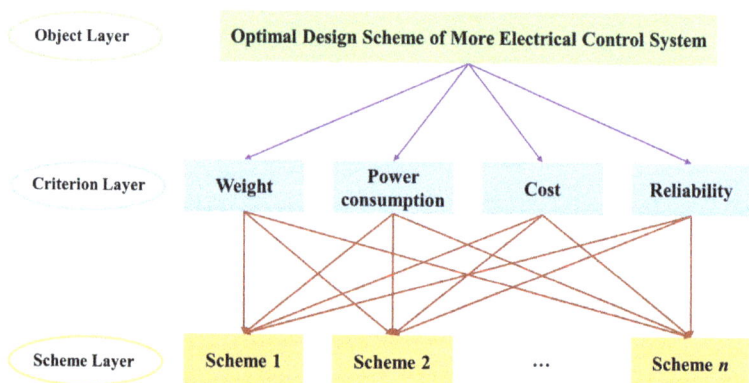

Figure 8. Three-level hierarchical framework of AHP.

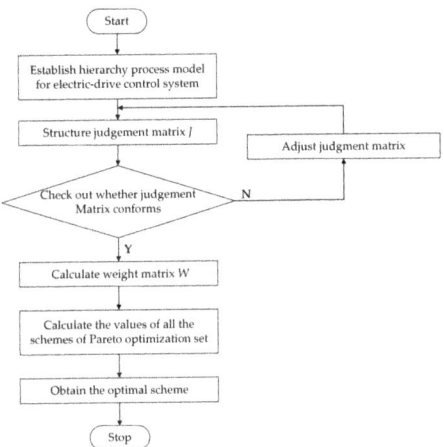

Figure 9. Flowchart of AHP.

1. Construct the decision-making model for AHP according to Figure 8.
2. Structure the judgement matrix J. Judgement matrix is established as per the association between the goals in the criterion layer.

$$J = \begin{bmatrix} 1 & 3 & 9 & 1 \\ \frac{1}{3} & 1 & \frac{1}{3} & \frac{1}{3} \\ \frac{1}{9} & \frac{1}{3} & 1 & \frac{1}{9} \\ 1 & 3 & 9 & 1 \end{bmatrix} \quad (31)$$

3. Validate the judgement matrix coherence. Coherence index is computed via:

$$\begin{cases} CI = \frac{\lambda_{max} - n}{n-1} \\ CR = \frac{CI}{RI} \end{cases} \quad (32)$$

where CI denotes the coherence index of the judgment matrix, RI denotes the average stochastic coherence index of the matrix (specific values were presented by Table 2), CR denotes the stochastic coherence ratio of the matrix, λ_{max} denotes the maximal value of the matrix characteristic value, and n denotes the matrix order.

Table 2. Values of the average stochastic coherence index.

n	1	2	3	4	5	6	7	8	9
RI	0.00	0.00	0.58	0.90	1.12	1.24	1.32	1.41	1.45

By computation, if λ_{max} = 3.7638 and CI = (3.7638 − 4)/(4 − 1) = −0.0787, then CR = −0.0787/0.90 = −0.0875 < 0.1, which suggests that the weight matrix is coherent. When the consistence test is not met, return to step 2 and reconstruct the judgement matrix.

4. Compute the weighted coefficient between the contrasted elements with the relevant standards. Compute the continued product M_i of each row element in A, the product of every row element, and its n-th root $\overline{w}_1$.

$$\begin{cases} M_i = \prod_{i=1}^{n} x_{ij} = \begin{bmatrix} 27 & 0.0370 & 0.0041 & 27 \end{bmatrix}^T, i = 1, \cdots, n \\ \overline{w}_1 = \sqrt[n]{M_i} = \begin{bmatrix} 2.2795 & 0.4387 & 0.2533 & 2.2795 \end{bmatrix}^T, i = 1, \cdots, n \end{cases} \quad (33)$$

Normalize $\overline{w}_1$ as $W_i = \overline{w}_1 / \sum_{i=1}^{n} \overline{w}_1$, where W_i denotes the weighted coefficient of every factor.

$$W = \begin{bmatrix} 0.4341 & 0.0835 & 0.0482 & 0.4341 \end{bmatrix} \quad (34)$$

5. Speculate the design in the Pareto frontier as per the weighted coefficients of every standard, and afterwards get the optimum design of MECS.

4. Case Study and Discussion

The diagram of an MECS with three fluid power supplies and two electric power supplies (3H2E) was presented by Figure 10. The tendency toward more electricity causes the transformation from hydraulic power into standby energy, or the replacement of hydraulic power with electrical energy, which produces more options for all control surfaces. For every actuator, it can be an HA linked to hydraulic source or an EHA linked to electrical source. In addition, each actuator is controlled by at least one controller. Therefore, the number of alternative configurations of the MECS is exceedingly high.

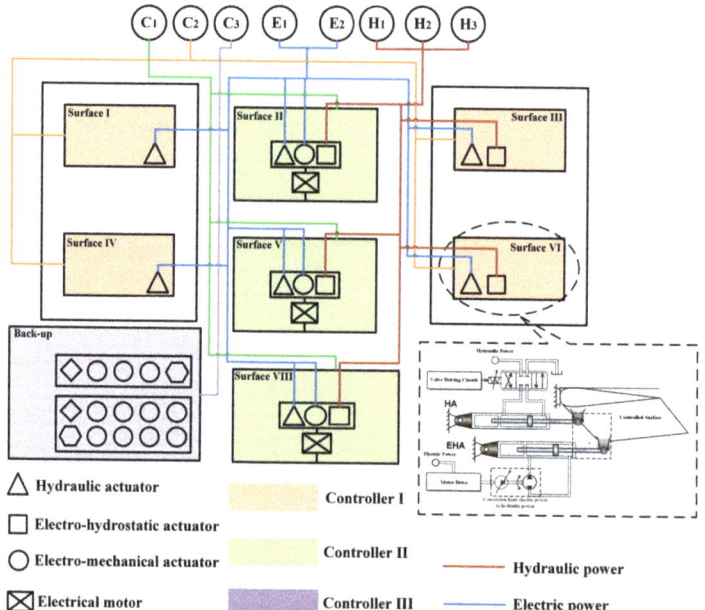

Figure 10. A typical 3H2E more electric control system (MECS).

In this section, we introduced the proposed optimization method into the case study of the MECS illustrated in Figure 10. By setting the maximum optimization iterations of NSGA-II as 200, and population size as 500, we obtained the Pareto frontier of the optimization result illustrated in Figure 11. W refers to the weight of the MECS, P is the power consumption of the MECS, and C represents the expense of the MECS. The red-round-dot-curved surface is a 3D Pareto optimum front curved surface. Every dot denotes an actuation design. The green line, marked by hollow pentagram; blue line, marked by cross symbol; and yellow line, marked by triangle were projected on three planes. The Pareto front was utilized to realize multi-objective optimization and to achieve a certain decrease in configuration quantity.

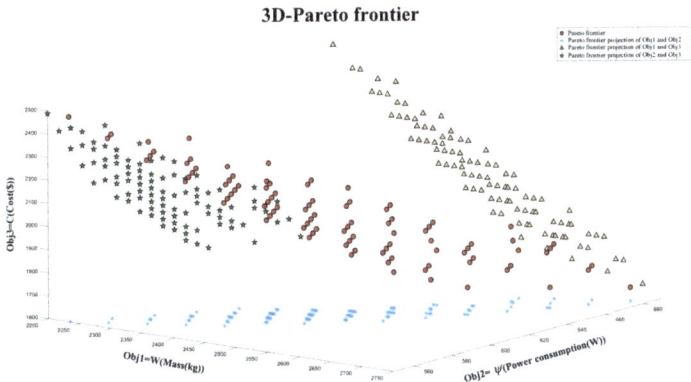

Figure 11. 3D-Pareto frontier of multi-objective optimization of MECS.

Moreover, the graph in Figure 11 presents the relationship among the three objectives. The line in x, y plane presents that weight and power have the relation of promotion. The line in x, z plane presents that weight and expense are contradictory. The line in y, z plane displays contradictory power and expense of an MECS. When the power decreases, the expense is elevated. The design objective is to minimize power and expense; hence, those two functions conflict with one another. Due to the conflicts between design indexes, it is hard to make each index optimum simultaneously.

In addition, to better illustrate the relationship between these three objectives and system reliability, we projected Figure 11 onto three 2-D planes, respectively, and obtained the 2D-pareto frontiers in Figures 12–14. Reliability was used as the scale parameter to color different system configurations, i.e., the brighter the color, the higher the reliability.

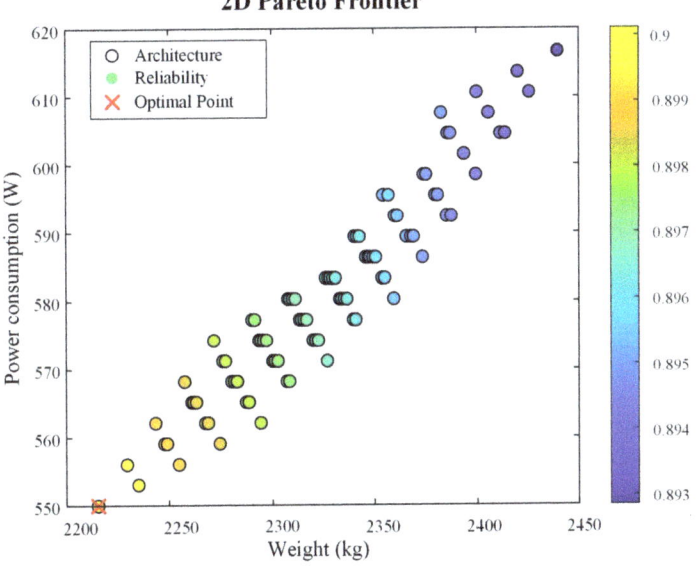

Figure 12. 2D Pareto frontier between weight, power consumption, and reliability.

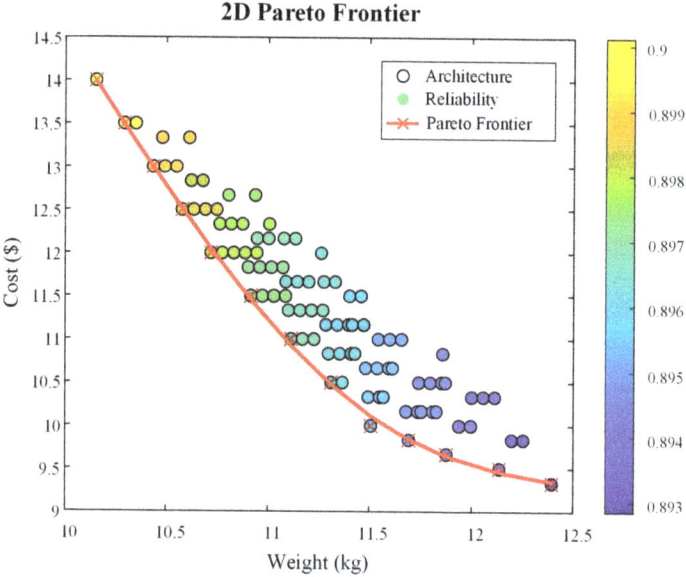

Figure 13. 2D Pareto frontier among weight, cost, and reliability.

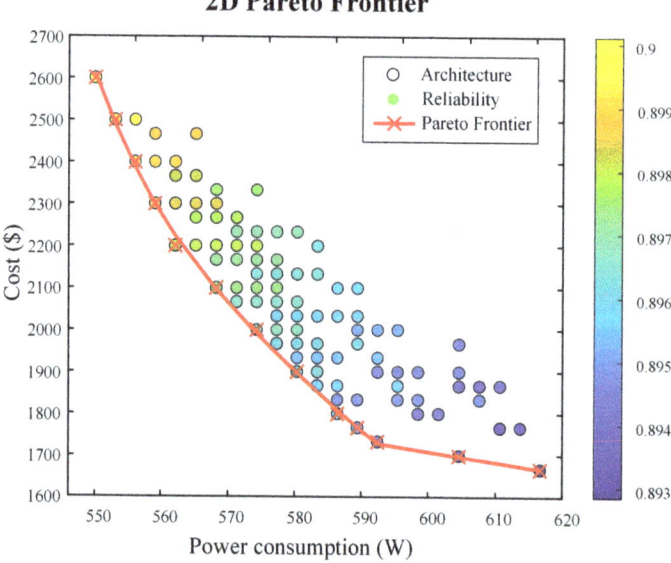

Figure 14. 2D Pareto frontier among power consumption, cost, and reliability.

Through multi-objective optimization, the Pareto optimization frontier was acquired. Nevertheless, finding the best way to identify the optimal solution in the Pareto frontier remains a daunting challenge. Thus, AHP was introduced to solve this problem. According to Section 3.3, diverse optimum solutions can be realized via structuring diverse judgment matrices when we use AHP to perform decision analyses. Thus, to better observe the association between the goals more, the judgment matrix was presented by Table 3, and the eventual result was illustrated in Figures 15 and 16.

Table 3. The association between the objectives.

Objectives	Weight	Power Dissipation	Cost	Reliability
Weight	1	3	9	1
Power dissipation	1/3	1	1/3	1/3
Cost	1/9	1/3	1	1/9
Reliability	1	3	9	1

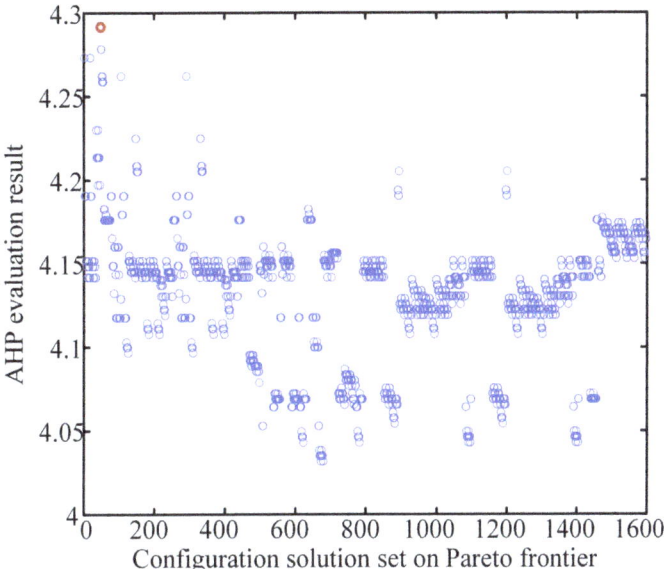

Figure 15. AHP assessment outcomes of Pareto frontier.

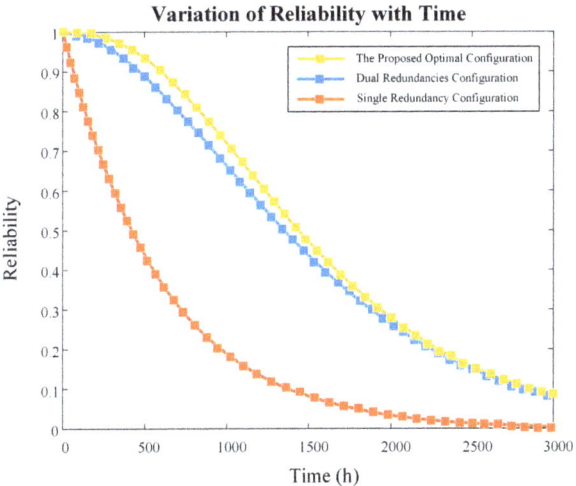

Figure 16. Reliability comparison among different configurations.

Through a comprehensive analysis of AHP, the score of each solution on the Pareto frontier was acquired. The solution with the greatest score is the best one. As presented by Figures 15 and 16, the optimal design strategy is number 96 of the Pareto frontiers, with triple redundancy configuration used in the MCES. The reliability of our designed configuration increased by 5.89% and 55.34% respectively, compared with the dual redundancy and single redundancy configurations. The results indicated that our approach combining multi-objective optimization and decision-making could realize the multi-objective optimization design of an MECS. The solutions on the Pareto frontier acquired by NSGA-II arithmetic were uniformly distributed, which suggested that we could offer designers more diversified choices. Subsequently, AHP was employed for the eventual decision analyses, which enabled designers to introduce predilections and experiences for different goals.

5. Conclusions and Future Work

The present research demonstrated the multi-objective optimization for an MECS. By designing the level length, the objectives are optimized. However, it is hard to minimize every goal simultaneously due to the conflicts between these objectives. Thus, an appropriate multi-objective optimization algorithm, which can compromise multiple objectives, is required to acquire the Pareto optimum solution set. Besides, since the Pareto frontier solution is not a solution but a solution set, an approach of multi-objective decision-making analysis is required to identify the optimum solution in the Pareto frontier. Herein, a combination of multi-objective optimization and multi-objective decision-making was utilized. The NSGA-II was employed to determine the optimum solution set, i.e., the Pareto frontier. Subsequently, the optimum design was acquired via an AHP. The outcome revealed the practicability of our optimization approach. In addition, the present research discovered that the multi-objective optimization and multi-objective decision-making approaches could be utilized in the optimization design of an MECS, and that this approach suited other components and systems as well.

Future work should encompass the improvement of real-time performance for an optimization algorithm, with more evaluation indicators considered.

Author Contributions: Conceptualization, Z.L. and R.C.; methodology, Z.L. and R.C.; software, D.L.; validation, Z.L. and D.L.; writing—original draft preparation, Z.L.; writing—review and editing, S.W.; supervision, S.W. and J.S.; project administration, J.S. All authors have read and agreed to the published version of the manuscript.

Funding: This research was funded by the National Natural Science Foundation of China (No. 51620105010) and the program of China Scholarship Council (No. 202106020106).

Institutional Review Board Statement: Not applicable.

Informed Consent Statement: Not applicable.

Data Availability Statement: This study does not report any data.

Conflicts of Interest: The authors declare no conflict of interest.

References

1. Sun, K.; Gebre-Egziabher, D. Air data fault detection and isolation for small UAS using integrity monitoring framework. *Navig.-J. Inst. Navig.* **2021**, *68*, 577–600. [CrossRef]
2. Kotyegov, V.; Kutovoy, S. Transfer of aerospace technologies for ensuring safety of industrial and civil objects. In Proceedings of the Second IEEE International Workshop on Intelligent Data Acquisition and Advanced Computing Systems: Technology and Applications, 2003 Proceedings, Lviv, Ukraine, 8–10 September 2003; pp. 136–138.
3. Cossentino, M.; Lopes, S.; Renda, G.; Sabatucci, L.; Zaffora, F. Smartness and autonomy for shipboard power systems reconfiguration. In Proceedings of the International Conference on Modelling and Simulation for Autonomous Systems, Palermo, Italy, 29–31 October 2019; pp. 317–333.
4. Joung, E.; Lee, H.; Oh, S.; Kim, G. Derivation of railway software safety criteria and management procedure. In Proceedings of the 2009 International Conference on Information and Multimedia Technology, Jeju, Korea, 16–18 December 2009; pp. 64–67.
5. Cai, B.G.; Jin, C.M.; Ma, L.C.; Cao, Y.; Nakamura, H. Analysis on the application of on-chip redundancy in the safety-critical system. *IEICE Electron. Express* **2014**, *11*, 20140153. [CrossRef]

6. Wang, S.P.; Cui, X.Y.; Jian, S.; Tomovic, M.M.; Jiao, Z.X. Modeling of reliability and performance assessment of a dissimilar redundancy actuation system with failure monitoring. *Chin. J. Aeronaut.* **2016**, *29*, 799–813. [CrossRef]
7. Li, T.Y.; Chen, T.T.; Ye, K.; Jiang, X.X. Research and simulation of marine electro-hydraulic steering gear system based on EASY5. In Proceedings of the IEEE 3rd Information Technology, Networking, Electronic and Automation Control Conference (ITNEC), Chengdu, China, 15–17 March 2019; pp. 1972–1979.
8. Jiao, Z.; Yu, B.; Wu, S.; Shang, Y.; Huang, H.; Tang, Z.; Wei, R.; Li, C. An intelligent design method for actuation system architecture optimization for more electrical aircraft. *Aerosp. Sci. Technol.* **2019**, *93*, 105079. [CrossRef]
9. Sarlioglu, B.; Morris, C.T. More electric aircraft: Review, challenges, and opportunities for commercial transport aircraft. *IEEE Trans. Transp. Electrif.* **2015**, *1*, 54–64. [CrossRef]
10. Benzaquen, J.; He, J.; Mirafzal, B. Toward more electric powertrains in aircraft: Technical challenges and advancements. *CES Trans. Electr. Mach. Syst.* **2021**, *5*, 177–193. [CrossRef]
11. Arbaiy, N. Weighted linear fractional programming for possibilistic multi-objective problem. In *Computational Intelligence in Information Systems*; Springer: Cham, Switzerland, 2015; pp. 85–94.
12. Marler, R.T.; Arora, J.S. The weighted sum method for multi-objective optimization: New insights. *Struct. Multidiscip. Optim.* **2010**, *41*, 853–862. [CrossRef]
13. Du, Y.; Xie, L.; Liu, J.; Wang, Y.; Xu, Y.; Wang, S. Multi-objective optimization of reverse osmosis networks by lexicographic optimization and augmented epsilon constraint method. *Desalination* **2014**, *333*, 66–81. [CrossRef]
14. Yang, X.; Leng, Z.; Xu, S.; Yang, C.; Yang, L.; Liu, K.; Song, Y.; Zhang, L. Multi-objective optimal scheduling for CCHP microgrids considering peak-load reduction by augmented ε-constraint method. *Renew. Energy* **2021**, *172*, 408–423. [CrossRef]
15. Anastasopoulos, P.C.; Sarwar, M.T.; Shankar, V.N. Safety-oriented pavement performance thresholds: Accounting for unobserved heterogeneity in a multi-objective optimization and goal programming approach. *Anal. Methods Accid. Res.* **2016**, *12*, 35–47. [CrossRef]
16. De, P.; Deb, M. Using goal programming approach to solve fuzzy multi-objective linear fractional programming problems. In Proceedings of the 2016 IEEE International Conference on Computational Intelligence and Computing Research (ICCIC), Chennai, India, 15–17 December 2016; pp. 1–5.
17. Anukokila, P.; Radhakrishnan, B.; Anju, A. Goal programming approach for solving multi-objective fractional transportation problem with fuzzy parameters. *RAIRO-Oper. Res.* **2019**, *53*, 157–178. [CrossRef]
18. Chang, Y.F.; Xie, H.; Huang, W.C.; Zeng, L. An improved multi-objective quantum genetic algorithm based on cellular automaton. In Proceedings of the 9th IEEE International Conference on Software Engineering and Service Science (ICSESS), Beijing, China, 23–25 November 2018; pp. 342–345.
19. Altay, E.V.; Alatas, B. Differential evolution and sine cosine algorithm based novel hybrid multi-objective approaches for numerical association rule mining. *Inf. Sci.* **2021**, *554*, 198–221. [CrossRef]
20. Tang, Q.R.; Li, Y.H.; Deng, Z.Q.; Chen, D.; Guo, R.Q.; Huang, H. Optimal shape design of an autonomous underwater vehicle based on multi-objective particle swarm optimization. *Nat. Comput.* **2020**, *19*, 733–742. [CrossRef]
21. Su, S.J.; Han, J.; Xiong, Y.P. Optimization of unmanned ship's parametric subdivision based on improved multi-objective PSO. *Ocean Eng.* **2019**, *194*, 106617. [CrossRef]
22. Guria, C.; Bhattacharya, P.K.; Gupta, S.K. Multi-objective optimization of reverse osmosis desalination units using different adaptations of the non-dominated sorting genetic algorithm (NSGA). *Comput. Chem. Eng.* **2005**, *29*, 1977–1995. [CrossRef]
23. Srinivas, N.; Deb, K. Muiltiobjective optimization using nondominated sorting in genetic algorithms. *Evol. Comput.* **1994**, *2*, 221–248. [CrossRef]
24. Deb, K.; Pratap, A.; Agarwal, S.; Meyarivan, T. A fast and elitist multiobjective genetic algorithm: NSGA-II. *IEEE Trans. Evol. Comput.* **2002**, *6*, 182–197. [CrossRef]
25. Yu, H.L.; Castelli-Dezza, F.; Cheli, F.; Tang, X.L.; Hu, X.S.; Lin, X.K. Dimensioning and power management of hybrid energy storage systems for electric vehicles with multiple optimization criteria. *IEEE Trans. Power Electron.* **2021**, *36*, 5545–5556. [CrossRef]
26. Xia, G.; Liu, C.; Chen, X. Multi-objective optimization for AUV conceptual design based on NSGA-II. In Proceedings of the OCEANS 2016-Shanghai, Shanghai, China, 10–13 April 2016; pp. 1–6.
27. Alam, K.; Ray, T.; Anavatti, S. Design and construction of an autonomous underwater vehicle. *Neurocomputing* **2014**, *142*, 16–29. [CrossRef]
28. Liu, X.Y.; Yuan, Q.Q.; Zhao, M.; Cui, W.C.; Ge, T. Multiple objective multidisciplinary design optimization of heavier-than-water underwater vehicle using CFD and approximation model. *J. Mar. Sci. Technol.* **2017**, *22*, 135–148. [CrossRef]
29. Yu, B.; Wu, S.; Jiao, Z.X.; Shang, Y.X. Multi-objective optimization design of an electrohydrostatic actuator based on a particle swarm optimization algorithm and an analytic hierarchy process. *Energies* **2018**, *11*, 2426. [CrossRef]
30. Cheng, X.; He, X.Y.; Dong, M.; Xu, Y.F. Transformer fault probability calculation based on analytic hierarchy process. In Proceedings of the 2021 International Conference on Sensing, Measurement & Data Analytics in the era of Artificial Intelligence (ICSMD), Nanjing, China, 21–23 October 2021; pp. 1–6.
31. Medineckiene, M.; Zavadskas, E.K.; Björk, F.; Turskis, Z. Multi-criteria decision-making system for sustainable building assessment/certification. *Arch. Civ. Mech. Eng.* **2015**, *15*, 11–18. [CrossRef]
32. Altuzarra, A.; Gargallo, P.; Moreno-Jiménez, J.M.; Salvador, M. Influence, relevance and discordance of criteria in AHP-Global Bayesian prioritization. *Int. J. Inf. Technol. Decis. Mak.* **2013**, *12*, 837–861. [CrossRef]

33. Wu, S.; Yu, B.; Jiao, Z.X.; Shang, Y.X. Preliminary design and multi-objective optimization of electro-hydrostatic actuator. *Proc. Inst. Mech. Eng. Part G J. Aerosp. Eng.* **2017**, *231*, 1258–1268. [CrossRef]
34. Lampl, T.; Knigsberger, R.; Hornung, M. Design and evaluation of distributed electric drive architectures for high-lift control systems. In *Deutsche Luft- und Raumfahrtkongress*; Technische Universität München: Munich, Germany, 2017.
35. Zhao, T. Acquisition Cost Estimating Methodology for Aircraft Conceptual Design. Ph.D. Thesis, Cranfield University, Cranfield, UK, 2004.

Article

Novel Approach to Fault-Tolerant Control of Inter-Turn Short Circuits in Permanent Magnet Synchronous Motors for UAV Propellers

Aleksander Suti *[ ], Gianpietro Di Rito [ ] and Roberto Galatolo

Department of Civil and Industrial Engineering, University of Pisa, Largo Lucio Lazzarino 2, 56122 Pisa, Italy; gianpietro.di.rito@unipi.it (G.D.R.); roberto.galatolo@unipi.it (R.G.)
* Correspondence: aleksander.suti@dici.unipi.it; Tel.: +39-05-0221-7211

Abstract: This paper deals with the development of a novel fault-tolerant control technique aiming at the diagnosis and accommodation of inter-turn short circuit faults in permanent magnet synchronous motors for lightweight UAV propulsion. The reference motor is driven by a four-leg converter, which can be reconfigured in case of a phase fault by enabling the control of the central point of the motor Y-connection. A crucial design point entails the development of fault detection and isolation (FDI) algorithms capable of minimizing the failure transients and avoiding the short circuit extension. The proposed fault-tolerant control is composed of two sections: the first one applies a novel FDI algorithm for short circuit faults based on the trajectory tracking of the motor current phasor in the Clarke plane; the second one implements the fault accommodation, by applying a reference frame transformation technique to the post-fault commands. The control effectiveness is assessed via nonlinear simulations by characterizing the FDI latency and the post-fault performances. The proposed technique demonstrates excellent potentialities: the FDI algorithm simultaneously detects and isolates the considered faults, even with very limited extensions, during both stationary and unsteady operating conditions. In addition, the proposed accommodation technique is very effective in minimizing the post-fault torque ripples.

Keywords: all-electric propulsion; electric machines; fault diagnosis; fault-tolerant control; inter-turn short circuit; modelling; simulation

Citation: Suti, A.; Di Rito, G.; Galatolo, R. Novel Approach to Fault-Tolerant Control of Inter-Turn Short Circuits in Permanent Magnet Synchronous Motors for UAV Propellers. *Aerospace* **2022**, *9*, 401. https://doi.org/10.3390/aerospace9080401

Academic Editor: Cengiz Camci

Received: 1 July 2022
Accepted: 25 July 2022
Published: 26 July 2022

Publisher's Note: MDPI stays neutral with regard to jurisdictional claims in published maps and institutional affiliations.

Copyright: © 2022 by the authors. Licensee MDPI, Basel, Switzerland. This article is an open access article distributed under the terms and conditions of the Creative Commons Attribution (CC BY) license (https://creativecommons.org/licenses/by/4.0/).

1. Introduction

The design of next-generation long-endurance UAVs is undoubtedly moving towards the full-electric propulsion. Though immature today, full-electric propulsion systems are expected to obtain large investments in the forthcoming years, to replace conventional internal combustion engines, as well as hybrid or hydrogen-based solutions. Full-electric propulsion would guarantee smaller CO_2-emissions, lower noise, a higher efficiency, a reduced thermal signature (crucial for military applications), and simplified maintenance. However, several reliability and safety issues are still open, especially for long-endurance UAVs flying in unsegregated airspaces.

In this context, the Italian Government and the Tuscany Regional Government cofunded the project TERSA (*Tecnologie Elettriche e Radar per Sistemi aeromobili a pilotaggio remoto Autonomi*) [1], led by Sky Eye Systems (Italy) in collaboration with the University of Pisa and other Italian industries. The TERSA project aims to develop an Unmanned Aerial System (UAS) with a fixed-wing UAV, Figure 1, having the following main characteristics:

- Take-off weight: from 35 to 50 kg;
- Endurance: >6 h;
- Range: >3 km;
- Take-off system: pneumatic launcher;

- Landing system: parachute and airbags;
- Propulsion system: Permanent Magnet Synchronous Motor (PMSM) powering a twin-blade fixed-pitch propeller;
- Innovative sensing systems:
  ○ Synthetic aperture radar, to support surveillance missions in adverse environmental conditions;
  ○ Sense-and-avoid system, integrating a camera with a miniaturised radar, to support autonomous flight capabilities in emergency conditions.

Figure 1. Rendering of the TERSA UAV.

With reference to the activities related to the TERSA UAV propulsion system development (which this work refers to), a special attention has been devoted to the architecture definition for obtaining fault-tolerant capabilities. In particular, propulsion systems based on three-phase PMSMs driven by conventional three-leg converters typically have failure rates from 100 to 200 per million flight hours [2], which are not compatible with the reliability and safety levels required for the airworthiness certification [3]. The failure rate of PMSMs is mostly driven by motor phase faults and converter faults (open-switch in a converter leg, open-phase, inter-turn, phase-to-leg, phase-to-ground, or capacitor short-circuits [4]), which cover from 60% to 70% of the total fault modes [5]. Provided that the weight and envelopes required by UAV applications impede the extensive use of hardware redundancy (e.g., redundant motors), the reliability enhancement of full-electric propulsion systems can be achieved only through the redundancy of phases or stator modules [6] or by unconventional converters. By reducing design complexity, weight and envelopes, the use of four-leg converters [7–9], Figure 2, could represent a suitable solution for UAV applications.

In this converter topology, a couple of power switches are added, as stand-by devices, to the conventional three-leg bridge, enabling the control of the central point of the "star" connection (often Y-connection) of the motor. Nevertheless, to benefit from their fault-tolerance capability, these devices must be promptly reconfigured when a fault occurs, and extremely fast Fault-Detection and Isolation (FDI) algorithms must be developed. Especially for PMSMs operating at high speeds (such as the ones used for aircraft propellers), phase faults can generate abnormal torque ripples and the related failure transients can potentially cause unsafe conditions [10,11].

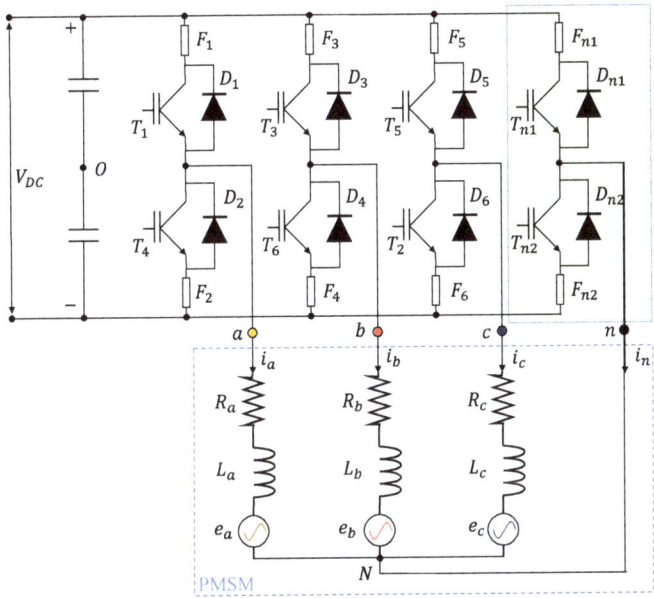

Figure 2. Four-leg converter driving a three-phase PMSM with access to the central point.

Many pieces of research on the development of FDI algorithms for PMSMs have been carried out in the last decades, and a special attention has been recently addressed to the Inter-Turn Short Circuit (ITSC) fault [12–15], which typically contributes to 10% of the PMSM failure rate [5,9,16].

Faiz [12] carried out a comprehensive literature review on FDI methods for ITSC faults in PMSMs. Depending on the measurements processed by the algorithms, they can be categorized as torque-based [17,18], flux-based [19,20], electromagnetic parameters-based [21], voltage-based, and current-based.

The most common strategy is given by the voltage-based methods. Immovilli and Bianchini [22] propose reconstructing harmonic patterns of the voltage components in the Clarke plane. The technique works very well in transient conditions, but it is not able to identify the shorted phase. The FDI algorithm proposed in [23], implements an estimator of back electromotive force and detects the ITSC by comparing the estimates to a reference model. The method is very robust against load disturbances, but it requires a very accurate model of the Back Electromotive Force (BEMF). Other approaches use as fault symptoms the harmonic components of the Zero Sequence Voltage Component (ZSVC) signal. Urresty et al. [24] proposes a Vold–Kalman Filtering Order Tracking (VKF-OT) algorithm to track the ZSVCs first harmonic component. The method offers reliable indicators, which are applicable in transient phases too, but it needs to be extended in terms of fault isolation. A robust algorithm is proposed by Hang et al., in [25], where the fault indicators are defined to remove the influence of rotor speed variation and a frequency tracking algorithm based on reference frame transformation is used to extract the fault symptoms. A possible drawback is the dependency of the fault indicator on the PMSM parameters, which may vary with environmental temperature. Boileau [26] considered as a fault indicator the amplitude of the positive voltage sequence while Meinguet [27] used as an indicator the ratio between the amplitudes of negative and positive voltage sequences. These methods are not able to locate the shorted phase, neither.

Concerning the current-based methods, the most relevant one is based on the so-called Park Vector Approach (PVA) [28–31]. When the PMSM is operating in normal conditions, the current phasor in the Clarke plane draws a circular trajectory, while it follows an elliptic one if a stator fault occurs. In PVAs, the fault index is thus defined by a geometrical distortion factor, obtained by currents measurements. Though demonstrating to be excellent for online ITSC FDI, the PVA is not able to locate the shorted phase.

All the current methods have in common the observation that in case of ITSCs, the stator currents are not balanced, and higher harmonic components in the signals can be used as fault symptoms. Different signal processing methods are employed to extract these symptoms (or patterns). Frequency-domain transforms such as Fast-Fourier Transforms (FFT) are applied in [13,32–34], while mixed frequency/time-domain transforms such as Short-Time Fourier Transform (STFT) are used in [14], Wavelet Transform (WT) in [15,35,36], and Hilbert–Huang Transform (HHT) in [37,38]. Frequency-domain transforms entail the loss of transient events such as speed or loads variations, so FFT analyses, though effective in stationary conditions, must be avoided if the FDI is required during transients. By requiring a higher computational effort, STFT analyses partially compensate the FFT limits, but FDI capabilities are still limited in unsteady conditions. On the other hand, WT methods permit to decompose time domain signals into stationary and non-stationary contributions, but the appropriate choice of wavelets is a drawback. Many limits of WT techniques are removed by HHT techniques, even if they are effective when applied to signals characterised by a narrow frequency content.

Other methods, based on artificial intelligence applications (Convolutional Neural Networks are used in [39]) have been also explored, but, though the results are remarkable, these techniques are not preferred for airworthiness certification [40].

Within this research context, this paper aims to develop an innovative Fault-Tolerant Control (FTC), capable of prompt FDI and accommodation of ITSC faults in three-phase PMSMs for lightweight UAVs propulsion, in both stationary and transient conditions. The proposed FTC is composed of two sections: the first one addresses the FDI problem via an original current-based method (here referred as Advanced-PVA, APVA), the second is dedicated to the fault accommodation, obtained by activating the stand-by leg of a four-leg converter and controlling the central point of the motor Y-connection. It is worth noting that the ITSC accommodation is developed by applying the Rotor Current Frame Transformation Control (RCFTC) technique, already applied by the authors in [9] for the FTC of open-circuit faults.

The paper is articulated as follows: in the first part, the nonlinear model of the propulsion system is presented; successively, the FDI algorithms and the fault accommodation technique are described by highlighting their basic design criteria; finally, a summary of the simulation results is proposed, by characterizing the failure transients and the post-fault behaviours after the injection of ITSC faults.

2. Materials and Methods

2.1. System Description

The reference propulsion system, designed for the full-electric propulsion of a lightweight UAV, is composed of (Figure 3):

- An electromechanical section, with:
    - Three-phase surface-mounted PMSM with phase windings in Y-connection;
    - Twin-blade fixed-pitch propeller [41];
    - Mechanical coupling joint.
- An Electronic Control Unit (ECU), including:
    - CONtrol/MONitoring (CON/MON) module, for the implementation of the closed-loop control and health-monitoring functions;
    - Four-leg converter;
    - Three current sensors (CSa, CSb, CSc), one per each motor phase;
    - One Angular Position Sensor (APS), measuring the motor angle;

- A Power Supply Unit (PSU), converting the power input coming from the UAV electrical power storage system to all components and sensors;
- Data and power connectors for the interface with the Flight Control Computer (FCC) and the UAV electrical system.

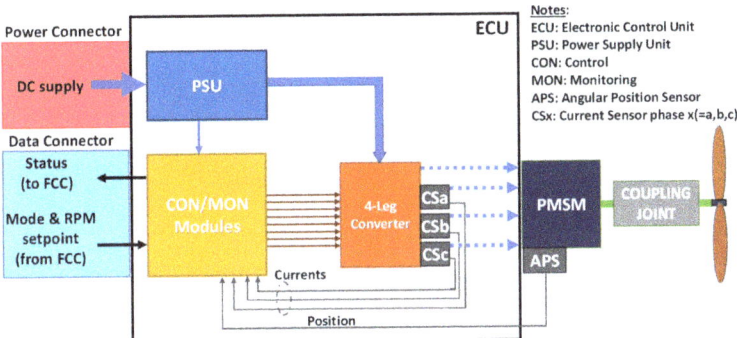

Figure 3. Schematics of the UAV propulsion system.

The CON module operates the closed-loop control of the system by implementing two nested loops on propeller speed and motor currents (via Field-Oriented Control, FOC). All the regulators implement proportional/integral actions on the tracking error signals, plus anti-windup functions with back-calculation algorithms to compensate for commands saturation. The MON module executes the health-monitoring algorithms, including the FTC proposed in this work.

2.2. Model of the Aero-Mechanical Section

The dynamics of the aero-mechanical section of the propulsion system, providing the UAV with the thrust is schematically depicted in Figure 4a and is modelled by [7,9]:

$$\begin{cases} J_p \ddot{\theta}_p = -Q_p - C_{gb}\left(\dot{\theta}_p - \dot{\theta}_m\right) - K_{gb}(\theta_p - \theta_m) + Q_d \\ J_m \ddot{\theta}_m = Q_m + C_{gb}\left(\dot{\theta}_p - \dot{\theta}_m\right) + K_{gb}(\theta_p - \theta_m) + Q_c \\ Q_p = C_{Q_p}\left(\dot{\theta}_p, AR\right) \rho D_p^5 \dot{\theta}_p^2 \\ AR = V_a / D_p \dot{\theta}_p \\ Q_c = Q_{cmax} \sin(n_h n_d \theta_m) \end{cases} \quad (1)$$

where J_p and θ_p, J_m, and θ_m are the inertia and the angular position of the propeller and the motor, respectively, Q_p is the propeller-resistant torque, Q_d is a gust-induced disturbance torque, Q_m is the motor torque, Q_c is the cogging torque, and Q_{cmax} is the maximum cogging torque, n_d is the pole pairs number, n_h is the harmonic index of the cogging disturbance, C_{Q_p} is the nondimensional torque coefficient of the propeller, AR is the propeller advance ratio, D_p is the propeller diameter, ρ is the air density, V_a is the UAV forward speed, while K_{gb} and C_{gb} are the stiffness and the damping of the mechanical coupling joint.

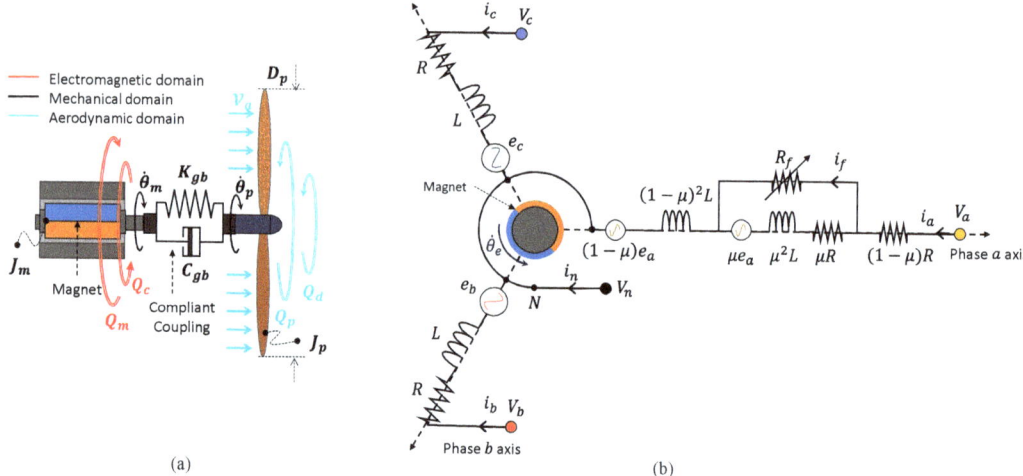

Figure 4. Full electric propulsion system: (**a**) mechanical scheme; (**b**) three-phase PMSM schematics with shorted turns in phase a (one pole pair and accessible neutral point).

2.3. *Model of the PMSM with ITSC Fault*

The mathematical modelling of PMSM with ITSC faults that will be used for this work has been previously presented and applied in literature, demonstrating it to be very accurate [13,42–44].

The occurrence of a short circuit in a stator phase causes an asymmetry in the motor magnetic flux, due to an additional circuit with insulation resistance R_f, in which the shorted current i_f flows, Figure 4b:

Once indicated, the fault extension along the winding with the parameter $\mu = N_f/N$ (defined as the ratio of the number of shorted turns to the total one), the winding affected by ITSC can be split into an undamaged and a damaged part, Figure 4b. Hence, the electrical equations can be written as:

$$V_{xn} = \mathbb{R}I + \mathbb{L}\frac{d}{dt}I + e. \quad (2)$$

In Equation (2), $V_{xn} = [V_a - V_n, V_b - V_n, V_c - V_n, 0]^T$ is the applied voltages vector, $I = [i_a, i_b, i_c, i_f]^T$ is the currents vector, e is the BEMF vector, which can be expressed as in Equation (3) (in Equation (3) and in the following equations, "s" and "c" briefly indicates "sine" and "cosine" functions):

$$e = \begin{bmatrix}(1-\mu)e_a \\ e_b \\ e_c \\ \mu e_a\end{bmatrix} = -\lambda_m \dot{\theta}_e \begin{bmatrix}(1-\mu)s(\theta_e) \\ s(\theta_e - 2/3\pi) \\ s(\theta_e + 2/3\pi) \\ \mu s(\theta_e)\end{bmatrix}, \quad (3)$$

where $\theta_e = n_d \theta_m$ is the electrical angle of the motor and λ_m is the rotor magnet flux linkage. The resistance and inductance matrixes are:

$$\mathbb{R} = \begin{bmatrix}(1-\mu)R & 0 & 0 & R_f \\ 0 & R & 0 & 0 \\ 0 & 0 & R & 0 \\ \mu R & 0 & 0 & -(\mu R + R_f)\end{bmatrix}, \mathbb{L} = \begin{bmatrix}(1-\mu)^2 L & 0 & 0 & 0 \\ 0 & L & 0 & 0 \\ 0 & 0 & L & 0 \\ \mu^2 L & 0 & 0 & -\mu^2 L\end{bmatrix}, \quad (4)$$

in which R is the phase resistance, L is the phase self-inductance, and R_f is the insulation resistance, given by:

$$R_f = k_{R_f} R(1-\mu), \qquad (5)$$

with k_{R_f} a factor depending on the insulation material.

By considering an inter-turn fault, the PMSM torque can be calculated as [44]:

$$Q_m = \left((1-\mu)e_a i_a + e_b i_b + e_c i_c - \mu e_a i_f\right)/\dot{\theta}_m, \qquad (6)$$

Since the neutral point voltage V_n is not null when a short circuit occurs, it is convenient to reformulate the electrical equations by extrapolating V_n. Hence, by applying Kirchhoff laws to the circuit in Figure 4b and by substituting in Equation (2), we have:

$$V = \mathcal{R} I + \mathcal{L}\frac{d}{dt} I + E, \qquad (7)$$

where the reformulated BEMF vector E is:

$$E = \frac{1}{3}\begin{bmatrix} 2e_{a_f} - (e_b + e_c) \\ 2e_b - \left(e_{a_f} + e_c\right) \\ 2e_c - \left(e_{a_f} + e_b\right) \\ 3e_f \end{bmatrix}, \qquad (8)$$

The new resistance and inductance matrixes are:

$$\mathcal{R} = \begin{bmatrix} (1-2/3\mu)R & 0 & 0 & 2/3R_f \\ \mu R/3 & R & 0 & -R_f/3 \\ \mu R/3 & 0 & R & -R_f/3 \\ \mu R & 0 & 0 & -\left(\mu R + R_f\right) \end{bmatrix}, \quad \mathcal{L} = \begin{bmatrix} (1-4/3\mu + 2/3\mu^2)L & 0 & 0 & 0 \\ (2-\mu)\mu L/3 & L & 0 & 0 \\ (2-\mu)\mu L/3 & 0 & L & 0 \\ \mu^2 L & 0 & 0 & -\mu^2 L \end{bmatrix}, \qquad (9)$$

while $V = [V_a, V_b, V_c, 0]^T$ is a vector defining the voltage commands sent by the converter to the motor terminals (first three components) and the voltage on the shorted turns. Finally, since the PMSM is controlled via FOC technique, it is convenient to express the voltage vector via the Clarke–Park transformation, as a function of direct and quadrature voltages V_d and V_q, generated by the closed-loop control algorithms:

$$\begin{bmatrix} V_a \\ V_b \\ V_c \\ V_n \end{bmatrix} = \sqrt{\frac{2}{3}} \begin{bmatrix} c(\theta_e) & -s(\theta_e) \\ c(\theta_e - 2/3\pi) & -s(\theta_e - 2/3\pi) \\ c(\theta_e + 2/3\pi) & -s(\theta_e + 2/3\pi) \\ 0 & 0 \end{bmatrix} \begin{bmatrix} V_d \\ V_q \end{bmatrix}, \qquad (10)$$

3. Fault-Tolerant Control System

3.1. FDI Algorithm Conceptualization

When a PMSM operates at constant speed without faults, the current phasor in the Clarke plane draws a circular trajectory, while it follows an elliptical trajectory when an ITSC fault occurs [28–31]. Torque oscillations appear, at the motor level, with amplitudes that depend on the short-circuit extension.

This section aims to demonstrate that the geometrical parameters of this elliptical trajectory (major, minor axes length, and axes inclination) are univocally related to the location and extension of the ITSC. In particular, we will demonstrate that:

- An ITSC fault can be detected by measuring the difference between the lengths of major and minor axes of the ellipse;
- An ITSC fault can be isolated by measuring the inclination of the major axis of the ellipse.

It is worth noting that, during constant speed operations, any elliptical trajectory of the motor current phasor in the Clarke plane (including the circular one, as a special case) can be reconstructed via the Fortescue decomposition [45], as the sum of three symmetrical and balanced rotating systems: *positive-*, *negative-*, and *zero-sequence* systems.

When an ITSC fault occurs, the central point of the Y-connection is isolated, so the *zero-sequence* current is zero, and the ellipse tracked by the current phasor ($\hat{I}$) can be considered as the sum of two counter-rotating phasors: *positive-* ($\hat{I}^+$) and *negative-sequence* phasors ($\hat{I}^-$), as given by:

$$\hat{I} = |\hat{I}(t)|e^{j\varphi(t)} = |\hat{I}^+|e^{j(\dot{\theta}_e t + \varphi^+)} + |\hat{I}^-|e^{-j(\dot{\theta}_e t - \varphi^-)}, \tag{11}$$

where φ, φ^+, and φ^- are the phase angles of the resultant, *positive-*, and *negative-sequence* current phasors, respectively. As illustrated in Figure 5, the semi major and semi minor axis lengths are obtained by:

$$s_{M,m} = |\hat{I}^+| \pm |\hat{I}^-|. \tag{12}$$

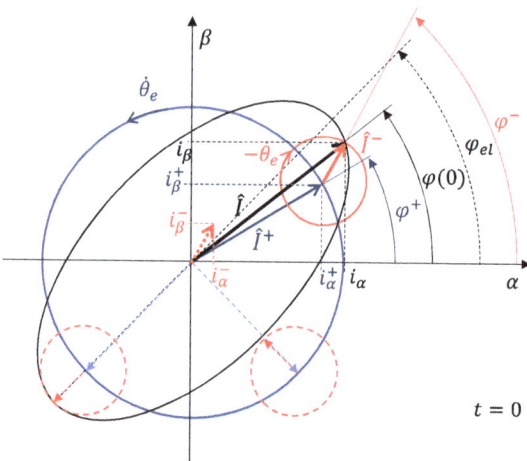

Figure 5. Decomposition of the current phasor in the Clarke plane into *positive-* and *negative-sequence* components.

While the inclination of the ellipse major axis corresponds to the angle that maximizes the current phasor amplitude, up to:

$$max(|\hat{I}|) = |\hat{I}^+| + |\hat{I}^-|. \tag{13}$$

The condition in Equation (13) is reached when the *positive-* and *negative-sequence* phasors have the same angular orientation:

$$\varphi_{el} = \dot{\theta}_e t|_{max(|\hat{I}|)} + \varphi^+ = -\dot{\theta}_e t|_{max(|\hat{I}|)} + \varphi^-, \tag{14}$$

which leads to:

$$\varphi_{el} = (\varphi^- + \varphi^+)/2. \tag{15}$$

To identify the ellipse inclination, it is thus crucial to evaluate the phase angles of the *positive-* and *negative-sequence* phasors, expressed by:

$$\varphi^* = \text{atan}\left(i_\beta^* / i_\alpha^*\right), \tag{16}$$

where "∗" stands for "+" or "−", and i_α^* and i_β^* are the projections of the phasor $\hat{I}^*$ on the α and β axes in the Clarke plane.

By introducing a complex analysis notation, the phasor $\hat{I}^*$ can be also represented in terms of a balanced three-phase space vector system, as:

$$\hat{I}^* = \sqrt{2/3}\left(Re(\hat{I}_a^*) + Re(\hat{I}_b^*)e^{j2\pi/3} + Re(\hat{I}_c^*)e^{j4\pi/3}\right), \tag{17}$$

Then, by expressing the space vector $\hat{I}^*$ into its real (i_α^*) and imaginary (i_β^*) components, we have:

$$\begin{aligned} i_\alpha^* &= \sqrt{2/3}\left[Re(\hat{I}_a^*) - 1/2(Re(\hat{I}_b^*) + Re(\hat{I}_c^*))\right] = \sqrt{3/2}Re(\hat{I}_a^*), \\ i_\beta^* &= \sqrt{2/3}\left[\sqrt{3}/2(Re(\hat{I}_b^*) - Re(\hat{I}_c^*))\right] = *-\sqrt{3/2}Im(\hat{I}_a^*), \end{aligned} \tag{18}$$

The phase angles of the *positive-* and *negative-sequence* phasors with ITSC faults can be thus calculated via the *positive-* and *negative-sequence* phasors related to phase a only ($\hat{I}_a^+$ and $\hat{I}_a^-$), given by the Fortescue transform:

$$\begin{bmatrix} \hat{I}_a^+ \\ \hat{I}_a^- \\ 0 \\ \hat{I}_f \end{bmatrix} = \frac{1}{3}\begin{bmatrix} 1 & e^{j2\pi/3} & e^{j4\pi/3} & 0 \\ 1 & e^{j4\pi/3} & e^{j2\pi/3} & 0 \\ 1 & 1 & 1 & 0 \\ 0 & 0 & 0 & 3 \end{bmatrix}\begin{bmatrix} \hat{I}_a \\ \hat{I}_b \\ \hat{I}_c \\ \hat{I}_f \end{bmatrix}, \tag{19}$$

The vector of phasors $\hat{I} = \begin{bmatrix} \hat{I}_a, & \hat{I}_b, & \hat{I}_c, & \hat{I}_f \end{bmatrix}$ can be obtained from:

$$\hat{I} = \left(\mathcal{R} + j\dot{\theta}_e\mathcal{L}\right)^{-1}(\hat{V} - \hat{E}), \tag{20}$$

where the voltage and BEMF phasors are:

$$\hat{V} = \sqrt{\frac{2}{3}}(V_d + jV_q)\begin{bmatrix} 1 \\ (-1/2 - j\sqrt{3}/2) \\ (-1/2 + j\sqrt{3}/2) \\ 0 \end{bmatrix}, \tag{21}$$

$$\hat{E} = j\lambda_m\dot{\theta}_e\begin{bmatrix} 1 - 2/3\mu \\ \mu - 1/2 - j\sqrt{3}/2 \\ \mu - 1/2 + j\sqrt{3}/2 \\ \mu \end{bmatrix}, \tag{22}$$

The solution of Equation (20) in terms of current phasors are presented by the polar-coordinate plot in Figure 6a, in which, given along the radius the ITSC fault extension ($\mu = N_f/N$), the inclination of the ellipse major axis of the current phasor trajectory is reported along the phase angle for each motor phase. In addition, Figure 6b shows the ratio of ellipse semi-axes lengths as a function of the ITSC fault extension parameter μ. It is worth noting that, if an ITSC fault occurs on phase a (b, c), the ellipse inclination will be 0° (240°, 120°) ±15°.

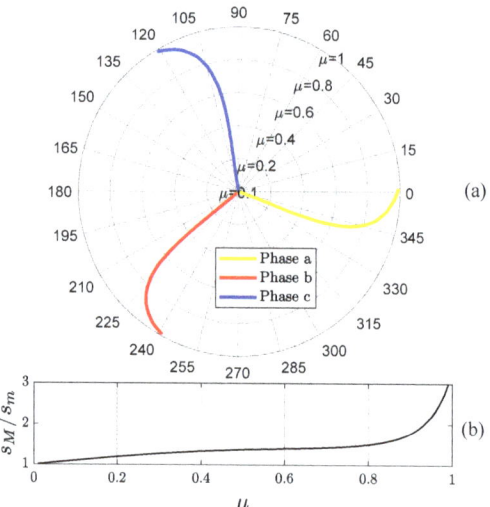

Figure 6. Inclination of the ellipse major axis (**a**) and ellipses axis ratio (**b**) under ITSC faults of different extensions and locations.

The above-mentioned discussion thus supports the development of an FDI algorithm based on the estimation of the geometrical characteristics of (generic) elliptical trajectories of current phasors in the Clarke plane, so that:

- The difference between the lengths of major and minor axes of the ellipse provides a symptom about the ITSC extension (fault detection);
- The inclination of the major axis of the ellipse provides a symptom about the ITSC location (fault isolation).

The FDI concept essentially entails the identification of geometrical properties of an ellipse. Clearly, at a single monitoring time, a single point of the ellipse is known (i_α, i_β Figure 5); therefore, the fitting problem is under-determined. On the other hand, the problem becomes over-determined if a sufficient number of measurements are used. The over-determined problem of fitting an ellipse to a set of data points arises in many applications such as computer graphics [46], hydraulic engineering [47], and statistics [48]. Many different methods are proposed in literature for fitting the geometric primitives of an ellipse: voting/clustering methods (e.g., the Hough transform [49] and the RANSAC technique [50]) are robust but they have poor accuracy and require large computing resources, optimization methods and direct least-square methods [51] are accurate, but, in case of nonlinear applications such as an ellipse fitting, they apply iterative techniques.

Here we instead use a direct least-square method, as proposed by Halíř and Flusser [52], who, starting from the quadratically-constrained least-squares minimization postulated by Fitzgibbon [53], reformulated the problem via partitioning techniques, by obtaining a very robust and computationally-effective algorithm.

Once given the vectorial conic definition of an ellipse,

$$\mathbf{\Gamma} \cdot \gamma = 0, \qquad (23)$$

in which $\mathbf{\Gamma} = [\alpha^2, \alpha\beta, \beta^2, \alpha, \beta, 1]$ and $\gamma = [A, B, C, D, E, F]^\mathrm{T}$ are vectors defining the Cartesian coordinates of the ellipse points and the ellipse coefficients, respectively, the over-determined fitting problem to a set of coordinate points α_i and β_i (where $i = 1, \ldots, n$, and

n is greater than the number of ellipse coefficients, i.e., $n > 6$) can be solved by the following eigenvalues problem (for more details, see Appendix A):

$$\begin{cases} \mathbb{M}\gamma_1 = \lambda\gamma_1 \\ \gamma_1^T \mathbb{C}_1 \gamma_1 = 1 \\ \gamma_2 = -\mathbb{S}_3^{-1}\mathbb{S}_2^T \gamma_1 \\ \gamma = (\gamma_1\ \gamma_2)^T \end{cases} \qquad (24)$$

where $\mathbb{M}$ is the reduced scatter matrix,

$$\mathbb{M} = \mathbb{C}_1^{-1}\left(\mathbb{S}_1 - \mathbb{S}_2\mathbb{S}_3^{-1}\mathbb{S}_2^T\right), \qquad (25)$$

$$\mathbb{S}_1 = \mathbb{D}_1^T\mathbb{D}_1, \quad \mathbb{S}_2 = \mathbb{D}_1^T\mathbb{D}_2, \quad \mathbb{S}_3 = \mathbb{D}_2^T\mathbb{D}_2, \qquad (26)$$

$$\mathbb{D}_1 = \begin{bmatrix} \alpha_1^2 & \alpha_1\beta_1 & \beta_1^2 \\ \vdots & \vdots & \vdots \\ \alpha_n^2 & \alpha_n\beta_n & \beta_n^2 \end{bmatrix},\ \mathbb{D}_2 = \begin{bmatrix} \alpha_1 & \beta_1 & 1 \\ \vdots & \vdots & \vdots \\ \alpha_n & \beta_n & 1 \end{bmatrix}, \qquad (27)$$

$$\mathbb{C}_1 = \begin{bmatrix} 0 & 0 & 2 \\ 0 & -1 & 0 \\ 2 & 0 & 0 \end{bmatrix}, \qquad (28)$$

$$\gamma = [\gamma_1\ \gamma_2]^T, \qquad (29)$$

with the ellipse coefficient vector segmented in $\gamma_1 = \begin{bmatrix} A & B & C \end{bmatrix}^T$ and $\gamma_2 = \begin{bmatrix} D & E & F \end{bmatrix}^T$. The solution of Equation (24) corresponds to the eigenvector γ yielding a minimal non-negative eigenvalue λ. Once obtained γ, the lengths of major and minor semi-axes s_M and s_m are given by [54]:

$$s_{M,m} = \frac{\sqrt{2(AE^2+CD^2-BDE+(B^2-4AC)F)\left(A+C\pm\sqrt{(A-C)^2+B^2}\right)^{-1}}}{4AC-B^2}, \qquad (30)$$

while the major axis inclination φ_{el} is [54]:

$$\varphi_{el} = \begin{cases} 0 & for\ B=0,\ A<C \\ \pi/2 & for\ B=0,\ A>C \\ 1/2\cot^{-1}((A-C)/B) & for\ B\neq 0,\ A<C' \\ \pi/2+1/2\cot^{-1}((A-C)/B) & for\ B\neq 0,\ A>C \end{cases} \qquad (31)$$

3.2. FDI Algorithm Design and Implementation

One of the basic limitations of the proposed FDI concept is that the technique is defined by assuming constant speed motor operations. During unsteady conditions, the current phasor changes its amplitude while rotating in the Clarke plane. Consequently, the ellipse identification does not work appropriately, and it could cause false alarms.

To overcome the problem, it is worth noting that, differently from the constant speed operations with ITSC faults, the axes of the ellipse reconstructed in unsteady conditions rotate. For this reason, the developed FDI algorithm detects an ITSC fault only if the value of the ellipse major axis is constant and stationary. This FDI logic has been implemented as represented by the flow chart in Figure 7: at each k-th monitoring sample, if the Boolean variable $mon^{(k)}$ is true (Equation (32)), a fault counter $n_c^{(k)}$ is increased by two, otherwise it is reduced by one. When the fault counter reaches a predefined maximum value (n_{th}), the algorithm outputs a true Boolean fault flag f_x, which indicates that a fault on the phase $x\ (= a,b,c)$ is detected and isolated.

$$mon^{(k)} = \left(\Delta_d^{(k)} \geq \varepsilon_{dth}\ \wedge\ \Delta_i^{(k)} \leq \varepsilon_{ith}\right), \qquad (32)$$

where:

$$\Delta_d^{(k)} = \left| s_M^{(k)} - s_m^{(k)} \right|, \tag{33}$$

$$\Delta_i^{(k)} = \min\left(\Delta_i^{(k)}\big|_a, \Delta_i^{(k)}\big|_c, \Delta_i^{(k)}\big|_b \right), \tag{34}$$

$$\Delta_i^{(k)}\big|_a = \left| \varphi_{el}^{(k)} \right|, \quad \Delta_i^{(k)}\big|_b = \left| \varphi_{el}^{(k)} \right| - \frac{2\pi}{3}, \quad \Delta_i^{(k)}\big|_c = \left| \varphi_{el}^{(k)} \right| + \frac{2\pi}{3}, \tag{35}$$

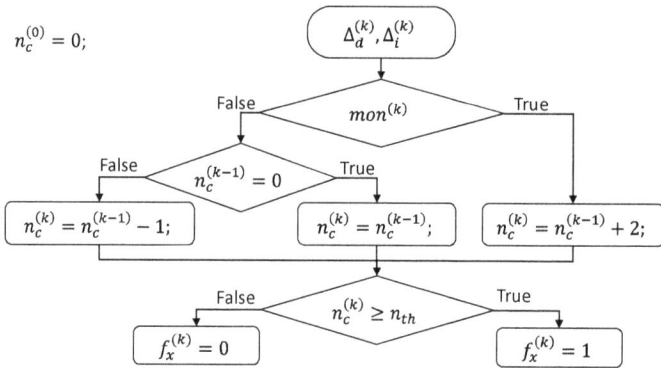

Figure 7. Flow chart for the FDI logic.

The details and a dedicated discussion about the FDI parameters design are proposed in Section 4, while tables reporting the FDI parameters are reported in Appendix B.

3.3. Fault Accommodation Algorithm

After the FDI algorithm detects and isolates the ITSC fault, the motor still has two active phases. To maintain the performances after the fault, the FTC should accommodate the system by restoring the operation of the current phasor in the Clarke plane as it was in healthy conditions. This is here achieved by enabling the control of the central point of the motor Y-connection, via the stand-by leg of the four-leg converter, and by applying an RCFTC technique [9,55].

To maintain the torque performances with only two phases, the currents amplitudes in the healthy phases in the rotor reference frame ($i_{xf}^\#$, $i_{yf}^\#$) must increase by $\sqrt{3}$ w.r.t. those before the fault, and they must shift by 60° along the electrical angle. In addition, the amplitude of the current flowing into the central point ($i_n^\#$) must be $\sqrt{3}$ times those in the healthy phases:

$$\begin{cases} i_{xf}^\# = \sqrt{2}\left(i_d^\# c\left[\theta_e + \frac{2\pi}{3}\left(m + \frac{7}{4}\right)\right] - i_q^\# s\left[\theta_e + \frac{2\pi}{3}\left(m + \frac{7}{4}\right)\right] \right) \\ i_{yf}^\# = \sqrt{2}\left(i_d^\# c\left[\theta_e + \frac{2\pi}{3}\left(m + \frac{5}{4}\right)\right] - i_q^\# s\left[\theta_e + \frac{2\pi}{3}\left(m + \frac{5}{4}\right)\right] \right) \\ i_{wf}^\# = 0 \\ i_n^\# = \sqrt{6}\left(i_d^\# c\left(\theta_e + \frac{2\pi}{3}m\right) - i_q^\# s\left(\theta_e + \frac{2\pi}{3}\right) \right) \end{cases} \tag{36}$$

In Equation (36), $i_d^\#$, $i_q^\#$ are the current demands in the Park frame before the fault. The subscripts xf, yf, and wf indicate the healthy and the isolated phases, while m is an integer number defined by the values in Table 1.

Table 1. Reconfiguration parameters.

Isolated Phase (w)	x	y	m
a	b	c	0
b	c	a	2
c	a	c	1

Since all the current demands in the stator frame are synchronous with the rotor motion, they can be expressed into a rotating frame by applying two new Clarke–Park transformations [9,55] (Figure 8):

- From the planar reference $\left(n_f, b_f, c_f\right)$ to a planar reference frame $(\alpha_f, \beta_f, \gamma_f)$, in which the α_f axis has an opposite direction w.r.t. the neutral current axis n_f;
- From the planar reference $(\alpha_f, \beta_f, \gamma_f)$ to a planar rotating frame (d_f, q_f, z_f) that maintains the same commands after the isolation ($i_{df} = i_d^*$ and $i_{qf} = i_q^*$).

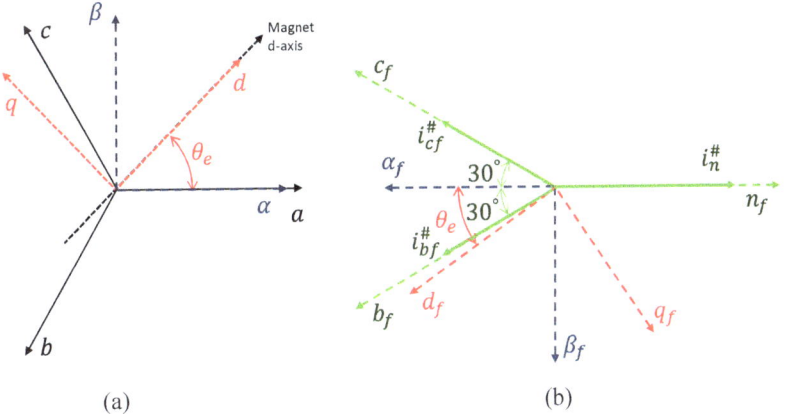

Figure 8. Reference frame transformation: (a) normal Clarke–Park transformation, (b) new transformation with phase a isolated.

$$\begin{bmatrix} i_{df}^\# \\ i_{qf}^\# \\ i_{zf}^\# \end{bmatrix} = R_{PCf}(m) T_{PCaf}(\theta_e) \begin{bmatrix} i_{xf}^\# \\ i_{yf}^\# \\ i_n^\# \end{bmatrix}, \tag{37}$$

where $R_{CPf}(m)$ is a rotation matrix that generalizes the transformation matrix $T_{CPaf}(\theta_e)$ related to the case of isolation of phase a [9]:

$$T_{CPaf}(\theta_e) = \begin{bmatrix} k_2 s(\theta_e) - k_1 c(\theta_e) & -k_2 s(\theta_e) - k_1 c(\theta_e) & k_3 c(\theta_e) \\ k_1 s(\theta_e) + k_2 c(\theta_e) & k_1 s(\theta_e) - k_2 c(\theta_e) & -k_3 s(\theta_e) \\ 0 & 0 & -1/\sqrt{3} \end{bmatrix},$$

$$R_{CPf}(m) = \begin{bmatrix} c(m 2\pi/3) & -s(m 2\pi/3) & 0 \\ s(m 2\pi/3) & c(m 2\pi/3) & 0 \\ 0 & 0 & 1 \end{bmatrix}, \tag{38}$$

$$k_1 = \sqrt{6}/\left(6 + 4\sqrt{3}\right), \quad k_2 = 1/\sqrt{2}, \quad k_3 = 2/\sqrt{3}, \tag{39}$$

The RCFTC accommodation is finally completed by inverting the direct transformation matrix to calculate the reference voltages for the converter:

$$\begin{bmatrix} V^{\#}_{xnf} \\ V^{\#}_{ynf} \\ V^{\#}_{nof} \end{bmatrix} = \left(T_{CPaf}(\theta_e)\right)^{-1} \left(R_{PCf}(m)\right)^{\mathrm{T}} \begin{bmatrix} V^{\#}_{df} \\ V^{\#}_{qf} \\ V^{\#}_{zf} \end{bmatrix}, \quad (40)$$

where

$$\left(T_{CPaf}(\theta_e)\right)^{-1} = \begin{bmatrix} \frac{s(\theta_e)}{2k_2} - \frac{c(\theta_e)}{2k_1} & \frac{s(\theta_e)}{2k_1} + \frac{c(\theta_e)}{2k_2} & -1 \\ -\frac{s(\theta_e)}{2k_2} - \frac{c(\theta_e)}{2k_1} & \frac{s(\theta_e)}{2k_1} - \frac{c(\theta_e)}{2k_2} & -1 \\ 0 & 0 & -\sqrt{3} \end{bmatrix}, \quad (41)$$

The integration of the proposed FTC strategy within the motor closed-loop system is schematically depicted in Figure 9.

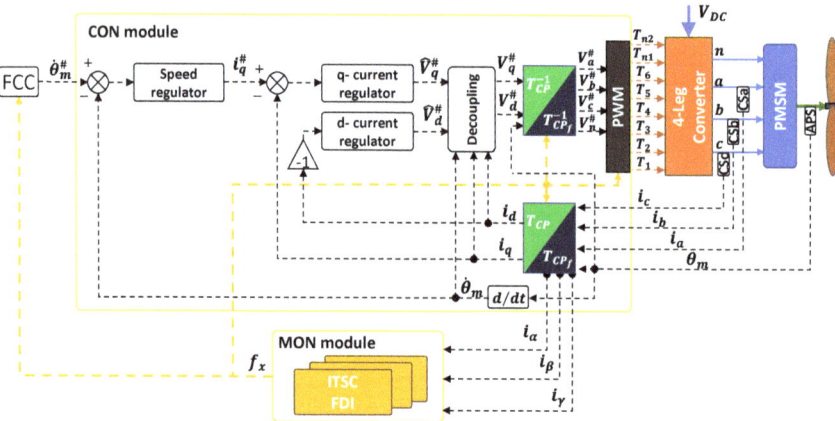

Figure 9. PMSM closed-loop architecture with FTC strategy.

When the system is in normal condition, the conventional Clarke–Parke transformations are employed (T_{CP}, T_{CP}^{-1}) and the central point of the Y-connection is isolated (T_{n1} and T_{n2} are not used). Once an ITSC is detected and isolated, the reference frame transformations of the RCFTC are employed (T_{CP_f}, $T_{CP_f}^{-1}$), and the reconfigured voltage references are sent to the four-leg converter to signal the central point (T_{n1} and T_{n2}).

4. Results and Discussion

4.1. Failure Transient Characterization

The effectiveness of the presented FTC strategy has been tested by using the nonlinear model of the propulsion system. The model is entirely developed in the MATLAB/Simulink environment, and its numerical solution is obtained via the fourth order Runge–Kutta method, using a 10^{-6} s integration step. It is worth noting that the choice of a fixed-step solver is not strictly related to the objectives of this work (in which the model is used for "off-line" simulations testing the FTC), but it has been selected for the next step of the project, when the FTC system will be implemented in the ECU boards via the automatic MATLAB compiler and executed in "real-time".

The closed-loop control is executed at a 20 kHz sampling rate and the maximum allowable fault latency has been set to 50 ms (for details, see Section 4.2). All the simulations started ($t = 0$ s) with a healthy PMSM, driving the propeller at 5800 rpm (UAV in straight-and-level flight at sea level Table A1).

The FTC strategy has been assessed by simulating the occurrence of an ITSC fault with $\mu = 0.5$ on phase a at $t = 150$ ms. The failure transient is characterized by applying or not the proposed FTC and by comparing the responses with those in healthy conditions. As shown by Figure 10, though its relevant extension, the ITSC fault implies minor impacts on the propeller speed response during stationary operations (Figure 10a), even if the FTC application assures a faster recovery of the pre-fault speed value. On the other hand, the failure transient during unsteady operations is much more limited with FTC, even if small-amplitude ripples (at approximately 100 Hz) appear immediately after the accommodation (Figure 10b). These responses highlight the importance of applying the FTC for ITSC faults: since the fault effects are minor during stationary operations, its detection is very difficult, but the ITSC is typically unstable, and it progressively spreads along the phase windings if the coil is not isolated.

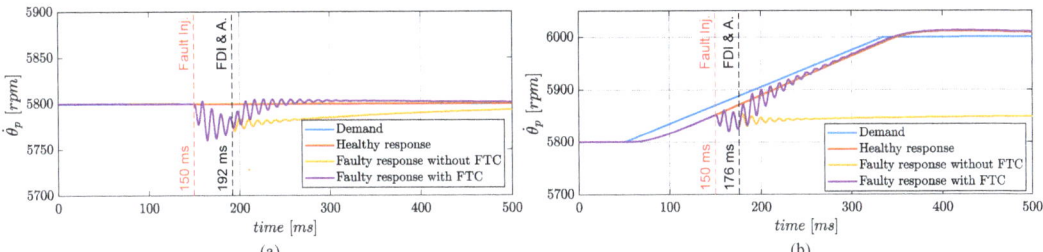

Figure 10. Propeller speed tracking with ITSC ($\mu = 0.5$) on phase a at $t = 150$ ms: (**a**) stationary and (**b**) unsteady operations.

The failure transient in terms of motor torque is then reported in Figure 11. It can be noted that, if the FTC is not applied, the post-fault behaviour is characterized, during both stationary and unsteady operations, by relevant high-frequency ripples (at approximately 1 kHz, i.e., twice the electrical frequency of the motor). In particular, during stationary operations, the FTC permits to rapidly restore the pre-fault torque level, by eliminating high-frequency loads that would inevitably cause damages at mechanical and electrical parts (Figure 11a). The failure transients in terms of phase currents are then reported in Figure 12, when the FTC is applied. The fault generates a short circuit current (i_f, Figure 12) causing the loss of symmetry of the three-phase system. Thanks to the RCFTC accommodation, the phase a is disengaged and the fourth leg of the converter is activated: this action stops the short circuit current and opens a current path through the central point (i_n, Figure 12).

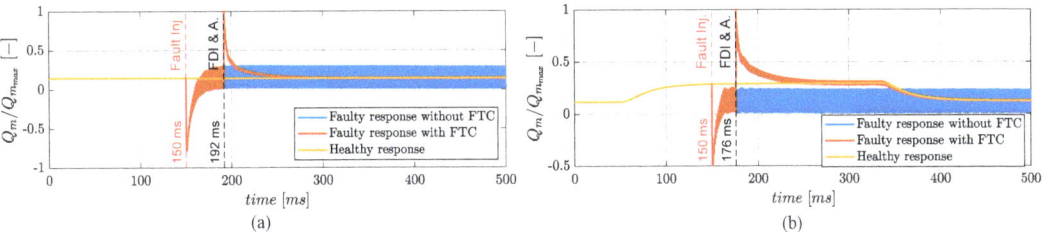

Figure 11. Normalized motor torque with ITSC ($\mu = 0.5$) on phase a at $t = 150$ ms: (**a**) stationary ($Q_{m_{max}} \approx 12$ Nm) and (**b**) unsteady operations ($Q_{m_{max}} \approx 15$ Nm).

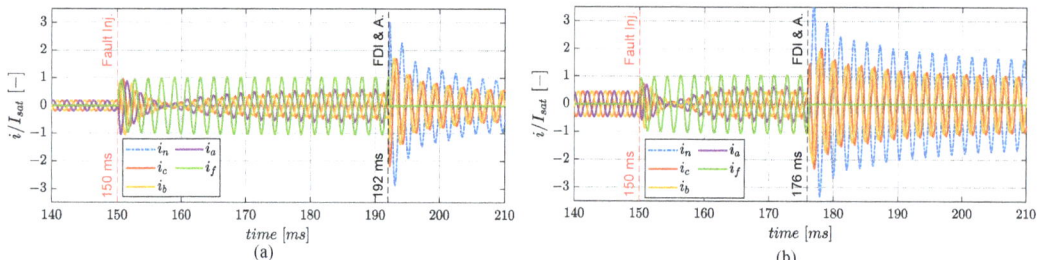

Figure 12. Normalized phase currents with ITSC ($\mu = 0.5$) on phase a at $t = 150$ ms ($I_{sat} = 80$ A): (**a**) stationary (**b**) unsteady operations.

The current phasor trajectories in the Clarke space are shown in Figure 13. It can be noted that, when an ITSC fault is injected in phase a, the trajectories are not strictly elliptical. This depends on the fact that the faulty currents are not perfectly sinusoidal, but they also contain higher harmonic contents. The phenomenon is caused by the phase voltages saturation. Despite the presence of these higher harmonic contents, the ellipse fitting technique successfully operates by demonstrating the relevant robustness of the FDI algorithm. On the other hand, when the fault is accommodated, the trajectory involves the neutral axis too, in such a way that, its projection on the α, β plane overlaps the healthy circular trajectory. Finally, the geometrical parameters of the current phasor elliptical trajectory (semi-axes lengths and major axis inclination) are plotted in Figure 14. It can be noted that, when the FTC intervenes, the projection of the current trajectory on the α, β plane becomes circular ($s_M/s_m = 1$) and the ellipse inclination returns to zero.

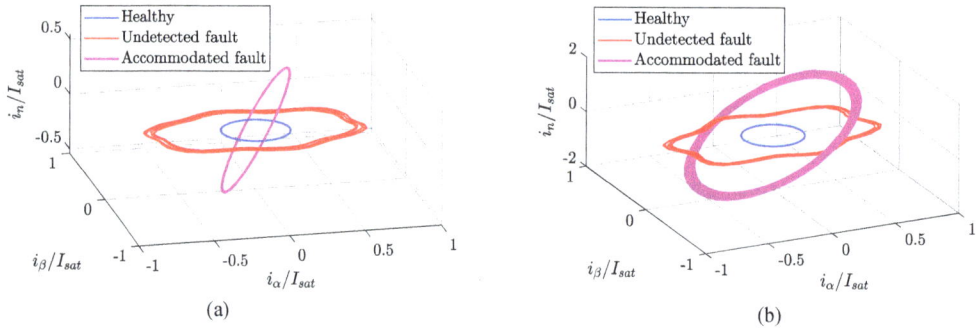

Figure 13. Current phasor trajectory in Clarke space with ITSC ($\mu = 0.5$) on phase a ($I_{sat} = 80$ A): (**a**) stationary (**b**) unsteady operations.

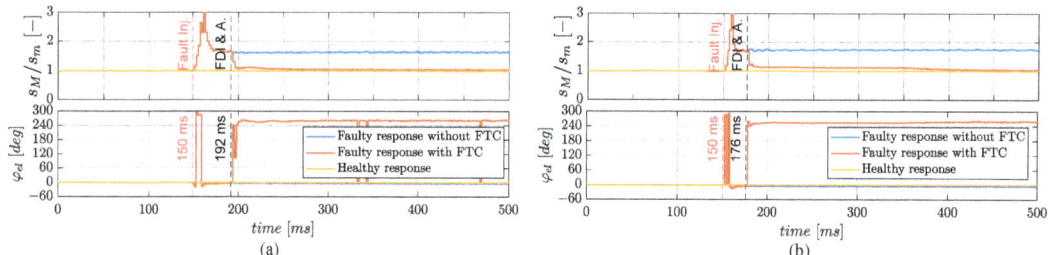

Figure 14. Ellipse parameters with ITSC ($\mu = 0.5$) on phase a at $t = 40$ ms: (**a**) stationary (**b**) unsteady operations.

4.2. FDI Parameters Definition

As described in Section 3.2, the design of the FDI algorithm requires the definition of four parameters, i.e., ε_{dth}, ε_{ith}, n_{th}, and n. Concerning the ellipse fitting samples n, this parameter has been defined by pursuing a good balance between the trajectory reconstruction accuracy and the FDI latency, both increasing when n increases. Considering that the propeller speed tracking bandwidth is approximately 15 Hz, the maximum allowable FDI latency has been set to 50 ms, which could be reasonably targeted by an equivalent monitoring frequency of 500 Hz. Being the sampling rate of the sensor system at 20 kHz, $n = 40$ has been imposed. The values of the remaining parameters (ε_{dth}, ε_{ith}, n_{th}) have been instead defined via time domain simulations, aiming to obtain correct ITSC FDI with very limited extension ($\mu = 0.1$). Two FDI design simulations have been performed, Figure 15:

- Simulation 1: cruise speed hold,
- Simulation 2: maximum speed ramp demand.

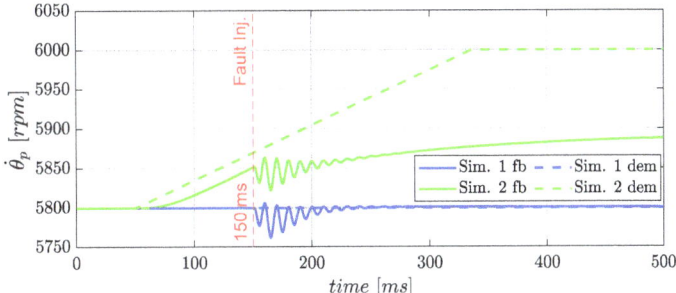

Figure 15. Propeller speed demands imposed for FDI design simulations (ITSC with $\mu = 0.1$ at $t = 150$ ms).

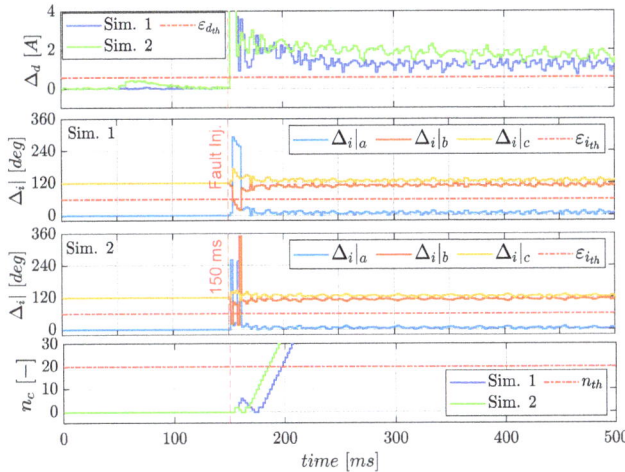

Figure 16. FDI signals during FDI design simulations (ITSC with $\mu = 0.1$ at $t = 150$ ms).

The FDI parameter plotted in Figure 16 shows that low values of ε_{dth} would clearly imply a too-sensitive algorithm, with high counts n_c during unsteady operations and in turn high values of n_{th} to avoid false alarms. On the other hand, high values of ε_{dth} will make the algorithm too robust to false alarms but it will reduce the effectiveness in detecting ITSC at an early stage. In the proposed case study, a good balance between the robustness against false alarms and the effectiveness in diagnosing ITSC higher than $\mu = 0.1$ can be obtained by selected $\varepsilon_{dth} = 0.6$ A and $n_{th} = 20$, which guarantees an FDI latency lower than the 50 ms among the two cases reported in Figure 15.

4.3. Critical Comparison with Other ITSC FDI Methods

A comparative analysis of the proposed FDI method with the most relevant ones developed in the literature (Table 2) has been carried out by using the list of capabilities given in Table 3. The results are reported in Table 4.

Table 2. Relevant FDI methods for ITSC faults.

Acronym	Method	Reference
M1	FFT	[13]
M2	HHT transforms	[38]
M3	PVA	[31]
M4	CNN	[39]
M5	APVA	Present work

Table 3. FDI capabilities for ITSC faults.

Acronym	Method
C1	Able to detect the faulty phase.
C2	Insensitive to operating loads.
C3	Robust against speed changes.
C4	Robust against current waveform.
C5	Minimum number of detected shorted turns.
C6	Electrical periods for FDI (*latency time*).
C7	Real-time computation.
C8	Tuning simplicity.

Table 4. Comparison of FDI methods for ITSC faults.

Capability	Method				
	M1	M2	M3	M4	M5
C1	Yes	No	No	Yes	Yes
C2	No	No	Yes	No	Yes
C3	No	Yes	No	Not provided	Yes
C4	Not provided	Not provided	Not provided	Not provided	Yes
C5	2	1	4	1	4
C6	Not provided	Not provided	Not provided	10 (*200* ms)	20 (*40* ms)
C7	No	No	No	Yes	Yes
C8	Yes	No	Yes	No	Yes

The FFT methods are robust against noise, but they are not applicable in unsteady operations. They can be enhanced by applying HHT transforms, but the capability to detect the faulty phase is lost. Similarly, the main disadvantage of the PVA methods is the inability to locate the fault. The CNN method is competitive in terms of the ITSC location, but their robustness in unsteady operations has not been proved; furthermore, the design of the neural detector requires a complex design and training process. On the other hand, the proposed APVA method demonstrates excellent capabilities in unsteady operations, it succeeds in fault location and detects ITSC faults with a very limited extension (10%, less than four turns) independently from the load level. All the methods considered are capable of detecting incipient faults, but many of them lack enough information for a comparison in terms of detection and isolation latency.

5. Conclusions

A novel FTC strategy for a high-speed PMSM with a four-leg converter employed for UAV propulsion is developed and characterized in terms of FDI and fault accommodation capabilities. The FTC performances are assessed via dynamic simulation by using a detailed nonlinear model of the electric propulsion system, which includes a physically based modelling of ITSC faults. For the proposed FTC, an original FDI algorithm is developed and applied, based on an innovative current signature technique, which uses as fault symptoms the geometrical parameters of the elliptical trajectory of the currents phasor in the Clarke plane (e.g., major and minor axes lengths, and major axis inclination). In addition, a theoretical analysis is carried out to support the FDI algorithms, by demonstrating that the major axis inclination can be used as a symptom for the faulty phase identification.

A comparative analysis with other ITSC FDI methods from the literature is also carried out. If compared with neural networks methods, which exhibit the best sensibility to incipient faults (a shorted turn is isolated within 10 electrical periods), the proposed technique behaves slightly worse (four shorted turns, corresponding to a 10% extension along the coil, and are isolated within 20 electrical periods). Nevertheless, the neural methods are more complicated to be tuned (due to the training process), while the proposed method only requires tuning four parameters. Finally, the paper demonstrates that a fault accommodation based on the RCFTC technique succeeds in minimizing the failure transient and eliminates the high-frequency torque ripples induced by the fault.

Author Contributions: Conceptualization, methodology and investigation, A.S. and G.D.R.; software, data curation and writing—original draft preparation, A.S.; validation, formal analysis and writing—review and editing, G.D.R.; resources, supervision and visualization, G.D.R. and R.G.; project administration and funding acquisition, R.G. All authors have read and agreed to the published version of the manuscript.

Funding: This research was co-funded by the Italian Government (*Ministero Italiano dello Sviluppo Economico*, MISE) and by the Tuscany Regional Government, in the context of the R&D project "*Tecnologie Elettriche e Radar per SAPR Autonomi* (TERSA)", Grant number: F/130088/01-05/X38.

Institutional Review Board Statement: Not applicable.

Informed Consent Statement: Not applicable.

Data Availability Statement: Not applicable.

Acknowledgments: The authors wish to thank Luca Sani, from the University of Pisa (*Dipartimento di Ingegneria dell'Energia, dei Sistemi, del Territorio e delle Costruzioni*), for the support in the definition of the PMSM model parameters, and Francesco Schettini, from Sky Eye Systems (Italy), for the support in the definition of the UAV propeller loads.

Conflicts of Interest: The authors declare no conflict of interest.

Appendix A.

The general ellipse expression is defined by an implicit second-order polynomial with specific constraints on coefficients.

$$\begin{cases} A\alpha^2 + B\alpha\beta + C\beta^2 + D\alpha + E\beta + F = 0 \\ B^2 - 4AC < 0 \end{cases}, \tag{A1}$$

in which A, B, C, D, E, and F are the ellipse coefficients, while α and β are the Cartesian coordinates of the ellipse points. The related vectorial conic definition is:

$$\mathbf{\Gamma} \cdot \mathbf{\gamma} = 0, \tag{A2}$$

where $\Gamma = [\alpha^2, \alpha\beta, \beta^2, \alpha, \beta, 1]$ and $\gamma = [A, B, C, D, E, F]^T$.

The ellipse fitting to a set of coordinate points (α_i, β_i), coming from i monitoring measurements (where $i = 1, \ldots, n$, and n is greater than the number of conic coefficients, i.e., $n > 6$) is an over-determined problem, which can be approached by minimizing the sum of distances of the points (α_i, β_i) to the conic represented by coefficients γ:

$$\begin{cases} \min_a \sum_{i=1}^{n} (\Gamma_i \cdot \gamma)^2 \\ B^2 - 4AC < 0 \end{cases} \tag{A3}$$

Due to the constraint, the problem cannot be solved directly with a conventional least-square approach. However, Fitzgibbon [53] showed that under a proper scaling, the inequality in Equation (A3) can be changed into an equality constraint as,

$$\begin{cases} \min_a \sum_{i=1}^{n} (\Gamma_i \cdot \gamma)^2 \\ 4AC - B^2 = 1 \end{cases} \tag{A4}$$

This minimization problem can be conveniently formulated as:

$$\begin{cases} \min_a \|\mathbb{D}\gamma\|^2 \\ \gamma^T \mathbb{C} \gamma = 1 \end{cases}, \tag{A5}$$

where $\mathbb{D}$ and $\mathbb{C}$ are known as the design and constraint matrices, respectively, defined as

$$\mathbb{D} = \begin{bmatrix} \alpha_1^2 & \alpha_1\beta_1 & \beta_1^2 & \alpha_1 & \beta_1 & 1 \\ \vdots & \vdots & \vdots & \vdots & \vdots & \vdots \\ \alpha_n^2 & \alpha_n\beta_n & \beta_n^2 & \alpha_n & \beta_n & 1 \end{bmatrix}, \quad \mathbb{C} = \begin{bmatrix} 0 & 0 & 2 & 0 & 0 & 0 \\ 0 & -1 & 0 & 0 & 0 & 0 \\ 2 & 0 & 0 & 0 & 0 & 0 \\ 0 & 0 & 0 & 0 & 0 & 0 \\ 0 & 0 & 0 & 0 & 0 & 0 \\ 0 & 0 & 0 & 0 & 0 & 0 \end{bmatrix} \tag{A6}$$

and it can be solved as a quadratically-constrained least-squares minimization by applying Lagrange multipliers, as:

$$\begin{cases} \mathbb{S}\gamma = \lambda \mathbb{C}\gamma \\ \gamma^T \mathbb{C} \gamma = 1 \end{cases}, \tag{A7}$$

where $\mathbb{S}$ is the *scatter matrix*, defined as:

$$\mathbb{S} = \mathbb{D}^T \mathbb{D}, \tag{A8}$$

The optimal solution of Equation (A7) is the eigenvector corresponding to the minimum positive eigenvalue λ_k. It is worth noting that the matrix $\mathbb{C}$ is singular, and $\mathbb{S}$ is also singular if all data points lie exactly on an ellipse. Because of that, the computation of eigenvalues is numerically unstable and it can produce wrong results (as infinite or complex numbers). To overcome the drawback, Halíř [52] suggested to partition the $\mathbb{C}$ and $\mathbb{S}$ matrices. The constraint matrix is defined as:

$$\mathbb{C} = \begin{bmatrix} \mathbb{C}_1 & 0 \\ 0 & 0 \end{bmatrix}, \tag{A9}$$

and

$$\mathbb{C}_1 = \begin{bmatrix} 0 & 0 & 2 \\ 0 & -1 & 0 \\ 2 & 0 & 0 \end{bmatrix}, \tag{A10}$$

The partition of matrix $\mathbb{S}$ is obtained by splitting the matrix $\mathbb{D}$ into its quadratic and linear parts:

$$\mathbb{D} = [\mathbb{D}_1 \ \mathbb{D}_2], \tag{A11}$$

where

$$\mathbb{D}_1 = \begin{bmatrix} \alpha_1^2 & \alpha_1\beta_1 & \beta_1^2 \\ \vdots & \vdots & \vdots \\ \alpha_n^2 & \alpha_n\beta_n & \beta_n^2 \end{bmatrix}, \ \mathbb{D}_2 = \begin{bmatrix} \alpha_1 & \beta_1 & 1 \\ \vdots & \vdots & \vdots \\ \alpha_n & \beta_n & 1 \end{bmatrix}. \tag{A12}$$

Then, the scatter matrix is constructed as:

$$\mathbb{S} = \begin{bmatrix} \mathbb{S}_1 & \mathbb{S}_2 \\ \mathbb{S}_2^T & \mathbb{S}_3 \end{bmatrix}, \tag{A13}$$

in which

$$\mathbb{S}_1 = \mathbb{D}_1^T \mathbb{D}_1, \ \mathbb{S}_2 = \mathbb{D}_1^T \mathbb{D}_2, \ \mathbb{S}_3 = \mathbb{D}_2^T \mathbb{D}_2. \tag{A14}$$

Similarly, the coefficients vector is partitioned as:

$$\gamma = [\gamma_1 \ \gamma_2]^T, \tag{A15}$$

where

$$\gamma_1 = \begin{bmatrix} A & B & C \end{bmatrix}^T, \ \gamma_2 = \begin{bmatrix} D & E & F \end{bmatrix}^T. \tag{A16}$$

Based on this decomposition, Equation (A7) can be written as:

$$\begin{cases} \mathbb{S}_1 \gamma_1 + \mathbb{S}_2 \gamma_2 = \lambda \mathbb{C}_1 a_1 \\ \mathbb{S}_2^T \gamma_1 + \mathbb{S}_3 \gamma_2 = 0 \\ \gamma_1^T \mathbb{C}_1 \gamma_1 = 1 \end{cases}, \tag{A17}$$

Considering that matrix $\mathbb{S}_3$ is singular only if all the points lie on a line [52], the second of equation in Equation (A17) can be solved to obtain γ_2. By substituting in Equation (A17), and by considering that $\mathbb{C}_1$ is not singular, we have:

$$\begin{cases} \mathbb{M} \gamma_1 = \lambda \gamma_1 \\ \gamma_1^T \mathbb{C}_1 \gamma_1 = 1 \\ \gamma_2 = -\mathbb{S}_3^{-1} \mathbb{S}_2^T \gamma_1 \\ \gamma = (\gamma_1 \ \gamma_2)^T \end{cases}, \tag{A18}$$

in which $\mathbb{M}$ is the reduced scatter matrix:

$$\mathbb{M} = \mathbb{C}_1^{-1} \left(\mathbb{S}_1 - \mathbb{S}_2 \mathbb{S}_3^{-1} \mathbb{S}_2^T \right). \tag{A19}$$

The optimal solution corresponds to the eigenvector γ that yields a minimal non-negative eigenvalue λ. Once obtained γ, the lengths of major and minor semi-axes s_M and s_m are [54]:

$$s_{M,m} = \frac{\sqrt{2(AE^2 + CD^2 - BDE + (B^2 - 4AC)F) \left(A + C \pm \sqrt{(A-C)^2 + B^2} \right)^{-1}}}{4AC - B^2}. \tag{A20}$$

while the major axis inclination φ_{el} is [54]:

$$\varphi_{el} = \begin{cases} 0 & \text{for } B = 0, \ A < C \\ \pi/2 & \text{for } B = 0, \ A > C \\ 1/2 \cot^{-1}((A-C)/B) & \text{for } B \neq 0, \ A < C \\ \pi/2 + 1/2 \cot^{-1}((A-C)/B) & \text{for } B \neq 0, \ A > C \end{cases}. \tag{A21}$$

Appendix B.

This section contains tables reporting the parameters of the UAV propeller (Table A1), the simulation model of the propulsion system (Table A2), and the design parameters of the FTC system (Table A3).

Table A1. APC 22 × 10E Propeller data.

Definition	Symbol	Value	Unit
Cruise speed	$\dot{\theta}_{p|cruise}$	5800	rpm
Cruise power	$P_{p|cruise}$	1100	W
Climb speed	$\dot{\theta}_{p|climb}$	7400	rpm
Climb power	$P_{p|climb}$	3238	W

Table A2. System model parameters.

Definition	Symbol	Value	Unit
Stator phase resistance	R	0.025	Ω
Stator phase inductance single module	L	1×10^{-5}	H
Pole pairs number	n_d	5	-
Total turns number per phase	N	36	-
Torque constant	k_t	0.12	Nm/A
Back-electromotive force constant	k_e	0.036	V/(rad/s)
Permanent magnet flux linkage	λ_m	0.008	Wb
Maximum current (continuous duty cycle)	I_{sat}	80	A
Voltage supply	V_{DC}	36	V
Rotor inertia	J_{em}	8.2×10^{-3}	kg·m^2
Propeller diameter	D_p	0.5588	m
Propeller inertia	J_p	1.62×10^{-2}	kg·m^2
Joint stiffness	K_{gb}	1.598×10^3	Nm/rad
Joint damping	$C_{gb\,t}$	0.2545	Nm/(rad/s)
Insulation resistance coefficient	k_{Rf}	11	-
Maximum cogging torque	Q_{cmax}	0.036	Nm
Harmonic index of the cogging disturbances	n_h	12	-

Table A3. FTC Algorithm parameters.

Definition	Symbol	Value	Unit
Control frequency	f_{CL}	20	kHz
Ellipse measurement points	n	40	—
Sampling frequency ($= f_{CL}/n$)	f_{FDI}	500	Hz
Detection index threshold	ε_{dth}	0.6	A
Isolation index threshold	ε_{ith}	60	deg
Fault counter threshold	n_{th}	20	—

References

1. Dipartimento di Ingegneria Civile e Industriale, Progetti istituzionali. TERSA (Tecnologie Elettriche e Radar per Sistemi aeromobili a pilotaggio remoto Autonomi). Available online: https://dici.unipi.it/ricerca/progetti-finanziati/tersa/ (accessed on 1 July 2022).
2. Nandi, S.; Toliyat, H.; Li, X. Condition Monitoring and Fault Diagnosis of Electrical Motors—A Review. *IEEE Trans. Energy Convers.* **2005**, *20*, 719–729. [CrossRef]
3. NATO Standardization Agency. *STANAG 4671—Standardization Agreement—Unmanned Aerial Vehicles Systems Airworthiness Requirements (USAR)*; NATO Standardization Agency (STANAG): Brussels, Belgium, 2009.
4. Kontarcek, A.; Bajec, P.; Nemec, M.; Ambrožic, V.; Nedeljkovic, D. Cost-Effective Three-Phase PMSM Drive Tolerant to Open-Phase Fault. *IEEE Trans. Ind. Electron.* **2015**, *62*, 6708–6718. [CrossRef]
5. Cao, W.; Mecrow, B.; Atkinson, G.; Bennett, J.; Atkinson, D. Overview of Electric Motor Technologies Used for More Electric Aircraft (MEA). *IEEE Trans. Ind. Electron.* **2012**, *59*, 3523–3531. [CrossRef]
6. Suti, A.; Di Rito, G.; Galatolo, R. Fault-Tolerant Control of a Dual-Stator PMSM for the Full-Electric Propulsion of a Lightweight Fixed-Wing UAV. *Aerospace* **2022**, *9*, 337. [CrossRef]
7. De Rossiter Correa, M.; Jacobina, C.; Da Silva, E.; Lima, A. An induction motor drive system with improved fault tolerance. *IEEE Trans. Ind. Appl.* **2001**, *37*, 873–879. [CrossRef]

8. Ribeiro, R.; Jacobina, C.; Lima, A.; Da Silva, E. A strategy for improving reliability of motor drive systems using a four-leg three-phase converter. In Proceedings of the APEC 2001. Sixteenth Annual IEEE Applied Power Electronics Conference and Exposition (Cat. No. 01CH37181), Anaheim, CA, USA, 4–8 March 2001. [CrossRef]
9. Suti, A.; Di Rito, G.; Galatolo, R. Fault-Tolerant Control of a Three-Phase Permanent Magnet Synchronous Motor for Lightweight UAV Propellers via Central Point Drive. *Actuators* **2021**, *10*, 253. [CrossRef]
10. Khalaief, A.; Boussank, M.; Gossa, M. Open phase faults detection in PMSM drives based on current signature analysis. In Proceedings of the XIX International Conference on Electrical Machines-ICEM 2010, Rome, Italy, 6–8 September 2010. [CrossRef]
11. Li, W.; Tang, H.; Luo, S.; Yan, X.; Wu, Z. Comparative analysis of the operating performance, magnetic field, and temperature rise of the three-phase permanent magnet synchronous motor with or without fault-tolerant control under single-phase open-circuit fault. *IET Electr. Power Appl.* **2021**, *15*, 861–872. [CrossRef]
12. Faiz, J.; Nejadi-Koti, H.; Valipour, Z. Comprehensive review on inter-turn fault indexes in permanent magnet motors. *IET Electr. Power Appl.* **2017**, *11*, 142–156. [CrossRef]
13. Krzysztofiak, M.; Skowron, M.; Orlowska-Kowalska, T. Analysis of the Impact of Stator Inter-Turn Short Circuits on PMSM Drive with Scalar and Vector Control. *Energies* **2021**, *14*, 153. [CrossRef]
14. Arabaci, H.; Bilgin, O. The Detection of Rotor Faults By Using Short Time Fourier Transform. In Proceedings of the 2007 IEEE 15th Signal Processing and Communications Applications, Eskisehir, Turkey, 11–13 June 2007. [CrossRef]
15. Mohammed, O.A.; Liu, Z.; Liu, S.; Abed, N.Y. Internal Short Circuit Fault Diagnosis for PM Machines Using FE-Based Phase Variable Model and Wavelets Analysis. *IEEE Trans. Magn.* **2007**, *43*, 1729–1732. [CrossRef]
16. Mazzoleni, M.; Di Rito, G.; Previdi, F. Fault Diagnosis and Condition Monitoring Approaches. In *Electro-Mechanical Actuators for the More Electric Aircraft*; Springer: Cham, Switzerland, 2021; pp. 87–117.
17. Awadallah, M.; Morcos, M.; Gopalakrishnan, S.; Nehl, T. A neuro-fuzzy approach to automatic diagnosis and location of stator inter-turn faults in CSI-fed PM brushless DC motors. *IEEE Trans. Energy Convers.* **2005**, *20*, 253–259. [CrossRef]
18. Awadallah, M.; Morcos, M.; Gopalakrishnan, S.; Nehl, T. Detection of stator short circuits in VSI-fed brushless DC motors using wavelet transform. *IEEE Trans. Energy Convers.* **2006**, *21*, 1–8. [CrossRef]
19. Penman, J.; Sedding, H.; Lloyd, B.; Fink, W. Detection and location of interturn short circuits in the stator windings of operating motors. *IEEE Trans. Energy Convers.* **1994**, *9*, 652–658. [CrossRef]
20. Ebrahimi, B.M.; Faiz, J. Feature Extraction for Short-Circuit Fault Detection in Permanent-Magnet Synchronous Motors Using Stator-Current Monitoring. *IEEE Trans. Power Electron.* **2010**, *25*, 2673–2682. [CrossRef]
21. Aubert, B.; Régnier, J.; Caux, S. Kalman-filter-based indicator for online inter turn short circuits detection in permanent-magnet synchonous generator. *IEEE Trans. Ind. Electron.* **2014**, *62*, 1921–1930. [CrossRef]
22. Immovilli, F.; Bianchini, C.; Lorenzani, E.; Bellini, A.; Fornasiero, E. Evaluation of Combined Reference Frame Transformation for Interturn Fault Detection in Permanent-Magnet Multiphase Machines. *IEEE Trans. Ind. Electron.* **2014**, *62*, 1912–1920. [CrossRef]
23. Sarikhani, A.; Mohammed, O.A. Inter-Turn Fault Detection in PM Synchronous Machines by Physics-Based Back Electromotive Force Estimation. *IEEE Trans. Ind. Electron.* **2012**, *60*, 3472–3484. [CrossRef]
24. Urresty, J.C.; Riba, J.R.; Romeral, L. Diagnosis of Interturn Faults in PMSMs Operating Under Nonstationary Conditions by Applying Order Tracking Filtering. *IEEE Trans. Power Electron.* **2012**, *28*, 507–515. [CrossRef]
25. Hang, J.; Zhang, J.; Cheng, M.; Huang, J. Online Interturn Fault Diagnosis of Permanent Magnet Synchronous Machine Using Zero-Sequence Components. *IEEE Trans. Power Electron.* **2015**, *30*, 6731–6741. [CrossRef]
26. Boileau, T.; Leboeuf, N.; Nahid-Mobarakeh, B.; Meibody-Tabar, F. Synchronous Demodulation of Control Voltages for Stator Interturn Fault Detection in PMSM. *IEEE Trans. Power Electron.* **2013**, *28*, 5647–5654. [CrossRef]
27. Meinguet, F.; Semail, E.; Kestelyn, X.; Mollet, Y.; Gyselinck, J. Change-detection algorithm for short-circuit fault detection in closed-loop AC drives. *IET Electr. Power Appl.* **2012**, *8*, 165–177. [CrossRef]
28. Cardoso, A.J.M.; Cruz, A.M.A.; Fonseca, D.S.B. Inter-turn stator winding fault diagnosis in three-phase induction motors, by Park's vector approach. *IEEE Trans. Energy Convers.* **1999**, *14*, 595–598. [CrossRef]
29. Cruz, S.M.A.; Cardoso, A.J.M. Stator winding fault diagnosis in three-phase synchronous and asynchronous motors, by the extended Park's vector approach. *IEEE Trans. Ind. Appl.* **2001**, *37*, 1227–1233. [CrossRef]
30. Abitha, M.; Rajini, V. Park's vector approach for online fault diagnosis of induction motor. In Proceedings of the 2013 International Conference on Information Communication and Embedded Systems (ICICES), Chennai, India, 21–22 February 2013. [CrossRef]
31. Goh, Y.-J.; Kim, O. Linear Method for Diagnosis of Inter-Turn Short Circuits in 3-Phase Induction Motors. *Appl. Sci.* **2019**, *9*, 4822. [CrossRef]
32. Kim, K.H. Simple Online Fault Detecting Scheme for Short-Circuited Turn in a PMSM Through Current Harmonic Monitoring. *IEEE Trans. Ind. Electron.* **2010**, *58*, 2565–2568. [CrossRef]
33. Jung, J.H.; Lee, J.J.; Kwon, B.-H. Online Diagnosis of Induction Motors Using MCSA. *IEEE Trans. Ind. Electron.* **2006**, *53*, 1842–1852. [CrossRef]
34. Haddad, R.Z.; Strangas, E.G. On the Accuracy of Fault Detection and Separation in Permanent Magnet Synchronous Machines Using MCSA/MVSA and LDA. *IEEE Trans. Energy Convers.* **2016**, *31*, 924–934. [CrossRef]
35. Khan, M.A.S.K.; Rahman, M.A. Development and Implementation of a Novel Fault Diagnostic and Protection Technique for IPM Motor Drives. *IEEE Trans. Ind. Electron.* **2008**, *56*, 85–92. [CrossRef]

36. Park, C.H.; Lee, J.; Ahn, G.; Youn, M.; Youn, B.D. Fault Detection of PMSM under Non-Stationary Conditions Based on Wavelet Transformation Combined with Distance Approach. In Proceedings of the 2019 IEEE 12th International Symposium on Diagnostics for Electrical Machines, Power Electronics and Drives (SDEMPED), Toulouse, France, 27–30 August 2019. [CrossRef]
37. Huang, N.E.; Shen, Z.; Long, S.R.; Wu, M.C.; Shih, H.H.; Zheng, Q.; Yen, N.-C.; Tung, C.C.; Liu, H.H. The empirical mode decomposition and the Hilbert spectrum for nonlinear and non-stationary time series analysis. *Proc. R. Soc. London. Ser. A Math. Phys. Eng. Sci.* **1998**, *454*, 903–995. [CrossRef]
38. Wang, C.; Liu, X.; Chen, Z. Incipient Stator Insulation Fault Detection of Permanent Magnet Synchronous Wind Generators Based on Hilbert–Huang Transformation. *IEEE Trans. Magn.* **2014**, *50*, 11. [CrossRef]
39. Skowron, M.; Orlowska-Kowalska, T.; Wolkiewicz, M.; Kowalski, C.T. Convolutional Neural Network-Based Stator Current Data-Driven Incipient Stator Fault Diagnosis of Inverter-Fed Induction Motor. *Energies* **2020**, *13*, 1475. [CrossRef]
40. Bellamy, W., III. Aviation Today. 19 February 2020. Available online: https://www.aviationtoday.com/2020/02/19/easa-expects-certification-first-artificial-intelligence-aircraft-systems-2025/ (accessed on 5 June 2022).
41. APC Propellers TECHNICAL INFO. Available online: https://www.apcprop.com/technical-information/performance-data/ (accessed on 2 May 2021).
42. Romeral, L.; Urresty, J.C.; Ruiz, J.R.R.; Espinosa, A.G. Modeling of Surface-Mounted Permanent Magnet Synchronous Motors With Stator Winding Interturn Faults. *IEEE Trans. Ind. Electron.* **2010**, *58*, 1576–1585. [CrossRef]
43. Vaseghi, B.; Nahid-Mobarakeh, B.; Takorabet, N.; Meibody-Tabar, F. Experimentally Validated Dynamic Fault Model for PMSM with Stator Winding Inter-Turn Fault. In Proceedings of the 2008 IEEE Industry Applications Society Annual Meeting, Edmonton, AB, Canada, 5–9 October 2008. [CrossRef]
44. Jeong, I.; Hyon, B.J.; Kwanghee, N. Dynamic Modeling and Control for SPMSMs With Internal Turn Short Fault. *IEEE Trans. Power Electron.* **2012**, *28*, 3495–3508. [CrossRef]
45. Fortescue, C.L. Method of Symmetrical Co-Ordinates Applied to the Solution of Polyphase Networks. *Trans. Am. Inst. Electr. Eng.* **1918**, *37*, 1027–1140. [CrossRef]
46. Pratt, V. Direct least-squares fitting of algebraic surfaces. *ACM SIGGRAPH Comput. Graph.* **1987**, *21*, 145–152. [CrossRef]
47. Heidari, M.; Heigold, P. Determination of Hydraulic Conductivity Tensor Using a Nonlinear Least Squares Estimator. *J. Am. Water Resour. Assoc.* **1993**, *29*, 415–424. [CrossRef]
48. Macdonald, P.; Linnik, Y.; Elandt, R. Method of Least Squares and Principles of the Theory of Observation. *J. R. Stat. Soc. Ser. D Stat.* **1962**, *12*, 335–336. [CrossRef]
49. Leavers, V. *Shape Detection in Computer Vision Using the Hough Transform*; Springer: London, UK, 1992. [CrossRef]
50. Bolles, R.; Fishler, M.A. A RANSAC-based approach to model fitting and its application to finding cylinders in range data. In Proceedings of the IJCAI'81: 7th International Joint Conference on Artificial Intelligence-Volume 2, Vancouver, Canada, 24 August 1981. [CrossRef]
51. Gander, W.; Golub, G.; Strebel, R. Least-squares fitting of circles and ellipses. *BIT Numer. Math.* **1994**, *34*, 558–578. [CrossRef]
52. Halir, R.; Flusser, J. Numerically Stable Direct Least Squares Fitting of Ellipses. In Proceedings of the International Conference in Central Europe on Computer Graphics, Visualization and Interactive Digital Media, Plzeň, Czech Republic, 9–13 February 1998.
53. Fitzgibbon, A.; Pilu, M.; Fisher, R. Direct least squares fitting of ellipses. In Proceedings of the 13th International Conference on Pattern Recognition, Vienna, Austria, 25–29 August 1996. [CrossRef]
54. Weisstein, E.W. "Ellipse," MathWorld—A Wolfram Web Resource, 17 December 2021. Available online: https://mathworld.wolfram.com/Ellipse.html (accessed on 2 June 2022).
55. Zhou, X.; Sun, J.; Li, H.; Song, X. High Performance Three-Phase PMSM Open-Phase Fault-Tolerant Method Based on Reference Frame Transformation. *IEEE Trans. Ind. Electron.* **2019**, *66*, 7571–7580. [CrossRef]

Article

Application of Deep Reinforcement Learning in Reconfiguration Control of Aircraft Anti-Skid Braking System

Shuchang Liu [1], Zhong Yang [1], Zhao Zhang [2,*], Runqiang Jiang [2], Tongyang Ren [2], Yuan Jiang [2], Shuang Chen [3] and Xiaokai Zhang [3]

[1] College of Automation Engineering, Nanjing University of Aeronautics and Astronautics, Nanjing 210016, China
[2] Changchun Institute of Optics, Fine Mechanics and Physics, Chinese Academy of Sciences, Changchun 130033, China
[3] Electronic Engineering Department, Aviation Key Laboratory of Science and Technology on Aero Electromechanical System Integration, Nanjing 211106, China
* Correspondence: zhangzhao@ciomp.ac.cn

Abstract: The aircraft anti-skid braking system (AABS) plays an important role in aircraft taking off, taxiing, and safe landing. In addition to the disturbances from the complex runway environment, potential component faults, such as actuators faults, can also reduce the safety and reliability of AABS. To meet the increasing performance requirements of AABS under fault and disturbance conditions, a novel reconfiguration controller based on linear active disturbance rejection control combined with deep reinforcement learning was proposed in this paper. The proposed controller treated component faults, external perturbations, and measurement noise as the total disturbances. The twin delayed deep deterministic policy gradient algorithm (TD3) was introduced to realize the parameter self-adjustments of both the extended state observer and the state error feedback law. The action space, state space, reward function, and network structure for the algorithm training were properly designed, so that the total disturbances could be estimated and compensated for more accurately. The simulation results validated the environmental adaptability and robustness of the proposed reconfiguration controller.

Keywords: aircraft anti-skid braking system; actuator faults; reconfiguration control; linear active-disturbance rejection control; deep reinforcement learning; twin delayed deep deterministic policy gradient algorithm

1. Introduction

The aircraft anti-skid braking system (AABS) is an essential airborne utilities system to ensure the safe and smooth landing of aircraft [1]. With the development of aircraft towards high speed and large tonnage, the performance requirements of AABS are increasing. Moreover, AABS is a complex system with strong nonlinearity, strong coupling, and time-varying parameters, and is sensitive to the runway environment [2]. These characteristics make AABS controller design an interesting and challenging topic.

The most widely used control method in practice is PID + PBM, which is a speed differential control law. However, it suffers from low-speed slipping and underutilization of ground bonding forces, making it difficult to meet high performance requirements. To this end, researchers have proposed many advanced control methods to improve the AABS performance, such as mixed slip deceleration PID control [3], model predictive control [4], extremum-seeking control [5], sliding mode control [6], reinforcement Q-learning control [7], and so on. Zhang et al. [8] proposed a feedback linearization controller with a prescribed performance function to ensure the transient and steady-state braking performance. Qiu et al. [9] combined backstepping dynamic surface control with an asymmetric barrier Lyapunov function to obtain a robust tracking response in the presence of disturbance and runway surface transitions. Mirzaei et al. [10] developed a fuzzy braking controller optimized by a genetic algorithm and introduced an error-based global optimization approach for fast convergence near the optimum point. The above-mentioned works provide an in-depth study on AABS control; however, the adverse effects caused by typical component faults such as actuator faults are neglected. Since most AABS are designed based on hydraulic control systems, the long hydraulic pipes create an enormous risk of air mixing with oil, and internal leakage. Without regular maintenance, it is easy to cause functional degradation or even failure, which raises many security concerns [11,12]. How to ensure the stability and the acceptable braking performance of AABS after actuator faults becomes a key issue.

In order to actually improve the safety and reliability of AABS, the fault probability can be reduced by reliability design and redundant technology on the one hand [13]. However, due to the production factors (cost/weight/technological level), the redundancy of aircraft components is so limited that the system reliability is hard to increase. On the other hand, fault-tolerant control (FTC) technology can be introduced into the AABS controller design, which is the future development direction of AABS and the key technology that needs urgent attention [14]. Reconfiguration control is a popular branch of FTC that has been widely used in many safety-critical systems, especially in aerospace engineering [15,16]. The essence of reconfiguration control is to consider the possible faults of the plant in the controller design process. When component faults occur, the fault system information is used to reconfigure the controller structure or parameters automatically [17]. In this way, the adverse effects caused by faults can be restrained or eliminated, thus realizing an asymptotically stable and acceptable performance of the closed-loop system. A number of common reconfiguration control methods can be classified as follows: adaptive control [18,19], multi-model switching control [20], sliding mode control [21], fuzzy control [22], other robust control [23], etc. In addition, the characteristics of AABS increase the difficulty of accurate modeling, and many nonlinear reconfiguration control methods are complex and relatively hard to apply in engineering. Therefore, it is crucial to design a reconfiguration controller with a clear structure, and which is model-independent, strong fault-perturbation resistant, and easy to implement.

Han retained the essence of PID control and proposed an active disturbance rejection control (ADRC) technique that requires low model accuracy and shows good control performance [24]. ADRC can estimate disturbances in internal and external systems and compensate for them [25]. Furthermore, ADRC has been widely used in FTC system design because of its obvious advantages in solving control problems of nonlinear models with uncertainty and strong disturbances [26–28]. Although the structure is not difficult to implement with modern digital computer technology, ADRC needs to tune a bunch of parameters which makes it hard to use in practice [29]. To overcome the difficulty, Gao proposed linear active disturbance rejection control (LADRC), which is based on linear extended state observer (LESO) and linear state error feedback (LSEF) [30,31]. The bandwidth tuning method greatly reduced the number of LADRC parameters. LADRC has been applied to solve various control problems [32–34].

However, it is well known that a controller with fixed parameters may not be able to maintain the acceptable (rated or degraded) performance of a fault system. For this reason, some advanced algorithms with parameter adaptive capabilities have been introduced by researchers that further improve the robustness and environmental adaptability of ADRC, such as neural networks [35,36], fuzzy logic [37,38], and the sliding mode [39,40]. With the development of artificial intelligence techniques, reinforcement learning has been applied to control science and engineering [41,42], and good results have been achieved. Yuan et al. proposed a novel online control algorithm for a thickener which is based on reinforcement learning [43]. Pang et al. studied the infinite-horizon adaptive optimal control of continuous-time linear periodic systems, using reinforcement learning techniques [44]. A Q-learning-based adaptive method for ADRC parameters was proposed by Chen et al. and has been applied to the ship course control [45].

Motivated by the above observations, in this paper, a reconfiguration control scheme via LADRC combined with deep reinforcement learning was developed for AABS which is subject to various fault perturbations. The proposed reconfiguration control method is a remarkable control strategy compared to previous methods for three reasons:

(1) AABS is extended with a new state variable, which is the sum of all unknown dynamics and disturbances not noticed in the fault-free system description. This state variable can be estimated using LESO. It indirectly simplifies the AABS modeling;

(2) Artificial intelligence technology is introduced and combined with the traditional control method to solve special control problems. By combining LADRC with the deep reinforcement learning TD3 algorithm, the selection of controller parameters is equivalent to the choice of agent actions. The parameter adaptive capabilities of LESO and LSEF are endowed through the continuous interaction between the agent and the environment, which not only eliminates the tedious manual tuning of the parameters, but also results in more accurate estimation and compensation for the adverse effects of fault perturbations;

(3) It is a data-driven robust control strategy that does not require any additional fault detection or identification (FDI) module, while the controller parameters are adaptive. Therefore, the proposed method corresponds to a novel combination of active reconfiguration control and FDI-free reconfiguration control, which makes it an interesting solution under unknown fault conditions.

The paper is organized as follows. Section 2 describes AABS dynamics with an actuator fault factor. The reconfiguration controller is presented in Section 3. The simulation results are presented to demonstrate the merits of the proposed method in Section 4, and conclusions are drawn in Section 5.

2. AABS Modeling

The AABS mainly consists of the following components: aircraft fuselage, landing gear, wheels, a hydraulic servo system, a braking device, and an anti-skid braking controller. The subsystems are strongly coupled and exhibit strong nonlinearity and complexity.

Based on the actual process and objective facts of anti-skid braking, the following reasonable assumptions can be made [46]:

(1) The aircraft fuselage is regarded as a rigid body with concentrated mass;
(2) The gyroscopic moment generated by the engine rotor is not considered during the aircraft braking process;
(3) The crosswind effect is ignored;
(4) Only the longitudinal deformation of the tire is taken into account and the deformation of the ground is ignored;
(5) All wheels are the same and controlled synchronously.

2.1. Aircraft Fuselage Dynamics

The force diagram of the aircraft fuselage is shown in Figure 1 and the specific parameters described in the diagram are shown in Table 1.

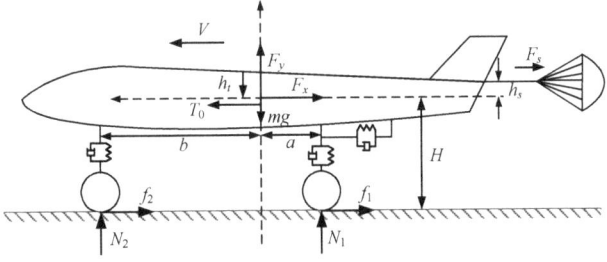

Figure 1. Force diagram of aircraft fuselage.

Table 1. Parameters of aircraft fuselage dynamics.

Name	Description	Value
H	Center of gravity height	
y	Center of gravity height variation	
V	Aircraft speed	
T_0	Engine force	
F_x	Aerodynamic drag	
F_y	Aerodynamic lift	
F_s	Parachute drag	
f_1	Braking friction force between main wheel and ground	
f_2	Braking friction force between front wheel and ground	
N_1	Main wheel support force	
N_2	Front wheel support force	
m	Mass of the aircraft	1761 kg
g	Gravitational acceleration	9.8 m/s^2
h_t	Distance between engine force line and center of gravity	0.1 m
h_s	Distance between parachute drag line and center of gravity	0.67 m
a	Distance between main wheel and center of gravity	1.076 m
b	Distance between front wheel and center of gravity	6.727 m
I	Fuselage inertia	4000 kg·s^2·m
S	Wing aera	50.88 m^2
S_s	Parachute area	20 m^2
C_x	Aerodynamic drag coefficient	0.1027
C_y	Aerodynamic lift coefficient	0.6
C_{xs}	Parachute drag coefficient	0.75
T_0'	Intimal engine force	426 kg
K_v	Velocity coefficient of engine	1 kg·s/m
ρ	Air density	4000 kg·s^2/m^4

The aircraft force and torque equilibrium equations are:

$$\begin{cases} m\dot{V} + F_x + F_s + f_1 + f_2 - T_0 = 0 \\ F_y + N_1 + N_2 - mg = 0 \\ N_2 b + F_s h_s - N_1 a - T_0 h_t - f_1 H - f_2 H = 0 \end{cases} \quad (1)$$

According to the influence of aerodynamic characteristics, we can obtain [46]:

$$\begin{cases} T_0 = T_0' + K_v V \\ F_x = \frac{1}{2}\rho C_x S V^2 \\ F_y = \frac{1}{2}\rho C_y S V^2 \\ F_s = \frac{1}{2}\rho C_{xs} S_s V^2 \\ f_1 = \mu_1 N_1 \\ f_2 = \mu_2 N_2 \end{cases} \quad (2)$$

2.2. Landing Gear Dynamics

The main function of the landing gear is to support and buffer the aircraft, thus improving the longitudinal and vertical forces. In addition to the wheel and braking device, the struts, buffers, and torque arm are also the main components of the landing gear. In this paper, it is assumed that the stiffness of the torque arm is large enough, and the torsional freedom of the wheel with respect to the strut and the buffer is ignored, so the torque arm is not considered.

The buffer can be reasonably simplified as a mass-spring-damping system [46], and the force acting on the aircraft fuselage by the buffer can be described as:

$$\begin{cases} N_1 = K_1 X_1 + C_1 \dot{X}_1^2 \\ N_2 = K_2 X_2 + C_2 \dot{X}_2^2 \end{cases} \quad (3)$$

$$\begin{cases} X_1 = a + y \\ X_2 = -b + y \end{cases} \quad (4)$$

whose parameters are shown in Table 2.

Table 2. Parameters of the buffer.

Name	Description	Value
X_1	Main buffer compression	
X_2	Front buffer compression	
K_1	Main buffer stiffness coefficient	42,529
K_2	Front buffer stiffness coefficient	2500
C_1	Main buffer damping coefficient	800
C_2	Front buffer damping coefficient	800

Due to the non-rigid connection between the landing gear and the aircraft fuselage, horizontal and angular displacements are generated under the action of braking forces. However, the struts are cantilever beams, and their angular displacements are very small and negligible. Therefore, the lateral stiffness model can be expressed by the following equivalent second-order equation:

$$\begin{cases} d_a = \dfrac{-\dfrac{f_1}{K_0}}{\dfrac{1}{W_n^2}s^2 + \dfrac{2\zeta}{W_n}s + 1} \\ d_V = \dfrac{d}{dt}(d_a) \end{cases} \quad (5)$$

whose parameters are shown in Table 3.

Table 3. Parameters of the landing gear lateral stiffness model.

Name	Description	Value
d_a	Navigation vibration displacement	Please see Equation (5)
d_V	Navigation vibration speed	Please see Equation (5)
K_0	Dynamic stiffness coefficient	536,000
ζ	Dynamic stiffness coefficient	0.2
W_n	Equivalent model natural frequency	60 Hz

2.3. Wheel Dynamics

The force diagram of the main wheel brake is shown in Figure 2.

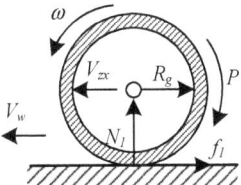

Figure 2. Force diagram of the main wheel.

It can be seen that during the taxiing, the main wheel is subjected to a combined effect of the braking torque M_s and the ground friction torque M_j. Due to the effect of the lateral stiffness, there is a longitudinal axle velocity V_{zx} along the fuselage, which is superimposed by the aircraft velocity V and the navigation vibration velocity d_V. The dynamics equation of the main wheel is [46]:

$$\begin{cases} \dot{\omega} = \dfrac{M_j - M_s}{J} + \dfrac{V_{zx}}{R_g} \\ V_w = \omega R_g \\ V_{zx} = V + d_V \\ R_g = R - Nk_\sigma \\ M_j = \mu N R_g n \end{cases} \quad (6)$$

whose parameters are shown in Table 4.

Table 4. Parameters of the main wheel.

Name	Description	Value
ω	Main wheel angular velocity	
$\dot{\omega}$	Main wheel angular acceleration	
V_w	Main wheel line speed	
R_g	Main wheel rolling radius	
N	Radical load	
J	Main wheel inertia	1.855 kg·s²·m
R	Wheel free radius	0.4 m
k_σ	Tire compression coefficient	1.07×10^{-5} m/kg
n	Equivalent model natural frequency	4

During the braking, the tires are subjected to the braking torque that keeps the aircraft speed always greater than the wheel speed, that is $V > V_w$. Thus, the slip ratio λ is defined to represent the slip motion ratio of the wheels relative to the runway. For the main wheel, using V_{zx} instead of V to calculate λ can avoid false brake release due to landing gear deformation, thus effectively reducing the landing gear walk situation [46]. The following equation is used to calculate the slip rate in this paper:

$$\lambda = \dfrac{V_{zx} - V_w}{V_{zx}} \quad (7)$$

The tire–runway combination coefficient is related to many factors, including real-time runway conditions, aircraft speed, slip rate, and so on. A simple empirical formula called 'magic formula' developed by Pacejka [47] is widely used to calculate and can be expressed as follows:

$$\mu(\lambda, \tau_j) = \tau_1 \sin(\tau_2 \text{arctg}(\tau_3 \lambda)) \quad (8)$$

where $\tau_j (j = 1, 2, 3)$, τ_1, τ_2, τ_3 are peak factor, stiffness factor, and curve shape factor, respectively. Table 5 lists the specific parameters for several different runway statuses [48].

Table 5. Parameters of the runway status.

Runway Status	τ_1	τ_2	τ_3
Dry runway	0.85	1.5344	14.5
Wet runway	0.40	2.0	8.2
Snow runway	0.28	2.0875	10

2.4. Hydraulic Servo System and Braking Device Modeling

Due to the complex structure of the hydraulic servo system, in this paper, some simplifications have been made so that only electro-hydraulic servo valves and pipes are considered. Their transfer functions are given as follows:

$$\begin{cases} M(s) = \dfrac{K_{sv}}{\dfrac{s^2}{\omega_{sv}^2} + \dfrac{2\xi_{sv}s}{\omega_{sv}} + 1} \\ L(s) = \dfrac{K_p}{T_p s + 1} \end{cases} \quad (9)$$

whose parameters are shown in Table 6.

It should be noted that the anti-skid braking controller should realize both braking control and anti-skid control. To this end, there is an approximately linear relationship between the brake pressure P and the control current I_c, which can be described as follows:

$$P = -I_c M(s) L(s) + P_0 \quad (10)$$

where $P_0 = 1 \times 10^7$ Pa.

The braking device serves to convert the brake pressure into brake torque, which is calculated as follows:

$$M_s = \mu_{mc} N_{mc} P R_{mc} \quad (11)$$

whose parameters are shown in Table 6.

Table 6. Parameters of the hydraulic servo system.

Name	Description	Value
K_{sv}	Servo valve gain	1
ω_{sv}	Servo valve natural frequency	17.7074 rad/s
ξ_{sv}	Servo valve damping ratio	0.36
K_p	Main wheel rolling radius	1
T_p	Pipe gain	0.01
μ_{mc}	Friction coefficient of brake material	0.23
N_{mc}	Number of friction surfaces	4
R_{mc}	Effective brake friction radius	0.142 m

The hydraulic servo system, as the actuator of AABS, is inevitably subject to some potential faults. Problems such as hydraulic oil mixing with air, internal leakage, and vibration seriously affect the efficiency of the hydraulic servo system [49]. Therefore, in this paper, the loss of efficiency (LOE) is introduced to represent a typical AABS actuator fault, which is characterized by a decrease in the actuator gain from its nominal value [26]. In the case of an actuator LOE fault, the brake pressure generated by the hydraulic servo system deviates from the commanded output expected by the controller. In other words, one instead has:

$$P_{fault} = k_{LOE} P \quad (12)$$

where P_{fault} represents the actuator actual output, and $k_{LOE} \in (0,1]$ refers the LOE fault factor.

Remark 1. *$n\%$ LOE is equivalent to the LOE fault gain $k_{LOE} = 1 - n/100$, $k_{LOE} = 1$ indicates that the actuator is fault-free.*

Remark 2. *Note that if the components do not always have the same characteristics as those of fault-free, it is necessary to establish the fault model. This not only provides an accurate model for the next reconfiguration on controller design, but also ensures that the adverse effects caused by fault perturbation can be effectively observed and compensated for.*

Thus, Equation (11) can be rewritten as follows:

$$M_s' = \mu_{mc} N_{mc} P_{fault} R_{mc} \tag{13}$$

where M_s' is the actual brake torque.

Remark 3. *As can be seen from the entire modeling process described above, AABS is nonlinear and highly coupled. The actuator fault leads to a sudden jump in the model parameters with greater internal perturbation compared to the fault-free case. Meanwhile, external disturbances such as the runway environment cannot be ignored.*

3. Reconfiguration Controller Design

3.1. Problem Description

Despite the aircraft having three degrees of freedom, only longitudinal taxiing is focused on in AABS. In this paper, AABS adopted the slip speed control type [48], that is, the braked wheel speed V_ω was used as the reference input, and the aircraft speed V was dynamically adjusted by the AABS controller to achieve anti-skid braking. According to Section 2, the AABS longitudinal dynamics model can be rewritten as follows:

$$\ddot{V} = f(V, \dot{V}, \omega_{out}, \omega_f) + b_v u \tag{14}$$

where $f(\cdot)$ is the controlled plant dynamics, ω_{out} represents the external disturbance, ω_f is an uncertain term including component faults, b_v is the control gain, and u is the system input.

Let $x_1 = V$, $x_2 = \dot{V}$. Set $f(V, \dot{V}, \omega_{out}, \omega_f)$ as the system generalized total perturbation and extend it to a new system state variable, i.e., $x_3 = f(V, \dot{V}, \omega_{out}, \omega_f)$. Then the state equation of System (14) can be obtained:

$$\begin{cases} \dot{x}_1 = x_2 \\ \dot{x}_2 = x_3 + b_v u \\ \dot{x}_3 = h(V, \dot{V}, \omega_{out}, \omega_f) \end{cases} \tag{15}$$

where x_1, x_2, x_3 are system state variables, and $h(V, \dot{V}, \omega_{out}, \omega_f) = \dot{f}(V, \dot{V}, \omega_{out}, \omega_f)$.

Assumption 1. *Both the system generalized total perturbation $f(V, \dot{V}, \omega_{out}, \omega_f)$ and its differential $h(V, \dot{V}, \omega_{out}, \omega_f)$ are bounded, i.e.,* $\begin{cases} |f(V, \dot{V}, \omega_{out}, \omega_f)| \leq \sigma_1 \\ |h(V, \dot{V}, \omega_{out}, \omega_f)| \leq \sigma_2 \end{cases}$, *where σ_1, σ_2 are two positive numbers.*

For System (14), affected by the total perturbation, a LADRC reconfiguration controller was designed next to restrain or eliminate the adverse effects, thus realizing the asymptotic stability and acceptable performance of the closed-loop system.

3.2. LADRC Controller Design

The control schematic of the LADRC is shown in Figure 3.

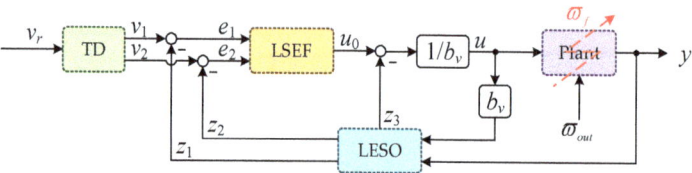

Figure 3. Control schematic of LADRC.

Firstly, the following tracking differentiator (TD) was designed:

$$\begin{cases} e(k) = v_1(k) - v_r(k) \\ \text{fh} = \text{fhan}(e(k), v_2(k), r, h) \\ v_1(k+1) = v_1(k) + hv_2(k) \\ v_2(k+1) = v_2(k) + h\text{fh} \end{cases} \quad (16)$$

where v_r is the desired input, v_1 is the transition process of v_r, v_2 is the derivative of v_1, and r and h are adjusted accordingly as filter coefficients. The function $\text{fhan}(\cdot)$ is defined as follows:

$$\text{fhan}(x_1, x_2, r, h) = -\begin{cases} r\text{sgn}(a), |a| > d_0 \\ r\frac{a}{d}, |a| \leq d_0 \end{cases} \quad (17)$$

We established the following form, LESO:

$$\begin{cases} \dot{z}_1 = z_2 - \beta_1(z_1 - v_1) \\ \dot{z}_2 = z_3 - \beta_2(z_1 - v_1) + b_v u \\ \dot{z}_3 = -\beta_3(z_1 - v_1) \end{cases} \quad (18)$$

Selecting the suitable observer gains $(\beta_1, \beta_2, \beta_3)$, LESO then enabled real-time observation of the variables in System (14) [50], i.e., $z_1 \to v_1$, $z_2 \to v_2$, $z_3 \to f(V, \dot{V}, \omega_{out}, \omega_f)$.

Set

$$u = \frac{u_0 - z_3}{b_v} \quad (19)$$

When z_3 can estimate $f(V, \dot{V}, \omega_{out}, \omega_f)$ without error, let LSEF be:

$$\begin{cases} e_1 = v_1 - z_1 \\ e_2 = v_2 - z_2 \\ u_0 = k_1 e_1 + k_2 e_2 \end{cases} \quad (20)$$

then the system (15) can be simplified to a double integral series structure:

$$\ddot{V} = (f(V, \dot{V}, \omega_{out}, \omega_f) - z_3) + u_0 \approx u_0 \quad (21)$$

Further, the bandwidth method [50] was used and we could obtain:

$$\begin{cases} \beta_1 = 3\omega_o \\ \beta_2 = 3\omega_o^2 \\ \beta_3 = \omega_o^3 \end{cases} \quad (22)$$

where ω_o is the observer bandwidth. The larger ω_o is, the smaller LESO observation errors are. However, the sensitivity of the system to noise may be increased, so the ω_o selection requires comprehensive consideration.

Similarly, according to the parameterization method and engineering experience [32], the LSEF parameters can be chosen as:

$$\begin{cases} k_1 = \omega_c^2 \\ k_2 = 2\xi\omega_c \end{cases} \quad (23)$$

where ω_c is the controller bandwidth, ξ is the damping ratio, and in this paper $\xi = 1$. Therefore, the parameter tuning problem of LADRC controller was simplified to the observer bandwidth ω_o and controller bandwidth ω_c configuration.

3.3. TD3 Algorithm

TD3 algorithm is an offline RL algorithm based on DDPG proposed in 2015 [51]. This approach adopted a similar method implemented in Double-DQN [52] to reduce the overestimation in function approximation, delaying the update frequency in the actor–network, and adding noises to target the actor–network to release the sensitivity and instability in DDPG. The structure of TD3 is shown in Figure 4.

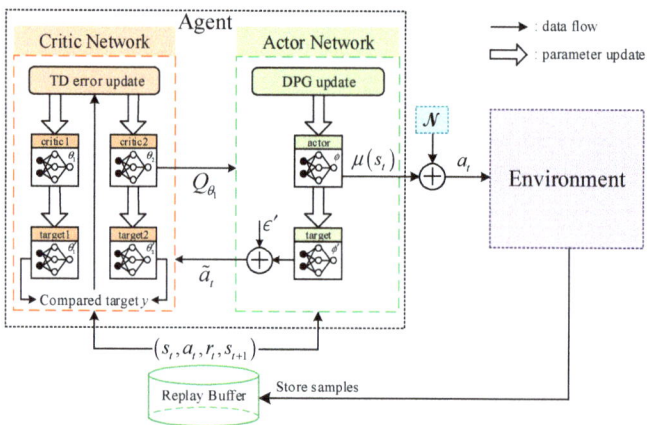

Figure 4. Structure of TD3.

Updating the parameters of critic networks by minimizing loss:

$$L = N^{-1} \sum (y - Q_{\theta_i}(s,a))^2 \tag{24}$$

where s is the current state, a is the current action, and $Q_{\theta_i}(s,a)$ stands for the parameterized state-action value function Q with parameter θ_i.

$$y = r + \gamma \min_{i=1,2} Q_{\theta'_i}(s', \tilde{a}) \tag{25}$$

is the target value of the function $Q_\theta(s,a)$, $\gamma \in [0,1]$ is the discount factor, and the target action is defined as:

$$\tilde{a} = \pi_{\phi'}(s) + \epsilon' \tag{26}$$

where noise ϵ' follows a clipped normal distribution clip $[\mathcal{N}(0,\sigma), -c, c]$, $c > 0$. This implies that ϵ' is a random variable with $\mathcal{N}(0,\sigma)$ and belongs to the interval $[-c,c]$.

The inputs of the actor network are both $Q_\theta(s,a)$ from the critic network and the minibatch form the memory, and the output is the action given by:

$$a_t = \pi_\phi(s_t) + \epsilon \tag{27}$$

where ϕ is the parameter of the actor network, and π_ϕ is the output form the actor network, which is a deterministic and continuous value. Noise ϵ follows the normal distribution $\mathcal{N}(0,\sigma)$, and is added for exploration.

Updating the parameters of the actor–network based on deterministic gradient strategy:

$$\nabla_\phi J(\phi) = N^{-1} \sum \nabla_a Q_{\theta_1}(s,a)\big|_{a=\pi_\phi(s)} \nabla_\phi \pi_\phi(s) \tag{28}$$

TD3 updates the actor–network and all three target networks every d steps periodically in order to avoid a too fast convergence. The parameters of the critic target networks and the actor–target network are updated according to:

$$\begin{cases} \theta'_i \leftarrow \tau\theta_i + (1-\tau)\theta'_i \\ \phi' \leftarrow \tau\phi + (1-\tau)\phi' \end{cases} \quad (29)$$

The pseudocode of the proposed approach is given in Algorithm 1.

Algorithm 1. TD3

1. Initialize critic networks Q_{θ_1}, Q_{θ_2} and actor network π_ϕ with random parameters θ_1, θ_2, ϕ;
2. Initialize target networks $Q_{\theta'_1}$, $Q_{\theta'_2}$ with $\theta'_1 \leftarrow \theta_1$, $\theta'_2 \leftarrow \theta_2$, and target actor network $\pi_{\phi'}$ with $\phi' \leftarrow \phi$;
3. Initialize replay buffer $\mathcal{R}$;
4. **For every episode:**
5. Initialize state s;
6. **Repeat**;
7. Select action with exploration noise $a \sim \pi(a) + \mathcal{N}(0, \sigma)$;
8. Observe reward r and new state s';
9. Store transition tuple (s, a, r, s') in $\mathcal{R}$;
10. Sample mini-batch of N transitions (s, a, r, s') from $\mathcal{R}$;
11. Attain $\tilde{a} \leftarrow \pi_{\phi'}(s) + \epsilon$, where $\epsilon \sim clip(\mathcal{N}(0, \sigma), -c, c)$;
12. Update critics $\theta_i \leftarrow \min_{\theta_i} N^{-1} \sum (y - Q_{\theta_i}(s, a))^2$;
13. Every d steps:
14. Update ϕ by the deterministic policy gradient:
15. $\nabla_\phi J(\phi) = N^{-1} \sum \nabla_a Q_{\theta_1}(s, a)|_{a=\pi_\phi(s)} \nabla_\phi \pi_\phi(s)$;
16. Update target network:
17. $\theta'_i \leftarrow \tau\theta_i + (1-\tau)\theta'_i$;
18. $\phi' \leftarrow \tau\phi + (1-\tau)\phi'$;
19. $s \leftarrow s'$;
20. **Until** s reaches terminal state s_T.

3.4. TD3-LADRC Reconfiguration Controller Design

Lack of environment adaptability, poor control performance, and weak robustness are the main shortcomings of parameter-fixed controllers [36]. When a fault occurs, it may not be possible to maintain the acceptable (rated or degraded) performance of the damaged system. Motivated by the above analysis, a reconfiguration controller called TD3-LADRC is proposed in this paper, and its control schematic is shown in Figure 5.

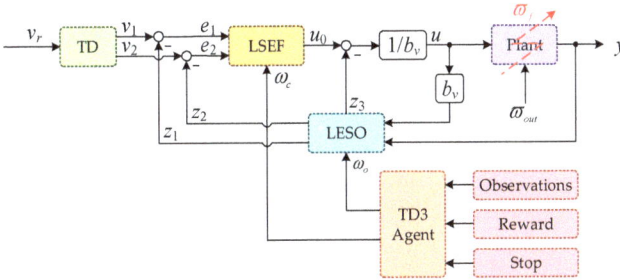

Figure 5. Control schematic of TD3-LADRC.

The deep reinforcement learning algorithm TD3 is introduced to realize the LADRC parameters adaption. The details of each part have been described above. The selection of control parameters is treated as the agent's action a_t, and the response result of the control system s_t is considered as the state, i.e., as follows:

$$\begin{cases} a_t = [\omega_o, \omega_c]^T \\ s_t = s_{obs} = \left[e, \dot{e}, V, \dot{V}\right]^T \end{cases} \qquad (30)$$

where $e = V - V_w$, and s_{obs} is the agent observations vector.

The range of each controller parameter is selected as follows:

$$\begin{cases} \omega_c \in [0, 4] \\ \omega_o \in [100, 200] \end{cases} \qquad (31)$$

The reward function plays a crucial role in the reinforcement learning algorithm. The appropriateness of the reward function design directly affects the training effect of the reinforcement learning, which in turn affects the effectiveness of the whole reconfiguration controller. According to the working characteristics of AABS, the following reward function is selected after several attempts to ensure stable and smooth braking:

$$r_t = 1\left(-6 \leq \dot{V} \leq -4\right) + 0\left(\dot{V} > -4 \parallel \dot{V} < -6\right) - 100(V < 2 \parallel e > 20) \qquad (32)$$

The stop conditions for each training episode are as follows, and one of the three will do:

(1) The aircraft speed $V < 2$;
(2) The error between main wheel speed and aircraft speed $e > 20$;
(3) Simulation time $t > 20$ s.

Remark 4. *TD3, TD, LESO, and LSEF together constitute the TD3-LADRC controller. Compared to normal LADRC, TD3-LADRC realizes the parameter adaption that makes the controller reconfigurable. The robustness and immunity are greatly improved. It can effectively compensate the adverse effects caused by the total perturbations including faults.*

3.5. TD3-LESO Estimation Capability Analysis

In order to prove the stability of the whole closed-loop system, the convergence of TD3-LESO is first analyzed in conjunction with Assumption 1 [53]. Let the estimation errors of TD3-LESO be $\tilde{x}_i = x_i - z_i, i = 1, 2, 3$, and the estimation error equation of the observer can be obtained as:

$$\begin{cases} \dot{\tilde{x}}_1 = \tilde{x}_2 - 3\omega_o \tilde{x}_1 \\ \dot{\tilde{x}}_2 = \tilde{x}_3 - 3\omega_o^2 \tilde{x}_1 \\ \dot{\tilde{x}}_3 = h(V, \dot{V}, \omega_{out}, \omega_f) - \omega_o^3 \tilde{x}_1 \end{cases} \qquad (33)$$

Let $\varepsilon_i = \dfrac{\tilde{x}_i}{\omega_o^{i-1}}, i = 1, 2, 3$, then Equation (33) can be rewritten as:

$$\dot{\varepsilon} = \omega_o A_3 \varepsilon + B \frac{h(V, \dot{V}, \omega_{out}, \omega_f)}{\omega_o^2} \qquad (34)$$

where $A_3 = \begin{bmatrix} -3 & 1 & 0 \\ -3 & 0 & 1 \\ -1 & 0 & 0 \end{bmatrix}, B = [0 \ 0 \ 1]^T$.

Based on Assumption 1 and Theorem 2 in Reference [54], the following theorem can be obtained:

Theorem 1. *Under the condition that $h(V, \dot{V}, \omega_{out}, \omega_f)$ is bounded, the TD3-LESO estimation errors are bounded and their upper bound decrease monotonically with the increase of the observer bandwidth ω_o.*

The proof is given in the Appendix A. Thus, it is clear that there are three positive numbers $v_i, i = 1, 2, 3$, such that the state estimation error $|\widetilde{x}_i| \leq v_i$ holds, i.e., the TD3-LESO estimation errors are bounded, which can effectively estimate the states of the controlled plant and the total perturbation.

3.6. Stability Analysis of Closed-loop System

The closed-loop system consisted of the control laws (19) and (20), and the controlled object (21) is:

$$\ddot{V} = f - z_3 + k_1 e_1 + k_2 e_2 \tag{35}$$

If we defined the tracking errors as $\varepsilon_i = v_i - x_i, i = 1, 2$, then we could attain:

$$\begin{cases} \dot{\varepsilon}_1 = \dot{r}_1 - \dot{x}_1 = r_2 - x_2 = \widetilde{e}_2 \\ \dot{\varepsilon}_2 = \dot{r}_2 - \dot{x}_2 = r_3 - \ddot{V} \\ \quad = -k_1 \varepsilon_1 - k_1 \widetilde{x}_1 - k_2 \varepsilon_2 - k_2 \widetilde{x}_2 - \widetilde{x}_3 \end{cases} \tag{36}$$

Let $\varepsilon = [\varepsilon_1, \varepsilon_2]^T, \widetilde{x} = [\widetilde{x}_1, \widetilde{x}_2, \widetilde{x}_3]^T$, then:

$$\dot{\varepsilon}(t) = A_\varepsilon \varepsilon(t) + A_{\widetilde{x}} \widetilde{x}(t) \tag{37}$$

where $A_e = \begin{bmatrix} 0 & 1 \\ -k_1 & -k_2 \end{bmatrix}, A_{\widetilde{x}} = \begin{bmatrix} 0 & 0 & 0 \\ -k_1 & -k_2 & -1 \end{bmatrix}$.

By solving Equation (37):

$$\varepsilon(t) = e^{A_\varepsilon t} \varepsilon(0) + \int_0^t e^{A_\varepsilon (t-\tau)} A_{\widetilde{x}} \widetilde{x}(\tau) d\tau \tag{38}$$

Combining Assumption 1, Theorem 1, Theorem 3, and Theorem 4 in the literature [54], the following theorem was proposed to analyze the stability of the closed-loop system:

Theorem 2. *Under the condition that the TD3-LESO estimation errors are bounded, there exists a controller bandwidth ω_c, such that the tracking error of the closed-loop system is bounded. Thus, for a bounded input, the output of the closed-loop system is bounded, i.e., the closed-loop system is BIBO-stable.*

See the Appendix A for proof.

4. Simulation Results

In order to verify the reconfiguration capability and disturbance rejection capabilities of the proposed method, the corresponding simulations are carried out in this section and compared with conventional PID + PBM and LADRC.

The initial states of the aircraft are set as follows:

(1) The initial speed of aircraft landing $V(0) = 72$ m/s;
(2) The initial height of the center of gravity $Hh = 2.178$ m.

To prevent deep wheel slippage as well as tire blowout, the wheel speed was kept following the aircraft speed quickly at first, and the brake pressure was applied only after 1.5 s. The anti-skid brake control was considered to be over when V was less than 2 m/s.

In the experiment, both the critic networks and the actor networks were realized by a fully connected neural network with three hidden layers. The number of neurons in the hidden layer was (50,25,25). The activation function of the hidden layer was selected as the ReLU function, and the activation function of the output layer of the actor network was selected as the tanh function. In addition, the parameters of the actor network and the critic network were tuned by an Adam optimizer. The remaining parameters of TD3-LADRC are shown in Table 7.

Table 7. Parameters of TD3-LADRC.

Name	Value
Control gain b_v	2
$TD - r$	0.001
$TD - h$	1
Discount factor γ	0.99
Actor learning rate	0.0001
Critic learning rate	0.001
Target update rate τ	0.001

Remark 5. *It is noted that the braking time t and braking distance x are selected as the criteria for braking efficiency, and the system stability is observed by slip rate λ.*

The model simulation was carried out in MATLAB 2022a, and the TD3 algorithm was realized through the reinforcement learning toolbox. The simulation time was 20 s, the sampling time was 0.001 s. The training stopped when the average reward reached 12,000. The training took about 6 h to complete. The learning curves of the reward obtained by the agent for each interaction with the environment during the training process are shown in Figure 6.

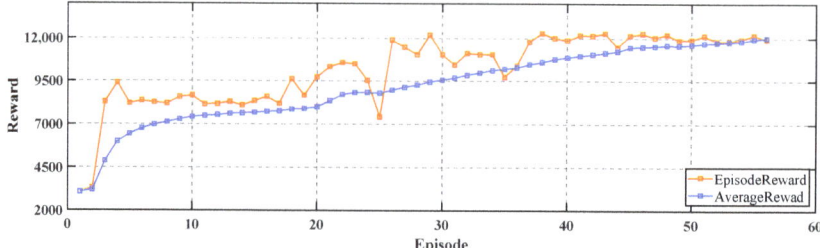

Figure 6. Learning curves.

It can be seen that at the beginning of the training, the agent was in the exploration phase and the reward obtained was relatively low. Later, the reward gradually increased, and after 40 episodes, the reward was steadily maintained at a high level and the algorithm gradually converges.

4.1. Case 1: Fault-Free and External Disturbance-Free in Dry Runway Condition

The simulation results of the dynamic braking process for different control schemes are shown in Figures 7 and 8 and Table 8.

As can be seen from Figure 7, PID + PBM leads to numerous skids during braking, which may cause serious loss to the tires. In contrast, LADRC and TD3-LADRC not only skid less frequently, but also have shorter braking time and braking distance. Moreover, the control effect of TD3-LADRC is better than LADRC. Figure 8 shows that TD3-LADRC can dynamically tune the controller parameters to accurately observe and compensate for the total disturbances, and thus improve the AABS performance.

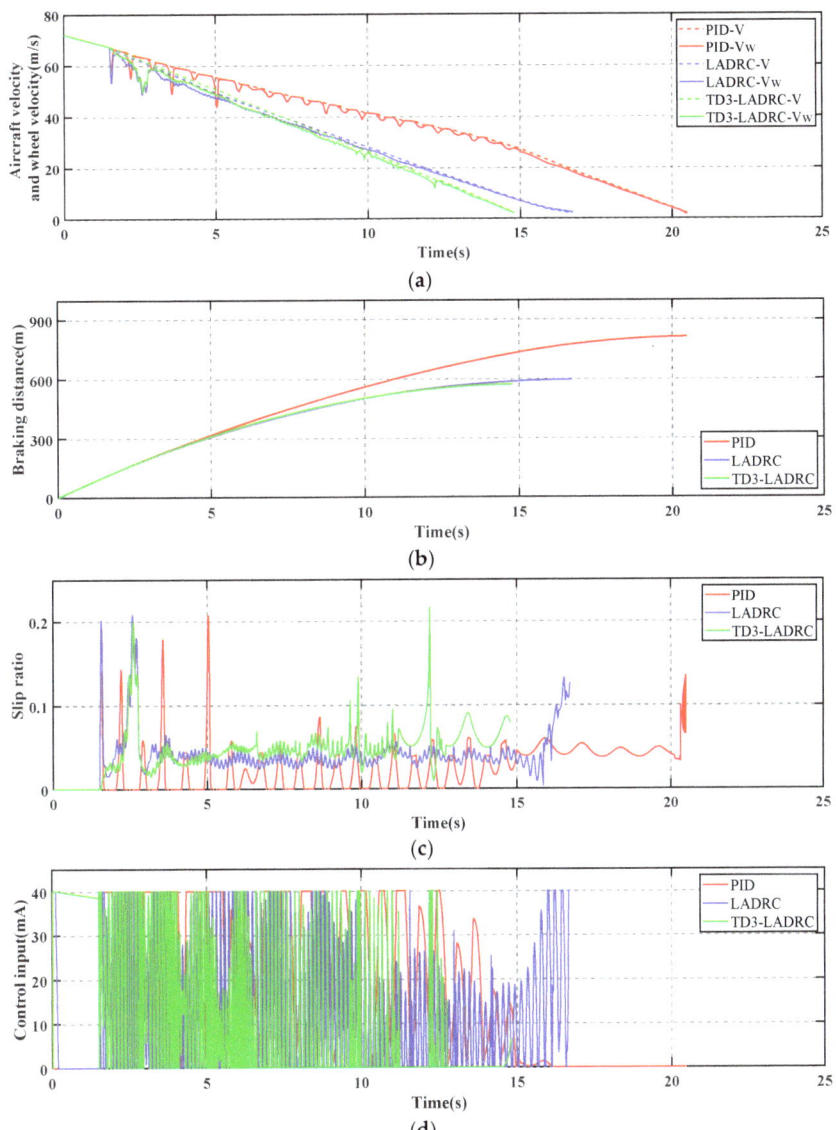

Figure 7. (**a**) Aircraft velocity and wheel velocity; (**b**) breaking distance; (**c**) slip ratio; (**d**) control input.

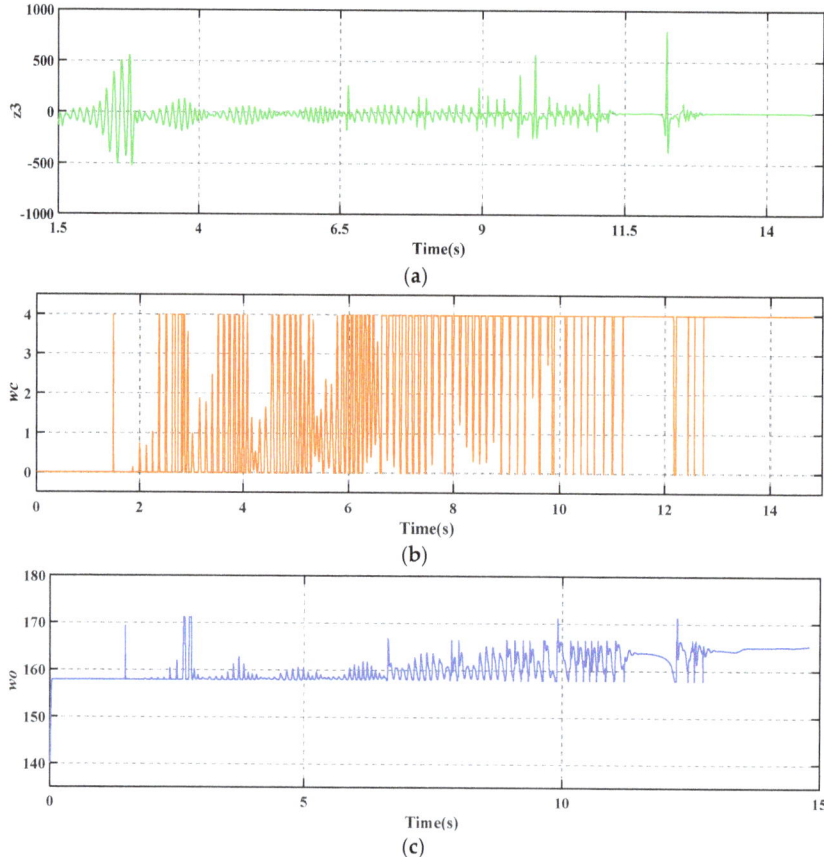

Figure 8. (a) Extended state of TD3-LADRC; (b) controller bandwidth ω_c; (c) controller bandwidth ω_o.

Table 8. AABS performance index.

Performance Index	PID + PBM	LADRC	TD3-LADRC
Braking time (s)	20.48	16.73	14.79
Braking distance (m)	811.9	595.46	571.18

Remark 6. *During the braking process, it is observed that in some instants $\omega_c = 0$. It may not affect the stability of the whole system. On the one hand, the value of ω_c does not change the fact that A_e is Hurwitz (see Proof of Theorem 2 for details). On the other hand, ω_c is constantly changed by the agent through a continuous interaction with the environment, and in these instants the agent considers $\omega_c = 0$ as optimal, i.e., no anti-skid braking control leads to better braking results.*

4.2. Case 2: Actuator LOE Fault in Dry Runway Condition

The fault considered here assumed a 20% actuator LOE at 5 s and escalated to 40% LOE at 10 s. The simulation results are shown in Figures 9 and 10 and Table 9.

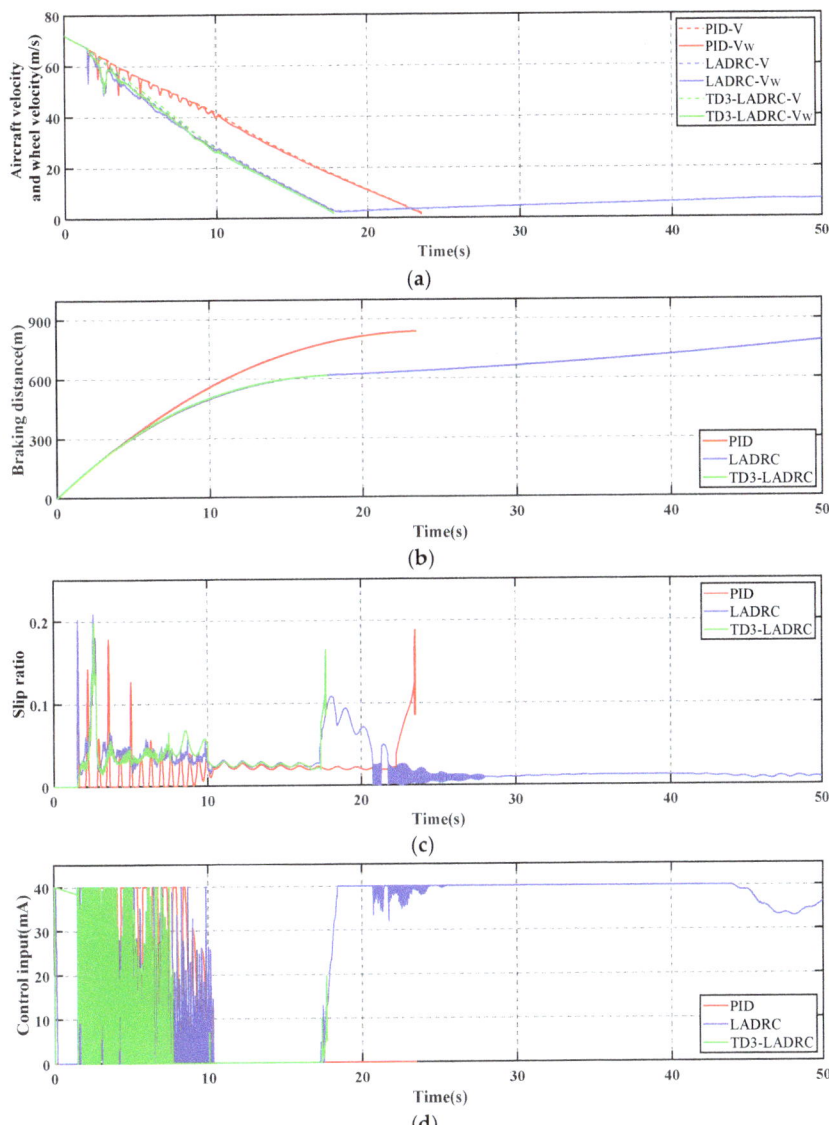

Figure 9. (**a**) Aircraft velocity and wheel velocity; (**b**) breaking distance; (**c**) slip ratio; (**d**) control input.

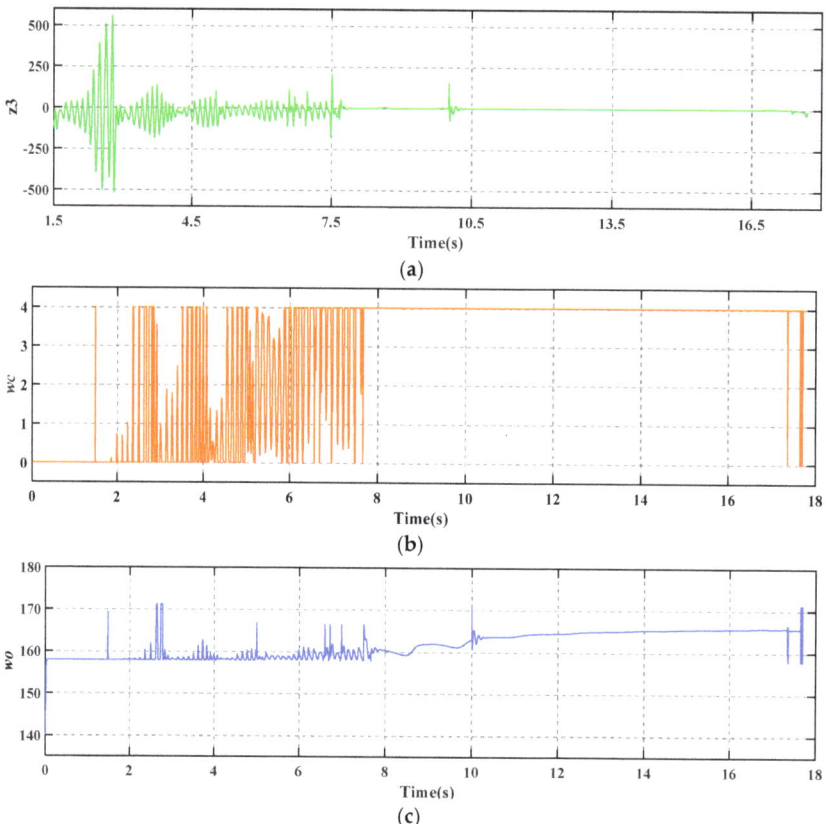

Figure 10. (a) Extended state of TD3-LADRC; (b) controller bandwidth ω_c; (c) controller bandwidth ω_0.

Table 9. AABS performance index.

Performance Index	PID + PBM	LADRC	TD3-LADRC
Braking time (s)	23.48	-	17.70
Braking distance (m)	838.46	-	618.12

As can be seen in Figure 9, PID + PBM continuously performed a large braking and releasing operation under the combined effect of fault and disturbance. This makes braking much less efficient and risks dragging and flat tires. In addition, LADRC cannot brake the aircraft to a stop which is not allowed in practice. Figure 9c shows that there is a high frequency of wheel slip in the low-speed phase of the aircraft. In contrast, TD3-LADRC retains the experience gained from the agent's prior training and continuously adjusts the controller parameters online based on the plant states, which ultimately allows the aircraft to brake smoothly. From Figure 10a, it can be seen that the total fault perturbations are estimated fast and accurately based on the adaptive LESO. Overall, TD3-LADRC not only improves the robustness and immunity of the controller in fault-perturbed conditions, but also greatly significantly improves the safety and reliability of AABS.

4.3. Case 3: Actuator LOE Fault in Mixed Runway Condition

The mixed runway structure is as follows: dry runway in the interval of 0–10 s, wet runway in the interval of 10–20 s, and snow runway after 20 s. The fault considered here assumed a 10% actuator LOE at 10 s. The simulation results are shown in Figures 11 and 12 and Table 10.

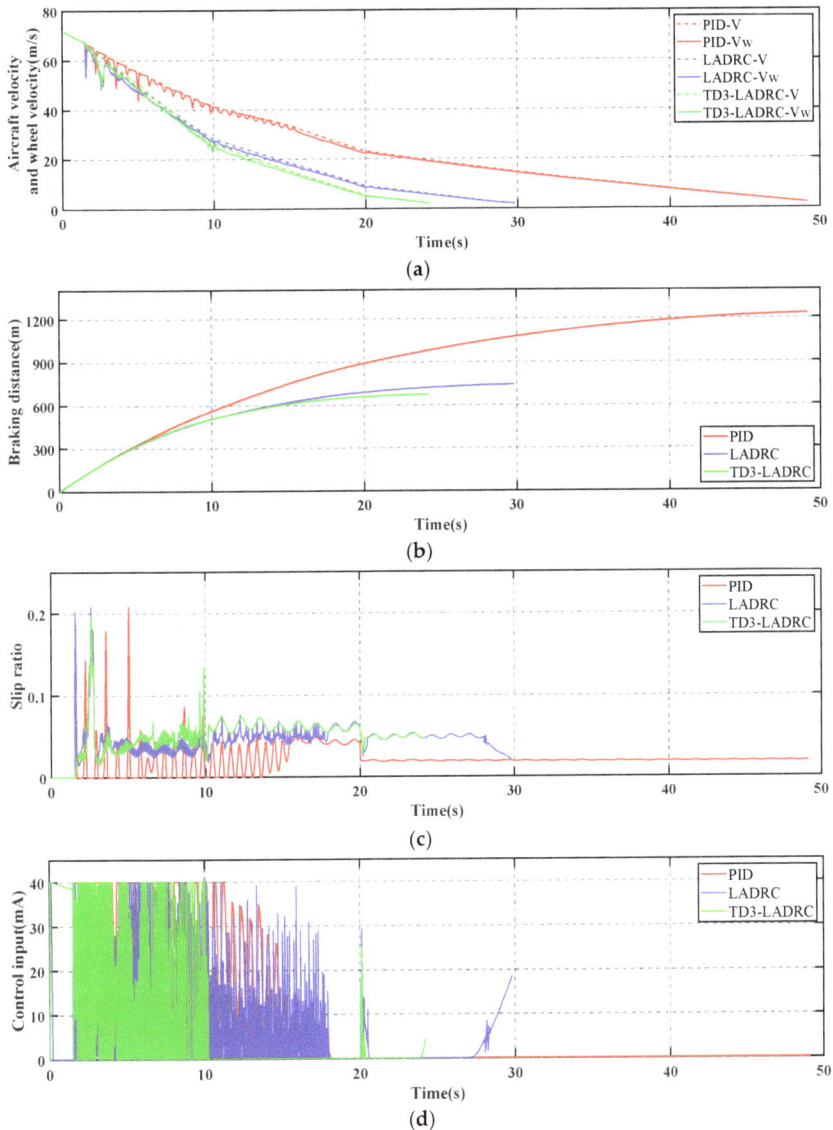

Figure 11. (a) Aircraft velocity and wheel velocity; (b) breaking distance; (c) slip ratio; (d) control input.

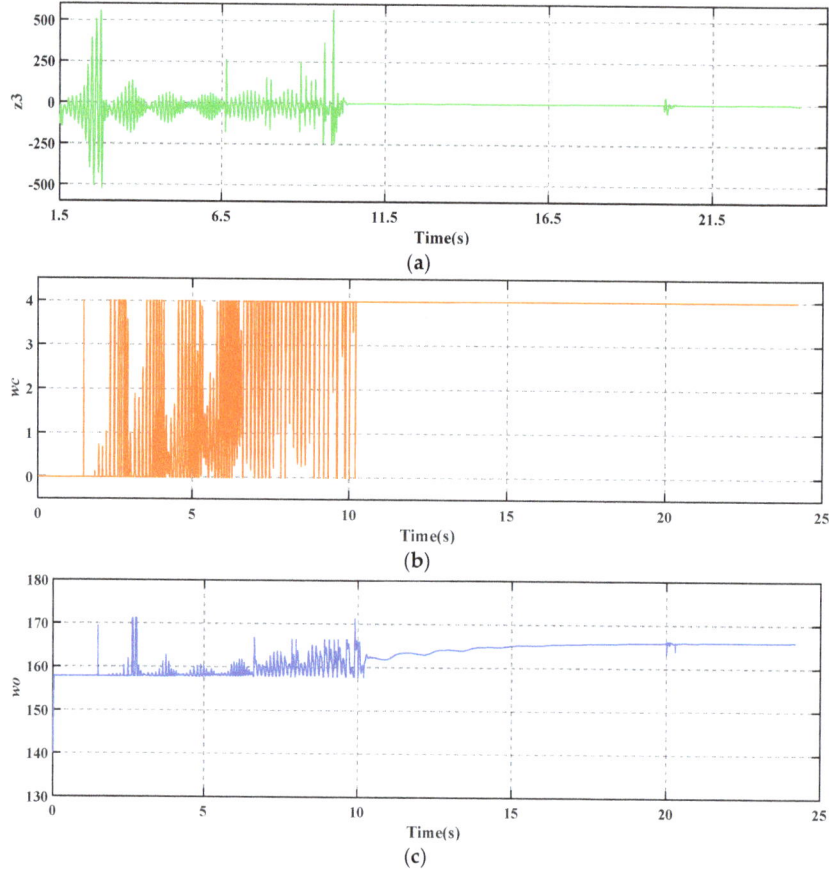

Figure 12. (a) Extended state of TD3-LADRC; (b) controller bandwidth ω_c; (c) controller bandwidth ω_0.

Table 10. AABS performance index.

Performance Index	PID + PBM	LADRC	TD3-LADRC
Braking time (s)	49.14	29.82	24.19
Braking distance (m)	1228.71	739.99	672.03

The deterioration of the runway conditions has resulted in a very poor tire–ground bond. It can be seen from Figure 11 that both braking time and braking distance have increased compared to the dry runway. Figure 12 shows that TD3-LADRC is still able to achieve controller parameters adaption, accurately observe the total fault perturbations, and effectively compensate for the adverse effects. The whole reconfiguration control system adapts well to runway changes. The environmental adaptability of AABS is improved.

5. Conclusions

A linear active disturbance rejection reconfiguration control scheme based on deep reinforcement learning was proposed to meet the higher performance requirements of AABS under fault-perturbed conditions. According to the composition structure and working principle, AABS mathematical model with an actuator fault factor is established.

A TD3-LADRC reconfiguration controller was developed, and the parameters of LSEF and LESO were adjusted online using the TD3 algorithm. The simulation results under different conditions verified that the designed controller can effectively improve the anti-skid braking performance even under faults and perturbations, as well as different runway environments. It successfully strengthened the robustness, immunity, and environmental adaptability of the AABS, thereby improving the safety and reliability of the aircraft. However, TD3-LADRC is so complex that its control effectiveness was verified only by simulations in this paper. The combined effect caused by various uncertainties in practical applications on the robustness of the controller cannot be completely considered. Therefore, in future work, an aircraft braking hardware-in-loop experimental platform is necessary to build, consisting of the host PC, the target CPU, the anti-skid braking controller, the actuators, and the aircraft wheel. The host PC and the target CPU are the software simulation part, while the other four parts are the hardware part.

Author Contributions: Conceptualization, S.L., Z.Y. and Z.Z.; methodology, S.L., Z.Y. and Z.Z.; software, S.L. and Z.Z.; validation, S.L., Z.Y., Z.Z., R.J., T.R., Y.J., S.C. and X.Z.; formal analysis, S.L., Z.Y. and Z.Z.; investigation, S.L., Z.Y., Z.Z., R.J., T.R., Y.J., S.C. and X.Z.; resources, Z.Y., Z.Z., R.J., T.R., Y.J., S.C. and X.Z.; data curation, S.L., Z.Y., Z.Z., R.J., T.R., Y.J., S.C. and X.Z.; writing—original draft, S.L. and Z.Z.; writing—review and editing, S.L. and Z.Z.; visualization, S.L. and Z.Z.; supervision, S.L., Z.Y., Z.Z., R.J., T.R., Y.J., S.C. and X.Z.; project administration, S.L., Z.Y. and Z.Z.; funding acquisition, Z.Y., Z.Z., R.J., T.R., Y.J., S.C. and X.Z. All authors have read and agreed to the published version of the manuscript.

Funding: This research was funded by the Key Laboratory Projects of Aeronautical Science Foundation of China, grant numbers 201928052006 and 20162852031, and Postgraduate Research & Practice Innovation Program of NUAA, grant number xcxjh20210332.

Data Availability Statement: Not applicable.

Conflicts of Interest: The authors declare no conflict of interest.

Appendix A. Proof of Theorems

Proof of Theorem 1. By solving Equation (34) we can attain:

$$\varepsilon(t) = e^{\omega_0 A_3 t}\varepsilon(0) + \int_0^t e^{\omega_0 A_3 (t-\tau)} B \frac{h(V(\tau), \dot{V}(\tau), \varpi_{out}, \varpi_f)}{\omega_0^2} d\tau \tag{A1}$$

Define $\zeta(t)$ as follows:

$$\zeta(t) = \int_0^t e^{\omega_0 A_3 (t-\tau)} B \frac{h(V(\tau), \dot{V}(\tau), \varpi_{out}, \varpi_f)}{\omega_0^2} d\tau \tag{A2}$$

From the fact that $h(V(\tau), \dot{V}(\tau), \varpi_{out}, \varpi_f)$ is bounded, we have:

$$|\zeta_i(t)| \leq \frac{\sigma}{\omega_0^3}\left[\left|\left(A_3^{-1}B\right)_i\right| + \left|\left(A_3^{-1}e^{\omega_0 A_3 t}B\right)_i\right|\right] \tag{A3}$$

Because $A_3^{-1} = \begin{bmatrix} 0 & 0 & -1 \\ 1 & 0 & -3 \\ 0 & 1 & -3 \end{bmatrix}$, we can attain:

$$\left|\left(A_3^{-1}B\right)_i\right| \leq 3 \tag{A4}$$

Considering that A_3 is Hurwitz, there is thus a finite time T_1 so that for any $t \geq T_1$, $i,j = 1,2,3$, the following formula holds [54]:

$$\left|\left[e^{\omega_o A_3 t}\right]_{ij}\right| \leq \frac{1}{\omega_o^3} \tag{A5}$$

Therefore, the following formula is satisfied:

$$\left|\left[e^{\omega_o A_3 t} B\right]_i\right| \leq \frac{1}{\omega_o^3} \tag{A6}$$

Finally, we can attain:

$$\left|\left(A_3^{-1} e^{\omega_o A_3 t} B\right)_i\right| \leq \frac{4}{\omega_o^3} \tag{A7}$$

From Equations (A3), (A4), and (A7) we can attain:

$$|\zeta_i(t)| \leq \frac{3\sigma}{\omega_o^3} + \frac{4\sigma}{\omega_o^6} \tag{A8}$$

Let $\varepsilon_{\text{sum}}(0) = |\varepsilon_1(0)| + |\varepsilon_2(0)| + |\varepsilon_3(0)|$, for all $t \geq T_1$, the following formula holds:

$$\left|\left[e^{\omega_o A_3 t} \varepsilon(0)\right]_i\right| \leq \frac{\varepsilon_{\text{sum}}(0)}{\omega_o^3} \tag{A9}$$

Form Equation (A1) we can attain:

$$|\varepsilon_i(t)| \leq \left|\left[e^{\omega_o A_3 t} \varepsilon(0)\right]_i\right| + |\zeta_i(t)| \tag{A10}$$

Let $\tilde{x}_{\text{sum}}(0) = |\tilde{x}_1(0)| + |\tilde{x}_2(0)| + |\tilde{x}_3(0)|$, from $\varepsilon_i = \frac{\tilde{x}_i}{\omega_o^{i-1}}$ and formulas (A8)–(A10), we can attain:

$$|\tilde{x}_i(t)| \leq \left|\frac{\tilde{x}_{\text{sum}}(0)}{\omega_o^3}\right| + \frac{3\sigma}{\omega_o^{4-i}} + \frac{4\sigma}{\omega_o^{7-i}} = v_i \tag{A11}$$

For all $t \geq T_1, i = 1,2,3$, the above formula holds. $\square$

Proof of Theorem 2. According to Equation (37) and Theorem 1, we can attain:

$$\begin{cases} [A_{\tilde{x}} \tilde{x}(\tau)]_1 = 0 \\ |[A_{\tilde{x}} \tilde{x}(\tau)]_2| \leq k_{\text{sum}} v_i = \gamma_l, \forall t \geq T_1 \end{cases} \tag{A12}$$

where $k_{\text{sum}} = 1 + k_1 + k_2$, bringing in the controller bandwidth $k_{\text{sum}} = 1 + \omega_c^2 + 2\omega_c$, and taking the parameters in this way ensures that A_ε is Hurwitz [54].

Define $\Theta = [0 \quad \gamma_l]^T$, let $\vartheta(t) = \int_0^t e^{A_\varepsilon(t-\tau)} A_{\tilde{x}} \tilde{x}(\tau) d\tau$, then we can attain:

$$|\vartheta_i(t)| \leq \left|\left(A_\varepsilon^{-1} \Theta\right)_i\right| + \left|\left(A_\varepsilon^{-1} e^{A_\varepsilon t} \Theta\right)_i\right|, i = 1,2 \tag{A13}$$

$$\begin{cases} |(A_\varepsilon^{-1} \Theta)_1| = \frac{\gamma_l}{k_1} = \frac{\gamma_l}{\omega_c^2} \\ |(A_\varepsilon^{-1} \Theta)_2| = 0 \end{cases} \tag{A14}$$

Consider that A_ε is Hurwitz; thus, there is a finite time T_2 so that for any $t \geq T_2$, $i,j = 1,2,3$, the following formula holds [54]:

$$\left|\left[e^{A_\varepsilon t}\right]_{ij}\right| \leq \frac{1}{\omega_c^3} \tag{A15}$$

Let $T_3 = \max\{T_1, T_2\}$, for any $t \geq T_3, i = 1, 2$, we can attain:

$$\left|(e^{A_\varepsilon t}\Theta)_i\right| \leq \frac{\gamma_l}{\omega_c^3} \tag{A16}$$

Then we can attain:

$$\left|\left(A_\varepsilon^{-1} e^{A_\varepsilon t}\Theta\right)_i\right| \leq \begin{cases} \dfrac{1+k_2}{\omega_c^2}\dfrac{\gamma_l}{\omega_c^3}, & i = 1 \\ \dfrac{\gamma_l}{\omega_c^3}, & i = 2 \end{cases} \tag{A17}$$

From Equations (A13), (A14), and (A17) we can attain that for any $t \geq T_3$:

$$|\vartheta_i(t)| \leq \begin{cases} \dfrac{\gamma_l}{\omega_c^2} + \dfrac{(1+k_2)\gamma_l}{\omega_c^5}, & i = 1 \\ \dfrac{\gamma_l}{\omega_c^3}, & i = 2 \end{cases} \tag{A18}$$

Let $\varepsilon_s(0) = |\varepsilon_1(0)| + |\varepsilon_2(0)|$, then for any $t \geq T_3$:

$$\left|\left[e^{A_\varepsilon t}\varepsilon(0)\right]_i\right| \leq \frac{\varepsilon_s(0)}{\omega_c^3} \tag{A19}$$

From Equation (A12), we can attain:

$$|\varepsilon_i(t)| \leq \left|\left[e^{A_\varepsilon t}\varepsilon(0)\right]_i\right| + |\vartheta_i(t)| \tag{A20}$$

From Equations (A12), (A18)–(A20), we can attain that for any $t \geq T_3, i = 1, 2$:

$$|\varepsilon_i(t)| \leq \begin{cases} \dfrac{\varepsilon_s(0)}{\omega_c^3} + \dfrac{k_{sum}v_i}{\omega_c^2} + \dfrac{(1+k_2)k_{sum}v_i}{\omega_c^5}, & i = 1 \\ \dfrac{k_{sum}v_i + \varepsilon_s(0)}{\omega_c^3}, & i = 2 \end{cases} \tag{A21}$$

□

References

1. Li, F.; Jiao, Z. Adaptive Aircraft Anti-Skid Braking Control Based on Joint Force Model. *J. Beijing Univ. Aeronaut. Astronaut.* **2013**, *4*, 447–452.
2. Jiao, Z.; Sun, D.; Shang, Y.; Liu, X.; Wu, S. A high efficiency aircraft anti-skid brake control with runway identification. *Aerosp. Sci. Technol.* **2019**, *91*, 82–95. [CrossRef]
3. Chen, M.; Liu, W.; Ma, Y.; Wang, J.; Xu, F.; Wang, Y. Mixed slip-deceleration PID control of aircraft wheel braking system. *IFAC-PapersOnLine* **2018**, *51*, 160–165. [CrossRef]
4. Chen, M.; Xu, F.; Liang, X.; Liu, W. MSD-based NMPC Aircraft Anti-skid Brake Control Method Considering Runway Variation. *IEEE Access* **2021**, *9*, 51793–51804. [CrossRef]
5. Dinçmen, E.; Güvenç, B.A.; Acarman, T. Extremum-seeking control of ABS braking in road vehicles with lateral force improvement. *IEEE Trans. Contr. Syst. Technol.* **2012**, *22*, 230–237. [CrossRef]
6. Li, F.B.; Huang, P.M.; Yang, C.H.; Liao, L.Q.; Gui, W.H. Sliding mode control design of aircraft electric brake system based on nonlinear disturbance observer. *Acta Autom. Sin.* **2021**, *47*, 2557–2569.
7. Radac, M.B.; Precup, R.E. Data-driven model-free slip control of anti-lock braking systems using reinforcement Q-learning. *Neurocomputing* **2018**, *275*, 317–329. [CrossRef]
8. Zhang, R.; Peng, J.; Chen, B.; Gao, K.; Yang, Y.; Huang, Z. Prescribed Performance Active Braking Control with Reference Adaptation for High-Speed Trains. *Actuators* **2021**, *10*, 313. [CrossRef]
9. Qiu, Y.; Liang, X.; Dai, Z. Backstepping dynamic surface control for an anti-skid braking system. *Control Eng. Pract.* **2015**, *42*, 140–152. [CrossRef]
10. Mirzaei, A.; Moallem, M.; Dehkordi, B.M.; Fahimi, B. Design of an Optimal Fuzzy Controller for Antilock Braking Systems. *IEEE Trans. Veh. Technol.* **2006**, *55*, 1725–1730. [CrossRef]

11. Xiang, Y.; Jin, J. Hybrid Fault-Tolerant Flight Control System Design Against Partial Actuator Failures. *IEEE Trans. Control Syst. Technol.* **2012**, *20*, 871–886.
12. Niksefat, N.; Sepehri, N. A QFT fault-tolerant control for electrohydraulic positioning systems. *IEEE Trans. Control Syst. Technol.* **2002**, *4*, 626–632. [CrossRef]
13. Wang, D.Y.; Tu, Y.; Liu, C. Connotation and research of reconfigurability for spacecraft control systems: A review. *Acta Autom. Sin.* **2017**, *43*, 1687–1702.
14. Han, Y.G.; Liu, Z.P.; Dong, Z.C. *Research on Present Situation and Development Direction of Aircraft Anti-Skid Braking System*; China Aviation Publishing & Media CO., LTD.: Xi'an, China, 2020; Volume 5, pp. 525–529.
15. Calise, A.J.; Lee, S.; Sharma, M. Development of a reconfigurable flight control law for tailless aircraft. *J. Guid. Control Dyn.* **2001**, *24*, 896–902. [CrossRef]
16. Yin, S.; Xiao, B.; Ding, S.X.; Zhou, D. A review on recent development of spacecraft attitude fault tolerant control system. *IEEE Trans. Ind. Electron.* **2016**, *63*, 3311–3320. [CrossRef]
17. Zhang, Y.; Jin, J. Bibliographical review on reconfigurable fault-tolerant control systems. *Annu. Rev. Control* **2008**, *32*, 229–252. [CrossRef]
18. Zhang, Z.; Yang, Z.; Xiong, S.; Chen, S.; Liu, S.; Zhang, X. Simple Adaptive Control-Based Reconfiguration Design of Cabin Pressure Control System. *Complexity* **2021**, *2021*, 6635571. [CrossRef]
19. Chen, F.; Wu, Q.; Jiang, B.; Tao, G. A reconfiguration scheme for quadrotor helicopter via simple adaptive control and quantum logic. *IEEE Trans. Ind. Electron.* **2015**, *62*, 4328–4335. [CrossRef]
20. Guo, Y.; Jiang, B. Multiple model-based adaptive reconfiguration control for actuator fault. *Acta Autom. Sinica* **2009**, *35*, 1452–1458. [CrossRef]
21. Gao, Z.; Jiang, B.; Shi, P.; Qian, M.; Lin, J. Active fault tolerant control design for reusable launch vehicle using adaptive sliding mode technique. *J. Frankl. Inst.* **2012**, *349*, 1543–1560. [CrossRef]
22. Shen, Q.; Jiang, B.; Cocquempot, V. Fuzzy Logic System-Based Adaptive Fault-Tolerant Control for Near-Space Vehicle Attitude Dynamics with Actuator Faults. *IEEE Trans. Fuzzy Syst.* **2013**, *21*, 289–300. [CrossRef]
23. Lv, X.; Jiang, B.; Qi, R.; Zhao, J. Survey on nonlinear reconfigurable flight control. *J. Syst. Eng. Electron.* **2013**, *24*, 971–983. [CrossRef]
24. Han, J. From PID to active disturbance rejection control. *IEEE Trans. Ind. Electron.* **2009**, *56*, 900–906. [CrossRef]
25. Huang, Y.; Xue, W. Active disturbance rejection control: Methodology and theoretical analysis. *ISA Trans.* **2014**, *53*, 963–976. [CrossRef] [PubMed]
26. Guo, Y.; Jiang, B.; Zhang, Y. A novel robust attitude control for quadrotor aircraft subject to actuator faults and wind gusts. *IEEE/CAA J. Autom. Sin.* **2017**, *5*, 292–300. [CrossRef]
27. Zhang, Z.; Yang, Z.; Zhou, G.; Liu, S.; Zhou, D.; Chen, S.; Zhang, X. Adaptive Fuzzy Active-Disturbance Rejection Control-Based Reconfiguration Controller Design for Aircraft Anti-Skid Braking System. *Actuators* **2021**, *10*, 201. [CrossRef]
28. Zhou, L.; Ma, L.; Wang, J. Fault tolerant control for a class of nonlinear system based on active disturbance rejection control and rbf neural networks. In Proceedings of the 2017 36th Chinese Control Conference (CCC), Dalian, China, 26–28 July 2017; IEEE: Piscataway, NJ, USA, 2017; pp. 7321–7326.
29. Tan, W.; Fu, C. Linear active disturbance-rejection control: Analysis and tuning via IMC. *IEEE Trans. Ind. Electron.* **2015**, *63*, 2350–2359. [CrossRef]
30. Gao, Z. Scaling and bandwidth-parameterization based controller tuning. In Proceedings of the 2003 American Control Conference, Denver, CO, USA, 4–6 June 2003; pp. 4989–4996.
31. Gao, Z. Active disturbance rejection control: A paradigm shift in feedback control system design. In Proceedings of the 2006 American Control Conference, Minneapolis, MN, USA, 14–16 June 2006; IEEE: Piscataway, NJ, USA, 2006; pp. 2399–2405.
32. Li, P.; Zhu, G.; Zhang, M. Linear active disturbance rejection control for servo motor systems with input delay via internal model control rules. *IEEE Trans. Ind. Electron.* **2020**, *68*, 1077–1086. [CrossRef]
33. Wang, L.X.; Zhao, D.X.; Liu, F.C.; Liu, Q.; Meng, F.L. Linear active disturbance rejection control for electro-hydraulic proportional position synchronous. *Control Theory Appl.* **2018**, *35*, 1618–1625.
34. Li, J.; Qi, X.H.; Wan, H.; Xia, Y.Q. Active disturbance rejection control: Theoretical results summary and future researches. *Control Theory Appl.* **2017**, *34*, 281–295.
35. Qiao, G.L.; Tong, C.N.; Sun, Y.K. Study on Mould Level and Casting Speed Coordination Control Based on ADRC with DRNN Optimization. *Acta Autom. Sin.* **2007**, *33*, 641–648.
36. Qi, X.H.; Li, J.; Han, S.T. Adaptive active disturbance rejection control and its simulation based on BP neural network. *Acta Armamentarii* **2013**, *34*, 776–782.
37. Sun, K.; Xu, Z.L.; Gai, K.; Zou, J.Y.; Dou, R.Z. Novel Position Controller of Pmsm Servo System Based on Active-disturbance Rejection Controller. *Proc. Chin. Soc. Electr. Eng.* **2007**, *27*, 43–46.
38. Dou, J.X.; Kong, X.X.; Wen, B.C. Attitude fuzzy active disturbance rejection controller design of quadrotor UAV and its stability analysis. *J. Chin. Inert. Technol.* **2015**, *23*, 824–830.
39. Zhao, X.; Wu, C. Current deviation decoupling control based on sliding mode active disturbance rejection for PMLSM. *Opt. Precis. Eng.* **2022**, *30*, 431–441. [CrossRef]

40. Li, B.; Zeng, L.; Zhang, P.; Zhu, Z. Sliding mode active disturbance rejection decoupling control for active magnetic bearings. *Electr. Mach. Control.* **2021**, *7*, 129–138.
41. Buşoniu, L.; de Bruin, T.; Tolić, D.; Kober, J.; Palunko, I. Reinforcement learning for control: Performance, stability, and deep approximators. *Annu. Rev. Control* **2018**, *46*, 8–28. [CrossRef]
42. Nian, R.; Liu, J.; Huang, B. A review on reinforcement learning: Introduction and applications in industrial process control. *Comput. Chem. Eng.* **2020**, *139*, 106886. [CrossRef]
43. Yuan, Z.L.; He, R.Z.; Yao, C.; Li, J.; Ban, X.J. Online reinforcement learning control algorithm for concentration of thickener underflow. *Acta Autom. Sin.* **2021**, *47*, 1558–1571.
44. Pang, B.; Jiang, Z.P.; Mareels, I. Reinforcement learning for adaptive optimal control of continuous-time linear periodic systems. *Automatica* **2020**, *118*, 109035. [CrossRef]
45. Chen, Z.; Qin, B.; Sun, M.; Sun, Q. Q-Learning-based parameters adaptive algorithm for active disturbance rejection control and its application to ship course control. *Neurocomputing* **2020**, *408*, 51–63. [CrossRef]
46. Zou, M.Y. Design and Simulation Research on New Control Law of Aircraft Anti-skid Braking System. Master's Thesis, Northwestern Polytechnical University, Xi'an, China, 2005.
47. Pacejka, H.B.; Bakker, E. The magic formula tyre model. *Veh. Syst. Dyn.* **1992**, *21*, 1–18. [CrossRef]
48. Wang, J.S. Nonlinear Control Theory and its Application to Aircraft Antiskid Brake Systems. Master's Thesis, Northwestern Polytechnical University, Xi'an, China, 2001.
49. Jiao, Z.; Liu, X.; Shang, Y.; Huang, C. An integrated self-energized brake system for aircrafts based on a switching valve control. *Aerosp. Sci. Technol.* **2017**, *60*, 20–30. [CrossRef]
50. Yuan, D.; Ma, X.J.; Zeng, Q.H.; Qiu, X. Research on frequency-band characteristics and parameters configuration of linear active disturbance rejection control for second-order systems. *Control Theory Appl.* **2013**, *30*, 1630–1640.
51. Lillicrap, T.P.; Hunt, J.J.; Pritzel, A.; Heess, N.; Erez, T.; Tassa, Y.; Silver, D.; Wierstra, D. Continuous control with deep reinforcement learning. *arXiv* **2015**, arXiv:1509.02971.
52. Fujimoto, S.; Hoof, H.; Meger, D. Addressing function approximation error in actor-critic methods. In Proceedings of the 35th International Conference on Machine Learning, Stockholm, Sweden, 10–15 July 2018; pp. 1587–1596.
53. Chen, Z.Q.; Sun, M.W.; Yang, R.G. On the Stability of Linear Active Disturbance Rejection Control. *Acta Autom. Sin.* **2013**, *39*, 574–580. [CrossRef]
54. Zheng, Q.; Gao, L.Q.; Gao, Z. On stability analysis of active disturbance rejection control for nonlinear time-varying plants with unknown dynamics. In Proceedings of the 2007 46th IEEE conference on decision and control, New Orleans, LA, USA, 12–14 December 2007; IEEE: Piscataway, NJ, USA, 2007; pp. 3501–3506.

Article

Multi-Mode Shape Control of Active Compliant Aerospace Structures Using Anisotropic Piezocomposite Materials in Antisymmetric Bimorph Configuration

Xiaoming Wang [1,*], Xinhan Hu [2], Chengbin Huang [2] and Wenya Zhou [2]

1 School of Mechanical and Electrical Engineering, Guangzhou University, Guangzhou 510006, China
2 Liaoning Provincial Key Laboratory of Aerospace Advanced Technology,
School of Aeronautics and Astronautics, Dalian University of Technology, Dalian 116024, China; huxinhan@mail.dlut.edu.cn (X.H.); 904368073@mail.dlut.edu.cn (C.H.); zwy@dlut.edu.cn (W.Z.)
* Correspondence: wangxm@gzhu.edu.cn

Abstract: The mission performance of future advanced aerospace structures can be synthetically improved via active shape control utilizing piezoelectric materials. Multiple work modes are required. Bending/twisting mode control receives special attention for many classic aerospace structures, such as active reflector systems, active blades, and compliant morphing wings. Piezoelectric fiber composite (Piezocomposite) material features in-plane anisotropic actuation, which is very suitable for multiple work modes. In this study, two identical macro-fiber composite (MFC) actuators of the F1 type were bonded to the base plate structure in an "antisymmetric angle-ply bimorph configuration" in order to achieve independent bending/twisting shape control. In terms of the finite element model and homogenization strategy, the locations of bimorph MFCs were determined by considering the effect of trade-off control capabilities on the bending and twisting shapes. The modal characteristics were investigated via both experimental and theoretical approaches. The experimental tests implied that the shape control accuracy was heavily reduced due to various uncertainties and nonlinearities, including hysteresis and the creep effect of the actuators, model errors, and external disturbances. A multi-mode feedback control law was designed and the experimental tests indicated that synthetic (independent and coupled) bending/twisting deformations were achieved with improved shape accuracy. This study provides a feasible multi-mode shape control approach with high surface accuracy, especially by employing piezocomposite materials.

Keywords: shape control; macro-fiber composites; bending; twisting; experimental validation; control system

1. Introduction

Owing to the increasingly stringent requirements for industrial equipment, particularly for the aerospace fields, smart devices and structures are receiving increasing attention to improve the synthetic or specific performances of systems [1–3]. Amongst these devices and structures, piezoelectric materials are the most widely used actuators or sensors for various applications, including shape control, vibration suppression, and health monitoring [4,5]. Active structures integrated with piezoelectric materials can change their shape or profile to enhance the accuracy and adaptability of the system via an active control approach. Some classic instances are active reflector systems, active blades, and compliant morphing wings [6–8].

Structural shape control utilizing piezoelectric materials is mainly implemented for two purposes: (1) To correct the shape error of the structure due to manufacturing error or external disturbance; and (2) to provide active shape control for morphing applications with a specific purpose, such as mechanically reconfigurable reflectors (MRRs) and compliant morphing wings. For reflector systems, particularly the large deployable antennas that will be used in the future, high surface precision is required to maintain position accuracy; however, surface errors are induced by many factors, including assembly error, deployment accuracy, environmental loads, or mechanical creep [9]. Active shape control has been proven as a feasible approach to correct reflector surface errors and ensure high shape accuracy. Bradford [6] investigated active reflector systems actuated by macro-fiber composite (MFC) arrays to correct thermally induced deformations and manufacturing errors. Hill [10] investigated the feasibility of using distributed polyvinylidene fluoride actuators in the active control of a large-scale reflector under thermal load. Wang [11] used PZT actuators to adjust the surfaces of flexible cable net antennas using quadratic criteria. Song [12] presented an experimental validation of a PZT-actuated CFRP reflector using a closed-loop iterative shape control method based on the influence coefficient matrix. In addition to modifying the shape error, piezoelectric actuators can also be used for mechanically reconfigurable reflectors that can actively reshape its surface according to the purpose, such as modifying the service coverage. Tanaka and Hiraku [13,14] developed MRR prototypes with six spherical piezoelectric actuators. Shao [7] designed a mechanically reconfigurable reflector using 30 piezoelectric inchworm actuators and presented a distributed time-sharing control strategy to minimize the size and power of the MRR system. In the aviation field, piezoelectric materials are often used for the active shape control of both fixed and rotating wings. Monner [15,16] developed "active twist blades" and used MFCs for twist actuation. Li [17] presented an experimental validation of the feasibility of using piezoelectric actuators to improve rolling power at all dynamic pressures via elastic wing twist. In the last decade, the use of piezocomposite materials, especially MFCs, in compliant morphing wing designs has been broadly investigated. Bilgen [8,18,19] devoted many studies towards the design, optimization, and wind-tunnel testing of MFC-actuated compliant morphing wings, and implemented the flight control of micro-air-vehicles (MAVs). LaCroix [20] used MFCs to deform the surface of the forward-swept thin, compliant composite wings of MAVs. Molinari [21] designed a three-dimensional adaptive compliant wing with embedded MFCs and presented aero-structural optimization. These prior works demonstrated that MFCs produce a large actuation effect to deform compliant structures. Compared with conventional hinged, discrete-control surfaces, MFC-actuated morphing wings perform with lower drag and offer more efficient production of control forces and moments.

For future smart structures, multiple work modes, such as bending, twisting, and expansion, will need to be controlled in many circumstances. Concerning some common structures, such as beams or plates, bending deformations are mostly considered due to their larger deformation magnitude and lower inherent frequencies. Twisting deformation and corresponding control issues receive specific attention in many fields, particularly for flexile wing structures. Twist morphing is one of the most popular categories of morphing wings and has resulted in a large number of wind-tunnel and flight tests in aircraft [1]. Some other instances are rotating blades, solar panels, and robot arms, whose behaviors also consist of both bending and twisting modes.

In a piezo-actuation context, in-plane polarized, anisotropic piezocomposite materials are the natural choices for multiple work modes including bending/twisting shape control. Conventional piezoceramics, such as lead zirconium titanate (PZT), are typically capable of large actuation forces; however, they have some limitations, such as small strains and low flexibility [22]. In addition, traditional actuators with through-the-thickness poling possess transverse isotropy in the plane and cannot supply sufficient twisting actuation moment [23]. Piezocomposite materials have emerged as the new class of hybrid materials; they consist of piezoelectric fiber reinforcements embedded in the epoxy matrix and interdigitated electrodes, so they can provide a wide range of effective material properties, good conformability, and strength integrity [24,25]. Piezocomposite materials can utilize the d_{33} piezoelectric effect in the direction of PZT fibers, which is larger than the d_{31} piezoelectric effect of conventional piezoelectric actuators [26]. In particular, piezocomposite actuator patches feature anisotropic actuation effects, which makes it possible to expand their work modes by designing specific PZF fiber orientations [15,23,27–29]. For the MFCs used in this study, three work modes (expansion, bending, and torsion), were realized and, of course, the appropriate actuator types and configurations were chosen for specific modes. Smart Material Corporation [30] provides two types of standard MFC actuator patch that utilize the d_{33} effect: P1 types with 0° fiber orientation and F1 types with 45° fiber orientation. The P1-type actuators are mainly used for the bending control of structures and the F1-type actuators are used for twisting control. In addition to unimorph-configuration actuators, piezocomposite actuators in bimorph configuration are commonly adopted to enhance control authority [21,31–34]. Bilgen [31] designed a lightweight high-voltage electronic circuit for MFC bimorphs and remedied the situations in which the MFCs had an asymmetry range in the positive and negative voltage directions. In the authors' previous studies, piezocomposite materials in "antisymmetric angle-ply bimorph configuration" were presented for the bending/twisting shape control of plate-like flexible wing structures [35,36]. The actuator optimizations (in unimorph or bimorph configuration) [37,38] and structure/actuator-integrated designs [39] for such piezocomposite-actuated structures were also presented. In addition to theoretical investigations, experimental validations are also presented in this paper to demonstrate the feasibility of bimorph MFCs for bending/twisting mode control and real-time control systems.

The primary aim of this paper is to present both theoretical and experimental investigations of the control performance of anisotropic piezocomposite actuators in synthetic (independent and coupled) bending/twisting shape control. In this study, two MFCs of the F1 type in bimorph configuration were used for the shape control of a cantilever aluminum plate. The two identical actuators were orientated at ±45° from the front and backside surfaces, respectively. Thus, in ideal situations, the same potential would induce pure twisting, whereas the opposite potential would cause pure bending. The finite element method was applied to model the system and optimize the actuators' locations. An experimental setup was built for the MFC-actuated flexible plate, whose deformation was measured using two laser displacement sensors. A feedback closed-loop control law was designed to improve the shape control accuracy when it was subjected to uncertainties and nonlinearities. Finally, the multi-mode control scheme was experimentally verified using pure bending, twisting, and coupled bending/twisting shape control.

2. Model Formulation
2.1. MFC-Actuated Plate Structures

The active structure in this study is characterized by means of a cantilever aluminum plate, as shown in Figure 1. Two identical MFC patches of M8557-F1 type, which were produced by Smart Material Corp., Sarasota, FL, US, were symmetrically glued to the front and backside surfaces of the base plate, respectively. This type of F1 MFC actuator features a 45° fiber orientation with respect to its length direction. Practically, due to the opposite surfaces, the actual fiber orientations for the two actuators were −45° and 45° with respect to the global x-axis, respectively. Note that the electrodes of the actuator always remained perpendicular to the fiber orientation. That is to say, the actuators were in the so-called "antisymmetric angle-ply bimorph configuration", which offers several unique advantages for active shape control. First and foremost, independent bending or twisting deformations could be produced using the opposite or the same potential, respectively; the detailed descriptions are given in the subsequent subsections. Compared with unimorph configuration, larger actuation ability and control authority could be produced. Furthermore, the elastic axis of the base plate is unchanged under bimorph configuration, which may be beneficial for flexible wings [35,36]. Table 1 lists the geometric and material properties (theoretical value) of the base plate and MFC actuators.

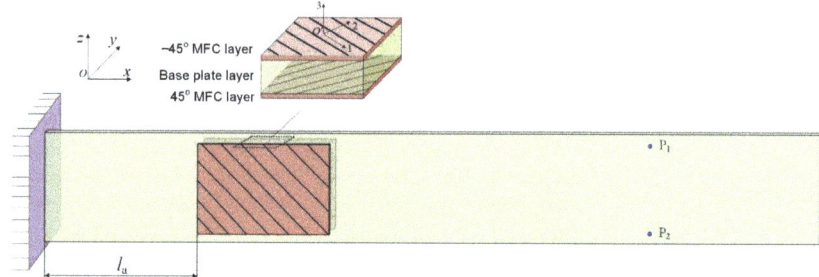

Figure 1. Schematics of the MFC-actuated compliant plate structure.

Table 1. The geometric and material properties (theoretical value) of plant.

Parameters	Values	
	Base Plate	MFC Actuator [30]
Length (mm)	500	85 (active area)
Width (mm)	64.5	57 (active area)
Thickness (mm)	1	0.3
Modulus of elasticity (GPa)	70.3	30.34, 15.86
Poisson's ratio	0.345	0.31, 0.16
Density (kg/m^3)	2700	5400
Actuator location		100 mm from the root
Measurement point locations		105 mm from the tip
Piezoelectric constants (m/V)		$400 \times 10^{-12}, -170 \times 10^{-12}$
Fiber orientations (deg)		$\pm 45°$
Electrode spacing		0.5 mm

Because the laser displacement sensors were used to measure the elastic deformation, two measurement points, which were symmetrical relative to the midline, as shown in the figure, were chosen to represent the bending and twisting deformations. The optimization of the locations of the MFC actuators is described in the following sections.

2.2. Finite Element Model

A mathematical model is commonly needed to predict the behaviors of systems and can also be used for model-based control system design. The finite element method was used in this study to model the piezocomposite-actuated plate. Figure 2 depicts the finite element model. Quadrilateral plate elements were adopted to discretize the structure. The key issue was to model the MFC actuator, which is a hybrid, layered material that consists of a rectangular cross-section, unidirectional piezoceramic fibers, the epoxy matrix, Kapton, and interdigitated electrodes [40]. Due to the complexity of MFCs, a homogenization strategy was adopted; thus, the actuator could be modeled in the form of homogenized orthotropic materials with arbitrary PZT fiber orientations, as well as composite materials [41,42]. Moreover, the local mass and stiffness change induced by the bonded patch was determined through composite laminate theory. Detailed descriptions of the FEM approach have been presented in [38,39]. For the sake of simplicity, the final governing equations are obtained and written as

$$M\ddot{x} + Kx = B_u u \quad (1)$$

where x is the vector of the nodal displacements. The matrices M and K denote the mass matrix and stiffness matrix, which are assembled as $M = M_b + M_p$ and $K = K_b + K_p$, respectively, where subscripts b and p denote the contribution of the base plate layer and piezocomposite layers, respectively. $u = \begin{bmatrix} u_1 & u_1 \end{bmatrix}^T$ is the vector of the applied voltages. B_u is the coefficient matrix, which depends on the actuator locations.

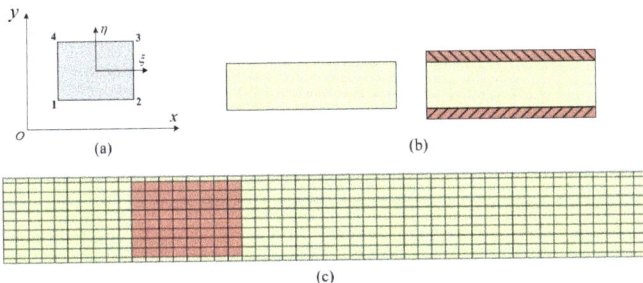

Figure 2. Finite element model of the piezo-actuated plate: (**a**) Quadrilateral plate element; (**b**) cross-sections of the elements; (**c**) meshes of the structure.

The deflections of the two measurement points P_1 and P_2 are given by

$$w_{P_1} = C_{P_1} x \\ w_{P_2} = C_{P_2} x \tag{2}$$

where C_{P_1} and C_{P_2} are the output matrices, which depend on the measurement locations of the laser sensors.

2.3. Actuator Position Optimization

It is an important issue to determine how to efficiently use piezoelectric capabilities to their fullest extent in active shape control. Accordingly, the locations of the actuators must be optimized before performing experiments. Extensive research on the optimal placement of piezoelectric actuators for structural control has also been carried out. Some detailed literature reviews on this topic can be found in [43–45]. The authors have also presented the optimization approach for the anisotropic piezocomposite actuators by considering the PZT fiber orientations as well as the distributed positions [37–39]. However, due to the fixed size and fiber orientation of M8557-F1-type MFC actuators, only the position of the actuators in the length direction can be designed. Thus, the best position can be chosen via traversal simulations, which is not time-consuming. Moreover, the influence of the actuator's position on shape control ability can also be reflected in this way. Therefore, only one design variable is concerned and given as

$$l_a \in [0, L-l] \tag{3}$$

where l_a is the distance from the plate root to the left edge of the active area of the MFC, as shown in Figure 1. Note that, to facilitate the theoretical simulation, only the active area of MFC is considered. The values L and l denote the length of the base plate and actuator, respectively.

Because both bending and twisting are concerned, two corresponding criteria are used to evaluate the shape control capabilities and given by

$$J_1 = |w_{P_1}|, \; u = [500, -500]^T \tag{4}$$

$$J_2 = |\alpha|, \; u = [500, 500]^T \tag{5}$$

$$\alpha = \arcsin \frac{w_{P_1} - w_{P_2}}{|P_1 P_2|} \approx \frac{w_{P_1} - w_{P_2}}{|P_1 P_2|} \tag{6}$$

where $|P_1 P_2|$ denotes the distance between two measurement points.

The variations of J_1 and J_2 with l_a are shown in Figure 3a, respectively. It was found that the bending deformation decreased monotonically with the length position of the actuator, i.e., the best position was the root area, as depicted in Figure 3(b.1). However, it was preferable to place the actuator in the area of 0.1 m~0.35 m so as to obtain enhanced twisting control capability, while the largest twisting deformation occurred in $l_a = 0.22$ m, as depicted in Figure 3(b.2). Hence, the area of the shaded part in Figure 3a is a kind of Pareto optimal area, in which any point could be viewed as an acceptable location. Therefore, after considering the trade-off in control authority between the bending and torsional modes, we chose a final position of $l_a = 0.1$ m. The above discussions explain why we placed the MFC actuators in this position.

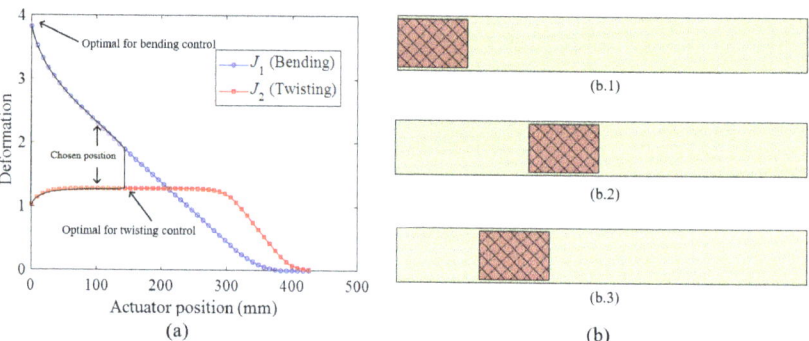

Figure 3. Variations in the bending and twisting deformations with the position of the actuator: (**a**) The variations in the bending and twisting deformations with actuator position; (**b.1**) the best position for bending control; (**b.2**) the best position for twisting control; (**b.3**) the chosen position for multi-mode shape control.

2.4. Theoretical Bending/Twisting Shape Control

As previously mentioned, the primary aim of this study was to achieve synthetic (independent and coupled) bending and twisting shape control. To theoretically demonstrate the bending/twisting control principle, the deformation configuration of the plate under the actuation of a single MFC is presented in Figure 4. The results demonstrate that both bending and twisting deformations were produced under the actuation of a single F1-type MFC. For the MFC1 that was bonded on the front side of the plate, the deflection amplitude of the lower measurement point was larger than the upper measurement $w_{P_1, \text{MFC1}}$. By contrast, the deformation trend produced by the MFC2 alone was opposite in not only the direction but also in the relations between w_{P_1} and w_{P_2}, i.e.,

$$\begin{vmatrix} w_{P_1,\text{MFC1}} \end{vmatrix} < \begin{vmatrix} w_{P_2,\text{MFC1}} \end{vmatrix} \\ \begin{vmatrix} w_{P_1,\text{MFC2}} \end{vmatrix} > \begin{vmatrix} w_{P_2,\text{MFC2}} \end{vmatrix} \tag{7}$$

In ideal situations, the following relation exists:

$$\begin{vmatrix} w_{P_1,\text{MFC1}} \end{vmatrix} = \begin{vmatrix} w_{P_2,\text{MFC2}} \end{vmatrix} \\ \begin{vmatrix} w_{P_1,\text{MFC2}} \end{vmatrix} = \begin{vmatrix} w_{P_2,\text{MFC1}} \end{vmatrix} \tag{8}$$

Consequently, employing this bimorph configuration, the two actuators could be polarized in the same direction for twisting deformation and polarized in opposite directions for bending deformation. The voltage input applied for the front and backside actuators are defined as u_1 and u_2, respectively. By applying the same voltages (i.e., $u_1 = u_2$), pure moments of torque are generated, while the bending moments are canceled out. On the other hand, pure bending deformation can be achieved by using the opposite voltages ($u_1 = -u_2$). Of course, a combination of bending and twisting deformations can be performed by designing the two voltages. In a practical context, for any two voltages, the decomposition is given as

$$u_1 = u_s + u_o \\ u_2 = u_s - u_o \tag{9}$$

where $u_s = \frac{u_1+u_2}{2}$, $u_o = \frac{u_1-u_2}{2}$ are the voltage components for twisting and bending deformation, respectively. The subscripts "s" and "o" correspond to the same and the opposite components. Figure 5 visualizes the voltage distribution for the two MFC actuators and Figure 6 shows the bending and twisting deformation obtained from the simulation by applying different voltages.

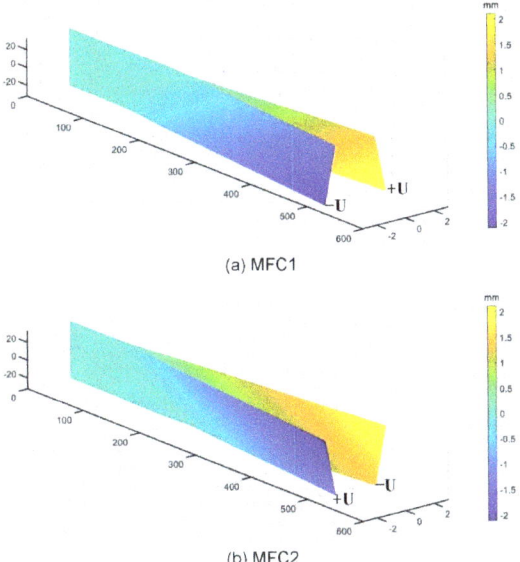

Figure 4. Deformation configuration of the substrate plate actuated by the single MFC. (**a**) Deformation configuration of the substrate plate actuated by the MFC1; (**b**) deformation configuration of the substrate plate actuated by the MFC2.

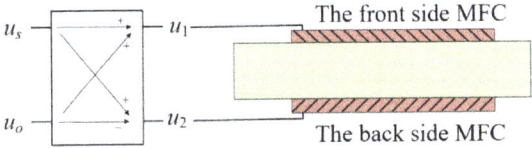

Figure 5. Voltage distributions for the two MFC actuators.

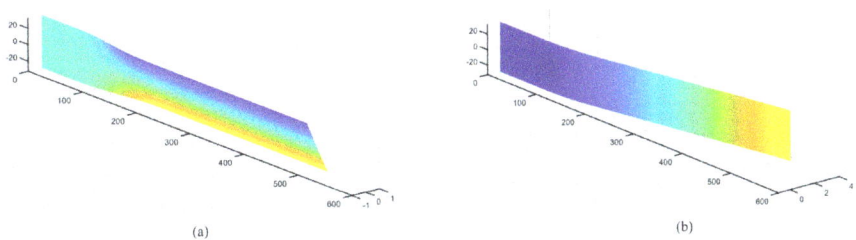

Figure 6. Bending (**a**) and twisting (**b**) shape control effect using different voltages for the actuators.

3. Experiment Implementation

3.1. Setup

An experimental setup was designed for validation according to the previous theoretical investigation, as shown in Figure 7. The base structure was a cantilever aluminum plate, whose dimensional and material properties are listed in Table 1. Two identical MFC patches of type M8557-F1, which were produced by Smart Material Corp., Sarasota, FL, US [30], were glued to the front and backside surfaces of the base plate, respectively. The actuators were located 0.1 m from the plate root, as determined previously. A PCI-1721 DAQ card, which was produced by Advantech Co., Ltd., Kunshan, China, was used to convert the digital signals from the computer to the analog signals. Since the operational voltage of the MFCs ranged from −500 to 1500 V, a HVA 1500-2 high-voltage amplifier, which was produced by Physical Instruments Corp., Berlin, Germany, was used to supply the high voltage for the MFC. The elastic deformation of the plate was measured by two OPTEX-CDX-30A-type laser displacement sensors, which was produced by Guangzhou Optex Industrial Automation Control Equipment Co., Ltd. Guangzhou, China; thus, the twisting angle could be computed using Equation (6). The measured data obtained from the laser sensors were conducted using an ADAM-USB card, which was produced by Advantech Co., Ltd., Kunshan, China; thus the signals could be received by the computer. A computer integrated with MATLAB (which was produced by MathWorks Corp, Natick, MA, USA) and LabVIEW (which was produced by NI Corp, Austin, TX, USA) code was used to generate the voltage signals, receive the sensor signals, and implement the feedback closed-loop control laws.

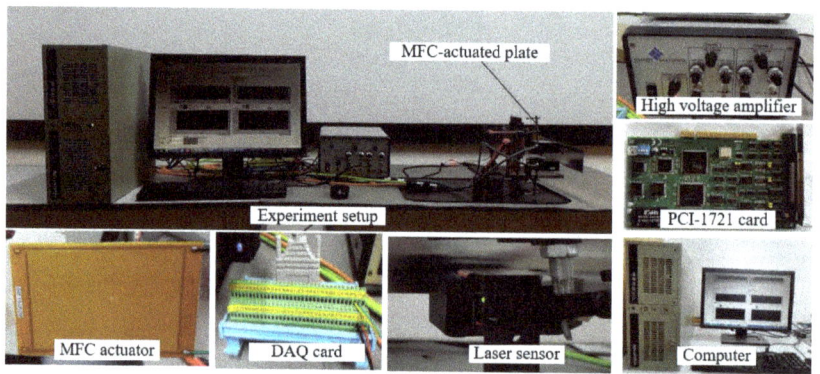

Figure 7. Experiment setup.

3.2. Modal Analysis

Before the active shape control tests, modal analysis was implemented first, for two purposes: (1) To investigate the dynamic characteristics of the compliant structure; and (2) to evaluate the influences of the bonded MFC actuators on the system. The modal frequencies and shapes can be determined by solving an eigenvalue problem of the FE model as follows:

$$\left[\mathbf{K} - \omega^2 \mathbf{M}\right] \mathbf{\Phi} = 0 \qquad (10)$$

where ω is the natural frequencies and $\mathbf{\Phi}$ denotes the corresponding modal shapes. Figure 8 shows the structural modal shapes of the first six modes, including four bending modes and two torsional modes (third and sixth modes). It was observed that the bonded MFCs had relatively little influence on the modal shapes of the structure.

The frequencies were recognized in terms of the free vibration data of the plate using fast Fourier transform. Table 2 lists the natural frequencies for the first three bending modes and the first torsional mode with and without the MFC actuators. In general, the theoretical values were in good agreement with the experimental results. The results showed that, after adding MFC actuators, the frequencies of the first bending and first torsional modes increased, while the frequencies of the second and third bending modes decreased. Note that the influences of piezoelectric actuators depend on many issues, including the size, position, and fiber orientation of anisotropic piezocomposite materials.

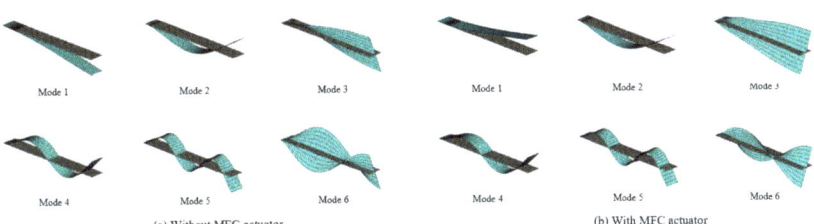

Figure 8. Structural modal shapes. (**a**) Structural modal shapes without MFC actuators; (**b**) structural modal shapes with MFC actuators.

Table 2. Natural frequencies of the structure with and without MFCs (Hz).

Mode	Without MFC			With MFCs		
	FEM	Experiment	Error	FEM	Experiment	Error
1st (bending)	3.34	3.27	2.14%	3.50	3.54	−1.13%
2nd (bending)	20.89	20.53	1.75%	19.57	19.51	0.30%
3rd (torsional)	49.90	46.63	5.80%	53.30	52.89	7.80%
4th (bending)	58.59	58.17	0.72%	54.40	57.45	−5.30%

3.3. Control Ability

In this study, the control ability of the MFCs in unimorph and bimorph was tested experimentally through static deformation analysis. To this end, the experimental steady-state deflection was obtained by directly applying a certain constant voltage for the MFC and recorded after a relative long time.

(a) Using single MFC

The elastic deflections of the two measurement points (i.e., w_{P_1}, w_{P_2}) under the single MFC are given in Figure 9. It can be observed that using a single MFC, both bending and twisting deformation were produced due to the off-line PZT fiber orientation. Moreover, as previously noted, the deflection amplitude of the lower-point P_2 produced by the MFC1 actuator was larger than the upper point P_1, i.e., $|w_{P_1,MFC1}| < |w_{P_2,MFC1}|$ for both positive and negative voltages. Conversely, the deflections produced by the MFC2 showed the opposite trend, i.e., $|w_{P_1,MFC2}| > |w_{P_2,MFC2}|$. That is to say, the remaker given by Equation (7) and Figure 4 was verified by experiments. However, it can be observed that the actuation abilities of the two actuators were not identical, since the deflection amplitude generated by the MFC1 is larger than the MFC2. This deviation was induced by a variety of uncertainties and errors that are discussed in Section 3.4. That is to say, the theoretical remaker given by Equation (8) was not strictly met in the experiments.

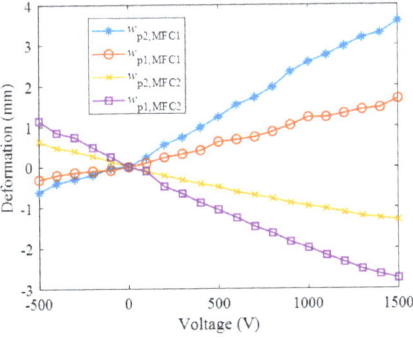

Figure 9. Deflection of the measurement points using a single MFC actuator.

(b) Using bimorph MFCs

The deflections of the plate under the actuation of the bimorph MFCs using the same and opposite voltages are shown in Figure 10, respectively. The x-axis in Figure 10b is labeled by the voltage of the MFC1, i.e., u_1, so the corresponding voltage of the MFC2 is $u_2 = -u_1$. Because of the asymmetrical operation voltages (i.e., $-500\sim1500$ V) for the MFCs, the operation voltage range is $-500\sim1500$ V for the same sign; however, it is $-500\sim500$ V for the opposite sign. It can be observed that the deflection amplitudes of the two points were generally equal; the deflection directions are opposite in Figure 10a (i.e., twisting deformation) and the same in Figure 10b (i.e., bending deformation), respectively. Moreover, the deflection amplitudes $|w_{P_1}|$ and $|w_{P_2}|$, which should be identical in terms of FE analysis, were still not the same in the two experimental cases. The above experiments imply that bending and twisting shape control can be qualitatively realized as theoretically predicted; however, they cannot be implemented with high control accuracy through an open-loop control approach alone.

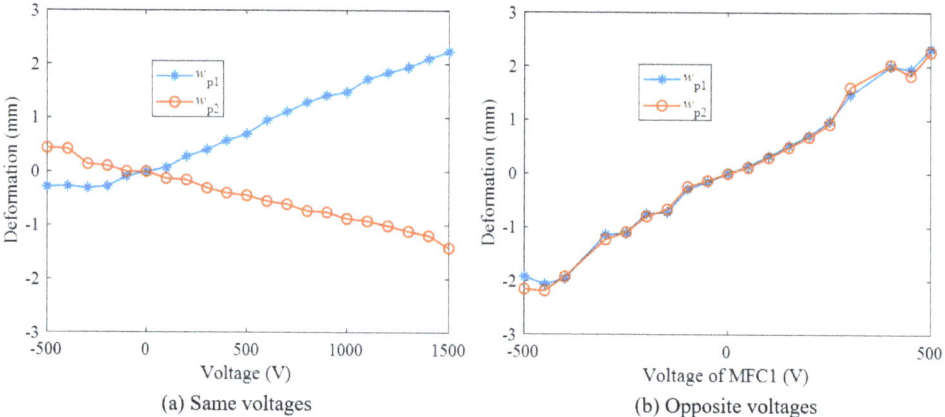

Figure 10. Deflection of the measurement points using bimorph MFC actuators. (**a**) Deflection of the measurement points using bimorph MFC actuators under the same voltages; (**b**) deflection of the measurement points using bimorph MFC actuators under the opposite voltages.

3.4. Uncertainty Analysis

The above experimental results demonstrate the independent bending/twisting shape control resulting from the use of MFCs in an antisymmetric angle-ply bimorph configuration. However, the open shape control accuracy is generally low due to a variety of uncertainties and nonlinearities.

Firstly, the shape control accuracy is also influenced by the non-uniform distribution of the substrate material of the plate, which is assumed as an isotropic aluminum plate in FE modeling. This non-uniform distribution induces changes in both the bending and torsional stiffness properties. Secondly, the MFC layer and the substrate plate are glued by using epoxy. Thus, the thickness and distribution of the epoxy layer also affect the actuation effect of the MFCs, especially the synchronization of the two actuators. Furthermore, the actuation performance is heavily affected by the nonlinearities of the MFCs, including hysteresis, creep, and varied piezoelectric coefficients [30]. Hysteresis and creep effects are the intrinsic nonlinear characteristics of piezoelectric actuators and can significantly affect the control performance [46], as shown in Figure 11. The hysteresis exhibited at a given time depends on the present input and the operational history of the system. Creep is related to the drift effect of output displacement with a constant applied voltage over extended periods. Furthermore, the piezoelectric coefficients of MFCs are also varied, depending on the situation; an example of this is the d_{33} and d_{31} piezoelectric constants in high electric field intensity compared with low field intensity [30]. Furthermore, the shape control accuracy is also affected by the placement of wires, external disturbance, and so on.

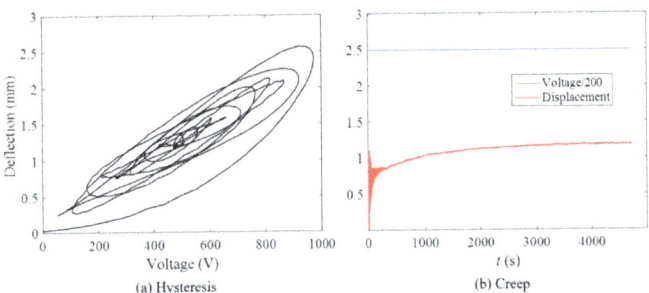

Figure 11. Hysteresis and creep effects of the MFC actuator. (**a**) Hysteresis effect of the MFC actuator; (**b**) creep effect of the MFC actuator.

Because of the above uncertainties and nonlinearities, accurate bending/twisting shape control cannot be ensured in experiments by using the voltage values estimated by the linear FE simulation. Some feedforward control schemes can be implemented to reduce or cancel out unwanted issues. For example, a feedforward inverse compensation control can be designed to cancel out the hysteresis and creep nonlinearities of MFCs in terms of a phenomenon-based model and experimental data [32]. However, it is still a difficult task to perform open-loop control given all the undesired effects. Therefore, a feedback control scheme is necessary to realize accurate bending/twisting shape control. It is easy to design a shape control law for a single-input–single-output (SISO) system, such as the bending shape control of a plate using M8528-P1-type MFCs in our previous study [47]. However, the plant in this study constitutes a two-input–two-output system, and the key point is how to design the control law to achieve both independent and coupled bending/twisting shape control.

4. Closed-Loop Multi-Mode Shape Control System

4.1. Feedback Control Law

To achieve the synthetic control of bending and twisting shape, the structural deformations are organized as follows:

$$B = \frac{w_{P_1} + w_{P_2}}{2} \\ T = \frac{w_{P_1} - w_{P_2}}{2} \quad (11)$$

where B and T denote the bending and twisting components in the deformation, respectively. On the other hand, the deflections of the two points can also be represented by the components, i.e., $w_{P_1} = B + T$, $w_{P_2} = B - T$.

Hence, the shape control error can be given as

$$\Delta B = B - B_d = \frac{\Delta w_{P_1} + \Delta w_{P_2}}{2} \\ \Delta T = T - T_d = \frac{\Delta w_{P_1} - \Delta w_{P_2}}{2} \quad (12)$$

where B_d and T_d denote the command bending and twisting requirements, respectively.

To adjust the shape control error, a feedback control law is designed as

$$\Delta u_s = K_s \Delta T + K_{sB} \Delta B \\ \Delta u_p = K_p \Delta B + K_{pT} \Delta T \quad (13)$$

where Δu_s and Δu_p are the incremental voltage values for the same and the opposite components of the actuators in each time step, respectively. The values K_s and K_p denote the primary control gains, which are designed according to ideal situations. The values K_{sB} and K_{pT} are used to compensate for the shape error associated with the uncertainties. Subsequently, the voltages for the two MFC actuators can be easily determined according to Equation (9).

4.2. Multi-Mode Shape Control Results

The multi-mode shape control was directly implemented based on the experiment by using B_c and T_c as the control requirements. According to Equation (11), the arbitrary deformation of the plate can be represented by a combination of bending and twisting components.

The pure twisting shape control results obtained by using $T_c = 0.8$ and $B_c = 0$ as the objectives are shown in Figure 12. Figure 12a gives the deflection histories of the two measurement points, i.e., w_{P_1} and w_{P_2}, which demonstrate that the same deformation amplitudes with opposite directions were achieved. The time histories of B and T shown in Figure 12b demonstrate that the pure twisting deformation of the plate was achieved, while the bending deformation was maintained at zero. It can be observed that the shape control accuracy was greatly improved, especially compared with the previous open-loop results. Figure 12c shows the time histories of the voltages for the two MFC actuators. It can be observed that the two voltages did not converge to the same value in terms of the ideal situation. Due to the influences of the creep effect, the voltages slowly varied with time. Furthermore, the deformations converged to the desired values. Such voltage profiles are difficult to determine by feedforward control due to complex uncertainties. The results again prove the necessity of closed-loop control in high-precision shape control.

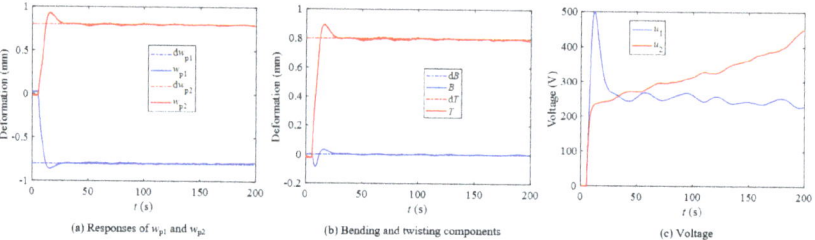

Figure 12. Pure twisting shape control performance. (**a**) The deflection histories of the two measurement points; (**b**) the time histories of B and T; (**c**) the time histories of the voltages for the two MFC actuators.

Similarly, Figure 13 shows the pure bending shape control results obtained by using $T_c = 0$ and $B_c = 0.8$ as the objectives. The displacement of the two measuring points was consistent, without producing twisting deformation. The voltage profiles of the MFC1 and MFC2 were generally opposite to each other, as theoretically predicted; however, they still varied with time to resist the uncertainties and nonlinearities.

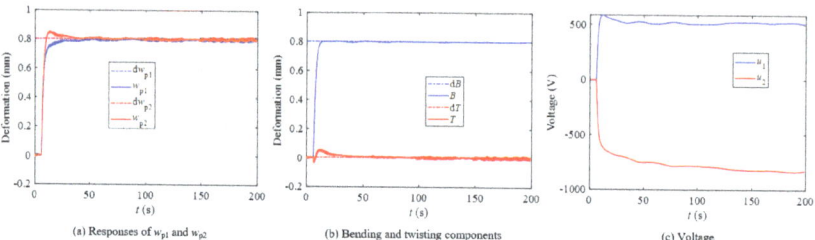

Figure 13. Pure bending shape control performance. (**a**) The deflection histories of the two measurement points; (**b**) the time histories of B and T; (**c**) the time histories of the voltages for the two MFC actuators.

Finally, the coupled bending/twisting shape control (or arbitrary shape control) results are presented by using $T_c = 0.3$ and $B_c = 0.8$ as the objectives, as shown in Figure 14. It can be observed that both the bending and twisting deformations reached the command values with high shape accuracy. The two voltage values were adjusted in time according to the feedback control law to achieve arbitrary deformation. The trend in the voltage profiles can also be explained by Figures 12 and 13. The above results imply that independent bending and twisting shape control with improved shape accuracy was achieved by employing the multi-mode feedback control approach.

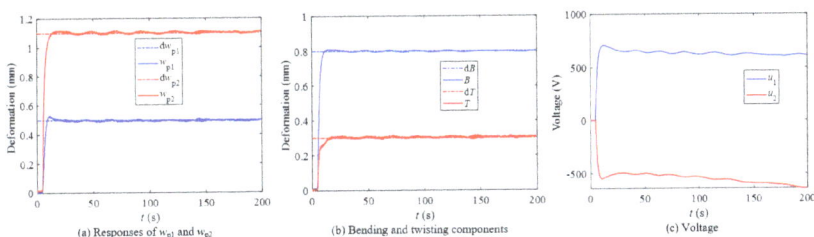

Figure 14. Arbitrary shape control performance. (**a**) The deflection histories of the two measurement points; (**b**) the time histories of B and T; (**c**) the time histories of the voltages for the two MFC actuators.

5. Conclusions

In this paper, the multi-mode shape control of piezo-actuated compliant morphing structures was achieved in both theoretical and experimental ways. Independent and coupled bending/twisting shape control of a plate structure was achieved by using F1-type MFCs in an antisymmetric angle-ply bimorph configuration. The optimal locations of the MFC actuators were determined by comprehensively considering the control of the bending and twisting deformations. The experimental tests implied that the shape control accuracy was heavily reduced due to various uncertainties and nonlinearities, including hysteresis and the creep effect of the actuators, model errors, and external disturbances. A multi-mode feedback control law was designed to cancel out the shape error. The experimental results implied that synthetic (independent and coupled) bending and twisting shape control with improved shape accuracy was achieved by employing the multi-mode feedback control approach.

Author Contributions: Conceptualization, X.W.; Data curation, X.H.; Formal analysis, C.H.; Funding acquisition, X.W.; Investigation, X.H.; Methodology, X.W. and W.Z.; Software, X.H. and C.H.; Supervision, W.Z.; Writing—original draft, X.W.; Writing—review & editing, W.Z. All authors have read and agreed to the published version of the manuscript.

Funding: This research was funded by the National Natural Science Foundation of China (11872381, 12102096) and the Guangdong Basic and Applied Basic Research Foundation (2019A1515010859).

Institutional Review Board Statement: Not applicable.

Informed Consent Statement: Not applicable.

Data Availability Statement: Not applicable.

Acknowledgments: The author would like to thank the reviewers and the editors for their valuable comments and constructive suggestions that helped to improve the paper significantly.

Conflicts of Interest: The authors declare no conflict of interest.

References

1. Barbarino, S.; Bilgen, O.; Ajaj, R.M.; Friswell, M.I.; Inman, D.J. A review of morphing aircraft. *J. Intell. Mater. Syst. Struct.* **2011**, *22*, 823–877. [CrossRef]
2. Sun, J.; Guan, Q.; Liu, Y.; Leng, J. Morphing aircraft based on smart materials and structures: A state-of-the-art review. *J. Intell. Mater. Syst. Struct.* **2016**, *27*, 2289–2312. [CrossRef]

3. Baier, H.; Datashvili, L. Active and morphing aerospace structures—A synthesis between advanced materials, structures and mechanisms. *Int. J. Aeronaut. Space Sci.* **2011**, *12*, 225–240. [CrossRef]
4. Giurgiutiu, V. Review of smart-materials actuation solutions for aeroelastic and vibration control. *J. Intell. Mater. Syst. Struct.* **2000**, *11*, 525–544. [CrossRef]
5. Ferreira, A.D.B.L.; Nóvoa, P.R.O.; Marques, A.T. Multifunctional material systems: A state-of-the-art review. *Compos. Struct.* **2016**, *151*, 3–35. [CrossRef]
6. Bradford, S.C.; Agnes, G.S.; Ohara, C.M.; Green, J.J.; Shi, F.; Zhou, H. Controlling wavefront in lightweight active reflector systems using piezocomposite actuator arrays. In Proceedings of the 54th AIAA/ASME/ASCE/AHS/ASC Structures, Structural Dynamics and Materials Conference, Boston, MA, USA, 8–11 April 2013; AIAA: Reston, VR, USA, 2013.
7. Shao, S.; Song, S.; Xu, M.; Jiang, W. Mechanically reconfigurable reflector for future smart space antenna application. *Smart Mater. Struct.* **2018**, *27*, 095014. [CrossRef]
8. Bilgen, O.; Kochersberger, K.B.; Inman, D.J.; Ohanian, O.J. Novel, Bidirectional, Variable-Camber Airfoil via Macro-Fiber Composite Actuators. *J. Aircr.* **2010**, *47*, 303–314. [CrossRef]
9. Echter, M.A.; Silver, M.J.; D'Elia, E.; Peterson, M.E. Recent Developments in Precision High Strain Composite Hinges for Deployable Space Telescopes. In Proceedings of the 2018 AIAA Spacecraft Structures Conference, Kissimmee, FL, USA, 8–12 January 2018; p. 14.
10. Hill, J.; Wang, K.W.; Fang, H. Advances of Surface Control Methodologies for Flexible Space Reflectors. *J. Spacecr. Rocket.* **2013**, *50*, 816–828. [CrossRef]
11. Wang, Z.; Li, T.; Cao, Y. Active shape adjustment of cable net structures with PZT actuators. *Aerosp. Sci. Technol.* **2013**, *26*, 160–168. [CrossRef]
12. Song, X.; Tan, S.; Wang, E.; Wu, S.; Wu, Z. Active shape control of an antenna reflector using piezoelectric actuators. *J. Intell. Mater. Syst. Struct.* **2019**, *30*, 2733–2747. [CrossRef]
13. Hiroaki, T.; Sakamoto, H.; Inagaki, A.; Ishimura, K.; Doi, A.; Kono, Y.; Kuratomi, T. Development of a smart reconfigurable reflector prototype for an extremely high-frequency antenna. *J. Intell. Mater. Syst. Struct.* **2016**, *27*, 764–773. [CrossRef]
14. Sakamoto, H.; Tanaka, H.; Ishimura, K.; Doi, A.; Kono, Y.; Matsumoto, N. Shape-control experiment of space reconfigurable reflector using antenna reception power. In Proceedings of the 3rd AIAA Spacecraft Structures Conference, San Diego, CA, USA, 4–8 January 2016; p. 0703.
15. Monner, H.P.; Riemenschneider, J.; Opitz, S.; Schulz, M. Development of active twist rotors at the German Aerospace Center (DLR). In Proceedings of the 52nd AIAA/ASME/ASCE/AHS/ASC Structures, Structural Dynamics and Materials Conference, Denver, CO, USA, 4–7 April 2011; AIAA: Reston, VR, USA, 2011.
16. Riemenschneider, J.; Keye, S.; Wierach, P.; Mercier Des Rochettes, H. Overview of the common DLR/ONERA project "Active Twist Blade" (ATB). In Proceedings of the 30th European Rotorcraft Forum, Marseilles, France, Confederation of European Aerospace Societies, Brussels, Belgium, 14–16 September 2005; pp. 273–281.
17. Li, M.; Chen, W.; Guan, D.; Li, W. Experimental validation of improving aircraft rolling power using piezoelectric actuators. *Chin. J. Aeronaut.* **2005**, *18*, 108–115. [CrossRef]
18. Bilgen, O.; Kochersberger, K.; Diggs, E.C.; Kurdila, A.J.; Inman, D.J. Morphing wing micro-air-vehicles via macro-fiber-composite actuators. In Proceedings of the 48th AIAA/ASME/ASCE/AHS/ASC Structures, Structural Dynamics, and Materials Conference, Honolulu, HI, USA, 23–26 April 2007; AIAA: Reston, VR, USA, 2007; pp. 1005–1020.
19. Bilgen, O.; Friswell, M.I. Piezoceramic composite actuators for a solid-state variable-camber wing. *J. Intell. Mater. Syst. Struct.* **2014**, *25*, 806–817. [CrossRef]
20. LaCroix, B.W.; Ifju, P.G. Aeroelastic model for macrofiber composite actuators on micro air vehicles. *J. Aircr.* **2016**, *54*, 199–208. [CrossRef]
21. Molinari, G.; Arrieta, A.F.; Ermanni, P. Aero-Structural Optimization of Three-Dimensional Adaptive Wings with Embedded Smart Actuators. *AIAA J.* **2014**, *52*, 1940–1951. [CrossRef]
22. Usher, T.D.; Ulibarri, K.R., Jr.; Camargo, G.S. Piezoelectric microfiber composite actuators for morphing wings. *ISRN Mater. Sci.* **2013**, *2013*, 1–9. [CrossRef]
23. Wetherhold, R.C.; Aldraihem, O.J. Bending and twisting vibration control of flexible structures using piezoelectric materials. *Shock. Vib. Dig.* **2001**, *33*, 187–197. [CrossRef]
24. Ray, M.C.; Reddy, J.N. Active damping of laminated cylindrical shells conveying fluid using 1–3 piezoelectric composites. *Compos. Struct.* **2013**, *98*, 261–271. [CrossRef]
25. Smith, W.A.; Auld, B.A. Modeling 1–3 composite piezoelectrics: Thickness-mode oscillations. *IEEE Trans. Ultrason. Ferroelectr. Freq. Control* **1991**, *38*, 40–47. [CrossRef]
26. Choi, S.C.; Park, J.S.; Kim, J.H. Vibration control of pre-twisted rotating composite thin-walled beams with piezoelectric fiber composites. *J. Sound Vib.* **2007**, *300*, 176–196. [CrossRef]
27. Bent, A.A.; Hagood, N.W.; Rodgers, J.P. Anisotropic actuation with piezoelectric fiber composites. *J. Intell. Mater. Syst. Struct.* **1995**, *6*, 338–349. [CrossRef]
28. Kwak, S.K.; Yedavalli, R.K. New modeling and control design techniques for smart deformable aircraft structures. *J. Guid. Control Dyn.* **2001**, *24*, 805–815. [CrossRef]

29. Wang, X.; Zhou, W.; Wu, Z.; Xing, J. Tracking control system design for roll maneuver via active wings using macro fiber composites. In Proceedings of the AIAA Modeling and Simulation Technologies Conference, Washington, DC, USA, 9–13 January 2017; AIAA: Reston, VR, USA, 2016.
30. Smart-Material-Corporation. 2022. Available online: https://www.smart-material.com/MFC-product-mainV2.html (accessed on 30 March 2022).
31. Bilgen, O.; Kochersberger, K.B.; Inman, D.J.; Ohanian, I.O.J. Lightweight high voltage electronic circuits for piezoelectric composite actuators. *J. Intell. Mater. Syst. Struct.* **2010**, *21*, 1417–1426. [CrossRef]
32. Schröck, J.; Meurer, T.; Kugi, A. Control of a flexible beam actuated by macro-fiber composite patches–Part II: Hysteresis and creep compensation, experimental results. *Smart Mater. Struct.* **2011**, *20*, 015016. [CrossRef]
33. Wickramasinghe, V.; Chen, Y.; Martinez, M.; Wong, F.; Kernaghan, R. Design and verification of a smart wing for an extreme-agility micro-air-vehicle. *Smart Mater. Struct.* **2011**, *20*, 125007. [CrossRef]
34. Ohanian, O.J.; David, B.M.; Taylor, S.L.; Kochersberger, K.B.; Probst, T.; Gelhausen, P.A. Piezoelectric morphing versus servo-actuated MAV control surfaces, part II: Flight testing. In Proceedings of the 51st AIAA Aerospace Sciences Meeting Including the New Horizons Forum and Aerospace Exposition, Grapevine, TX, USA, 7–10 January 2013; AIAA: Reston, VR, USA, 2013.
35. Wang, X.; Zhou, W.; Wu, Z. Feedback tracking control for dynamic morphing of piezocomposite actuated flexible wings. *J. Sound Vib.* **2018**, *416*, 17–28. [CrossRef]
36. Wang, X.; Zhou, W.; Xun, G.; Wu, Z. Dynamic shape control of piezocomposite-actuated morphing wings with vibration suppression. *J. Intell. Mater. Syst. Struct.* **2018**, *29*, 358–370. [CrossRef]
37. Zhou, W.; Wang, X.; Qian, W.; Wu, W. Optimization of Locations and Fiber Orientations of Piezocomposite Actuators on Flexible Wings for Aeroelastic Control. *J. Aerosp. Eng.* **2019**, *32*, 04019056. [CrossRef]
38. Wang, X.; Zhou, W.; Wu, Z.; Wu, W. Optimal unimorph and bimorph configurations of piezocomposite actuators for bending and twisting vibration control of plate structures. *J. Intell. Mater. Syst. Struct.* **2018**, *29*, 1685–1696. [CrossRef]
39. Wang, X.; Zhou, W.; Wu, Z.; Zhang, X. Integrated design of laminated composite structures with piezocomposite actuators for active shape control. *Compos. Struct.* **2019**, *215*, 166–177. [CrossRef]
40. Williams, R.B.; Grimsley, B.W.; Inman, D.J.; Wilkie, W.K. Manufacturing and mechanics-based characterization of macro fiber composite actuators. In Proceedings of the ASME 2002 International Mechanical Engineering Congress and Exposition, New Orleans, LA, USA, 17–22 November 2002; ASME: New York, NY, USA, 2002; pp. 79–89.
41. Zhang, S.-Q.; Li, Y.-X.; Schmidt, R. Modeling and simulation of macro-fiber composite layered smart structures. *Compos. Struct.* **2015**, *126*, 89–100. [CrossRef]
42. Zhang, S.-Q.; Zhao, G.-Z.; Rao, M.N.; Schmidt, R.; Yu, Y.-J. A review on modeling techniques of piezoelectric integrated plates and shells. *J. Intell. Mater. Syst. Struct.* **2019**, *30*, 1133–1147. [CrossRef]
43. Chee, C.Y.K.; Tong, L.Y.; Steven, G.P. A review on the modelling of piezoelectric sensors and actuators incorporated in intelligent structures. *J. Intell. Mater. Syst. Struct.* **1998**, *9*, 3–19. [CrossRef]
44. Gupta, V.; Sharma, M.; Thakur, N. Optimization criteria for optimal placement of piezoelectric sensors and actuators on a smart structure: A technical review. *J. Intell. Mater. Syst. Struct.* **2010**, *21*, 1227–1243. [CrossRef]
45. Frecker, M.I. Recent advances in optimization of smart structures and actuators. *J. Intell. Mater. Syst. Struct.* **2003**, *14*, 207–216. [CrossRef]
46. Cao, Y.; Chen, X.B. A Survey of Modeling and Control Issues for Piezo-Electric Actuators. *J. Dyn. Syst. Meas. Control Trans. ASME* **2015**, *137*, 010101. [CrossRef]
47. Wang, X.; Zhou, W.; Zhang, Z.; Jiang, J.; Wu, Z. Theoretical and experimental investigations on modified LQ terminal control scheme of piezo-actuated compliant structures in finite time. *J. Sound Vib.* **2021**, *491*, 115762. [CrossRef]

Article

Finite Element Method-Based Optimisation of Magnetic Coupler Design for Safe Operation of Hybrid UAVs

Sami Arslan [1,*], Ires Iskender [2] and Tuğba Selcen Navruz [3]

1 Department of Electrical and Electronics Engineering, Gazi University, Graduate School of Natural and Applied Sciences, 06500 Ankara, Turkey
2 Department of Electrical Electronics Engineering, Çankaya University, 06790 Ankara, Turkey
3 Department of Electrical and Electronics Engineering, Gazi University, 06560 Ankara, Turkey
* Correspondence: sami.arslan1@gazi.edu.tr; Tel.: +90-5300407511

Abstract: The integration of compact concepts and advances in permanent-magnet technology improve the safety, usability, endurance, and simplicity of unmanned aerial vehicles (UAVs) while also providing long-term operation without maintenance and larger air gap use. These developments have revealed the demand for the use of magnetic couplers to magnetically isolate aircraft engines and starter-generator shafts, allowing contactless torque transmission. This paper explores the design aspects of an active cylindrical-type magnetic coupler based on finite element analyses to achieve an optimum model for hybrid UAVs using a piston engine. The novel model is parameterised in Ansys Maxwell for optimetric solutions, including magnetostatics and transients. The criteria of material selection, coupler types, and topologies are discussed. The Torque-Speed bench is set up for dynamic and static tests. The highest torque density is obtained in the 10-pole configuration with an embrace of 0.98. In addition, the loss of synchronisation caused by the piston engine shaft locking and misalignment in the case of bearing problems is also examined. The magnetic coupler efficiency is above 94% at the maximum speed. The error margin of the numerical simulations is 8% for the Maxwell 2D and 4.5% for 3D. Correction coefficients of 1.2 for the Maxwell 2D and 1.1 for 3D are proposed.

Keywords: active cylindrical coupler; correction coefficient; finite element method; hybrid UAV; magnetic coupler; magnetic coupling; noncontact torque transmission

1. Introduction

Newly increased environmental apprehension, consciousness, and continuous development to improve the safety and reliability of all aircraft are some of the biggest challenges to be addressed in aviation. Such impressive and challenging issues require the development of more efficient and innovative hybrid systems, as shown in Figure 1.

In conventional systems, the propulsion system of small UAVs is provided only by fuel engines, usually piston engines (PEs). In hybrid systems, there is a high-speed, direct-drive, and highly efficient electric machine called starter/generator (S/G) [1] that provides the initial starting mechanism of the PE and charges the system battery group in generator mode while cruising or contributes to the propulsion in motor mode during climbing.

The modernisation of unmanned aerial vehicles (UAVs) under the concept of More Electric Aircraft (MEA) has been on the agenda. However, the challenge of isolating the shafts of the aircraft engine and the electrically driven system, typically the S/G, for more functional and stealthy operations [2] imposes a critical function on magnetic couplers (MCs). MCs provide both contactless torque transmission and hermetic separation using static seals or containment shrouds, which are essential for hybrid UAVs.

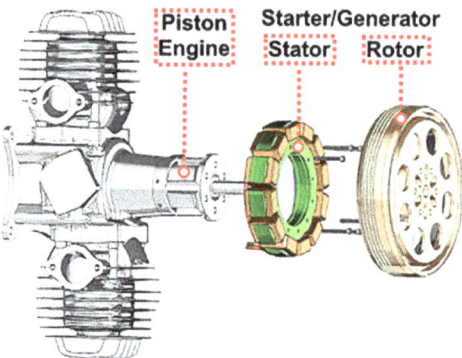

Figure 1. Simple hybrid UAV illustration with the Sullivan S676 Starter/Generator.

MCs have significant advantages such as overload protection, reduced maintenance, simple design, and highly tolerant shaft misalignment, vibration, and noise absorption. An MC consists of permanent magnets (PMs), rotor yokes, a protective cover to protect PMs from high speeds, a containment shroud for sealing, and shafts. Typically, it consists of two rotating parts, an inner and outer rotor, and is grouped as shown in Figure 2 [3].

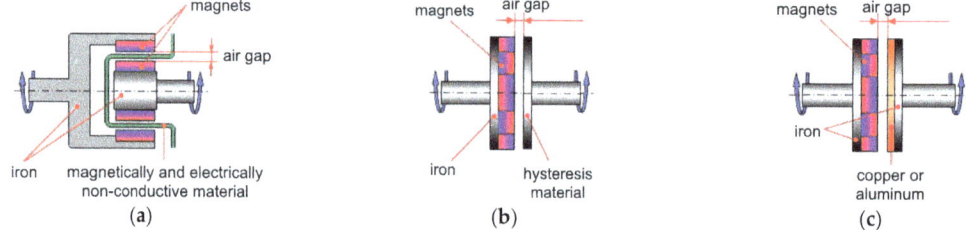

Figure 2. Type of MCs: (**a**) Active/Reactive coupler; (**b**) Hysteresis coupler; (**c**) Eddy-current coupler.

Active and reactive couplers have PMs inserted into both rotors. In reactive couplers, the PMs are mounted on only one-half of the rotor, while the other half is steel in the form of PMs. The inner and outer rotors rotate at synchronous speed. The hysteresis and eddy-current couplers have PMs on the half-rotor side, while the other side has hysteresis and conductive materials, respectively. The inner and outer rotor speeds are not the same.

Recently, substantial work has been concentrated on MCs using analytical and numerical approaches. Carpentier et al. [4] suggested implementing the analytical virtual work approach to the framework of the volume integral method to compute the magnetic forces. Li et al. [5] obtained 3D analytical torque equations with a closed form for an ideal radial MC. Ravaud et al. [6] performed a 3D semi-analytical study of the transmitted torque between uniformly magnetised PMs based on the Coulombian model. The theoretical aspects of these studies are predominant. Apart from this, the coupler parameters affecting the transferable torque have not been comparatively studied.

On the other hand, studies have accelerated with the development of powerful numerical analysers [7] using finite element methods (FEMs) [8]. Ziolkowski et al. [9] compared transient, quasi-static, and fast-quasi-static modelling techniques to calculate force profiles using FEMs. Ose-Zala et al. [10] investigated the influence of basic design parameters on the mechanical torque for cylindrical MCs with rounded PMs using QuickField software based on 2D FEMs. Kang et al. [11] showed the torque calculation and parametric analysis of synchronous PM couplers. The analytical results are compared with 2D FEMs. Nevertheless, different temperature and grade conditions of PMs have not been studied. Torque variations against different rotor materials have not yet been investigated.

Baiba Ose et al. [12] examined the influence of the PM width and the number of pole pairs on the mechanical torque of MCs. Meng et al. [13] performed transient magnetic field calculations for MCs by using Ansys Maxwell 3D software. However, the studies do not simultaneously examine multi-objective design parameters that affect each other.

In addition, different MC topologies [14] have been the subject of comparison. Kang et al. [15] compared the torque of synchronous PM couplers with parallel and Halbach-magnetised magnets by using field calculations. Recently, studies of an axially magnetised MC [16] have been reported. At the same time, magnetic gear concepts [17–19] inspired by mechanical gearboxes were studied. Structures combining magnetic gears and electrical machines [20,21] have begun to be developed. Moreover, hybrid coupler studies [22] have become widespread. Loss calculations [23,24] for MC efficiency studies are shown. However, static tests in response to torque angle variations have not been investigated.

MCs are safely used in many areas [25–27], such as automotive, marine, pump, and compressor applications. One of the practical benefits of MCs is to prevent mechanical faults [28,29] due to torque overloads in some critical applications with the help of slipping when excessive torque is applied. MCs are also impactful for use in hazardous or corrosive environments while transmitting torque through a containment shroud [30].

Optimisation studies [31,32] have been performed to achieve the optimum design. Furthermore, it has been investigated whether magnetic bearings [33] could be used instead of mechanical bearings to reduce maintenance and operating costs.

Although not in large numbers, MCs have started to be used in the aviation industry [34,35]. Benarous et al. [36] summarised all the findings and revealed test data from a magnetic gear coupler designed for an aerospace application. Finally, coreless design [37], which is demanded chiefly in aviation applications, has also been mentioned.

Since most systems traditionally have design limitations that directly affect output characteristics, the system-specific design of MCs is required where performance investigations against correlative system parameters are considered.

This paper clarifies the design aspects and implications of active cylindrical MCs, particularly for small-sized UAVs, to achieve the optimum design. The use of proposed MCs in hybrid UAVs comprising PE and S/G units is important because they provide a significant advantage in protecting the UAV, especially under severe conditions such as excessive loading and shaft lock-up [38,39]. In such catastrophic situations, the S/G is operated in motor mode, allowing the UAV to continue its cruise mission or land safely with the help of the loss of synchronisation between the inner and outer rotors of the MC, as shown in Figure 3. Although this loss of synchronisation may seem like a problem in ordinary machine designs, the use of MCs provides a great advantage in terms of protection against breakage in hybrid UAVs.

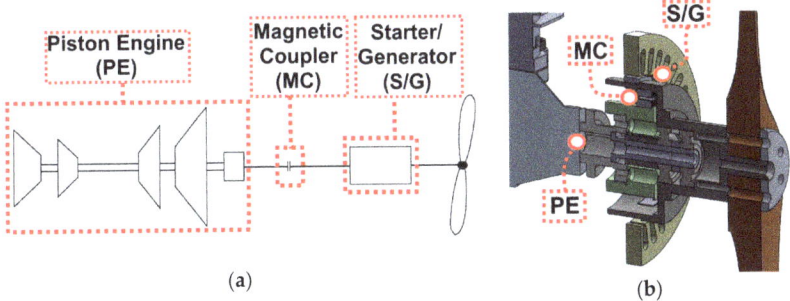

Figure 3. (a) Block diagram of the hybrid UAV; (b) 3D drawing of the proposed model.

The novel MC is part of the customised Bearcat F85F model aircraft, which is a 60% hybrid by replacing the conventional 3W-140i PE [40] with the 3W-55XI PE and S/G of 6 kW and 4500 rpm. The MCs are prototyped, and the given dimensions are verified.

The main contribution of this paper is to explore the effects of the design parameters of the MC by applying a multi-objective optimisation approach. Supporting the numerical analyses with experiments and systematically collecting the results under a unique study paves the way for researchers to facilitate the design process and validate the proof of concept. This work is distinguished from other studies with the following novelties:

- Performing transient analysis on the dynamically modelled state of the MC;
- Dynamic investigation of the effect of misalignments on the transmitted torque;
- Examination of the MC efficiency depending on the operating speed at a critical angle;
- Exploring the negative torque between the rotors in case of a loss of synchronisation;
- Proposing the correction coefficients to identify the error margin of simulations.

In addition, exploring the effect of PM temperature changes, orientations, and grades on pullout torque increases the novelty of the article. In previous studies, the multi-objective optimisation of the MC design was considered analytically [10–14]. However, nonlinear parameters affecting the performance, such as leakage flux, core losses, and end effects, are ignored in analytical methods to avoid complex and time-consuming calculations. In addition, the eddy-current losses induced in the PMs due to the continuously varying torque angle depending on the natural vibration of the piston engine are very difficult to handle analytically. Therefore, it requires more precise FEM analysis.

For this purpose, the design parameters are considered as a whole, and accordingly, the system is numerically optimised. Thus, the leakage flux, core losses, and end effects are evaluated with the FEM model. The efficiency of the optimised MC is considered an important performance indicator and is analysed together with the nonlinear effects of the materials. Furthermore, the experimental verification of an optimised FEM model in accordance with PE output parameters for hybrid UAVs also makes this study interesting for researchers. The experimental results are in agreement with the FEM outputs.

This study consists of four main frameworks. Section 2 covers design considerations such as dimension criteria, constraints, and rotor topologies. Analytical pre-dimensioning and FEMs by Ansys Maxwell are included in Section 3. The MC is dynamically modelled to improve the simulation accuracy. The effects of the air gap clearance, model length, pole numbers, PM thickness, and thickness of the rotor yokes are investigated in magnetostatics and transients. The torque ripple of the MC is explored. Section 4 comparatively presents and discusses the performance test results of the MCs with different design parameters carried out on the dynamic test bench. Locked-rotor and dynamic tests are performed with steps, full loads, and overloads [41]. Finally, the findings are reviewed in Section 5.

2. Design Considerations

The block diagrams of the hybrid UAV system and the proposed model are shown in Figure 3a,b, respectively. Numerous criteria are used for UAV classification [42], such as the mean take-off weight (MTOW), size, operating conditions, and capabilities.

The modernised Bearcat F85F Warbirds 1/4.2 scale aircraft with 22 kg, 256 cm wingspan, 204 cm length, and 150 m ceiling altitude is in the Open Category A3 (small size) based on European Union Aviation Safety Agency (EASA) regulations [43,44].

The maximum torque of the replaced 3W-55XI PE and, therefore, the minimum torque to be transmitted by the MC is 4.4 N·m. However, considering the load variations due to sudden manoeuvres, a safety factor of 1.2 is determined. In addition, the correction coefficient of 1.3 is initially chosen at the beginning of the design to account for the simulation errors and high starting kickback torque of the PE. Accordingly, in light of the MTOW, including PE and S/G, the allowable weight and length for the MC are set by the manufacturer at 375 g and 15 mm, respectively. The optimisation parameters of the MC design sought in the design reviews and given by the UAV manufacturer are summarised in Table 1.

Table 1. Design parameters.

Parameters	Value
Pullout torque, with safety factor and correction coefficient	6.9 N·m
Minimum torque density, required	18.4 N·m/kg
Rated speed	4500 rpm
Operation speed range	2500–6500 rpm

In the design of active couplers, an objective function such as torque per magnet volume, torque per coupler volume, or cost per weight should be considered to obtain the final design. The minimum weight that meets the requirements is often preferred for hybrid UAVs. However, the optimum design study is based on the achievable maximum torque within the manufacturable size and weight limitations to compensate for the unpredictable high kickback torque experienced during the initial start-up of the PE. The optimal design parameters are identified in Section 3.5 by comparing different topologies. MCs are also classified by the shape of PMs [45], such as star-type, cylindrical, ring-type, rectangular or sector shape, and toothed surface. In terms of practical use, the cylindrical type is more popular. Further classification can be performed according to the magnetisation direction of PMs as radial, axial, and linear orientations. The active cylindrical type is intended for synchronous speed and radial motion requirements.

2.1. Determination of the Minimum Outer Diameter of the Inner Rotor

The inner rotor of the MC is directly connected to the flange of the PE, as marked in red in Figure 4a, thus providing magnetic separation [46] between the shafts of the PE and S/G to improve the safe operation [47] of the hybrid UAV.

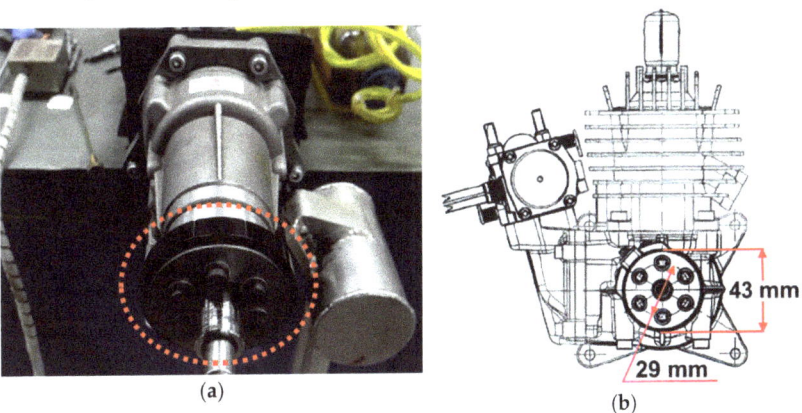

Figure 4. Mounting illustration of the (**a**) Inner rotor; (**b**) Inner flange.

The design of the MC should start from the inner rotor to the outer rotor, as opposed to the conventional method, due to the diameter limitations of the inner flange in the 3W-55XI PE, as shown in Figure 4b. It is ensured that the outer rotor of the MC is also the rotor of the S/G to take advantage of this design limitation.

2.2. Selection of Rotor Topology

Figure 5a–f illustrate the conventional rotor topologies selected depending on the objective function and the application area. Some disruptive topologies have also been applied, such as Halbach arrays to increase the field strength of PMs, as shown in Figure 5g [15], and enhanced hybrid couplers to increase the torque density, as shown in Figure 5h [48]. However, arc surface PMs are preferred due to the ease of fabrication and access.

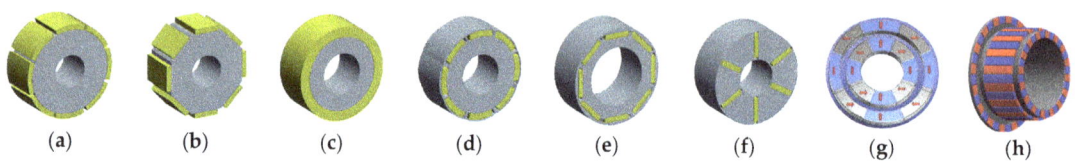

Figure 5. Rotor topologies: (**a**) Arc surface-mounted; (**b**) Rectangular surface-mounted; (**c**) Ring-type; (**d**) Buried arc type; (**e**) Buried type; (**f**) Inset type; (**g**) Halbach arrays; (**h**) Enhanced hybrid.

2.3. Materials Overview

Electric steel, carbon steel, and metals are used as MC rotor materials. NdFeB and SmCo stand out among ferrite, ceramic, and alnico magnets due to their high energy density. Epoxy, the most common type of coating for aerospace applications, has been preferred among coating types such as zinc, gold, plastic, nickel, and Teflon. The temperature assignment of PMs is made at 80 degrees Celsius, which is the most likely to be encountered in the system. In some special cases, a protective sleeve made of stainless steel, fibres, or plastics is used to prevent the PMs from leaving the rotor surface.

The containment shroud for sealing fixed to the stationary part of the MC hermetically separates the inner and outer rotors. There are several materials, such as nonferrous stainless steel, Hastelloy, carbon fibre peek, oxide ceramics, and nonmetallics.

3. Design Studies

Analytical approaches [49,50] are simple and fast methods for estimating preliminary design dimensions. The margins of error are high because the calculations are made under the assumptions that the magnetisation of PMs is homogeneous, the model length and the average air gap radius are very large compared to the PM thickness and air gap length, and the rotor materials are not saturated and have high permeability [51]. However, analytical calculations involving these effects are laborious and complicated.

3.1. Analytical Preliminary Sizing

The analytical subdomain method based on the Maxwell stress tensor and virtual work approach are accurate methods for the analytical calculations of the transmitted torque of MCs. The analytical subdomain method uses Laplace's and Poisson's equations [7] for the air gap and PM regions to find the flux density distribution by using the derivative of the vector potential equation in the air gap, as in Equations (1) and (2). Then, the transmitted torque is calculated from the Maxwell stress tensor method as in Equation (3).

$$B_{IIr}(r,\theta) = \frac{1}{r}\frac{\partial A_{II}}{\partial \theta} \qquad (1)$$

$$B_{II\theta}(r,\theta) = \frac{1}{r}\frac{\partial A_{II}}{\partial \theta} \qquad (2)$$

$$T = \frac{l_s r_{mean}}{\mu_0} \int_0^{2\pi} [B_{IIr}(r_{mean},\theta) B_{II\theta}(r_{mean},\theta)] d\theta \qquad (3)$$

where $B_{IIr}(r,\theta)$ and $B_{II\theta}(r,\theta)$ are the air gap flux density distributions depending on the radial distance (r) and angle (θ) according to polar coordinate adoption, respectively. ∂A_{II} and l_s are the vector potential in the air gap and the total model active length, respectively. r_{mean} is the radius of the middle of the air gap, as shown in Figure 6a. r_{oi} and r_{io} are the outer radius of the inner rotor and inner diameter of the outer rotor, respectively.

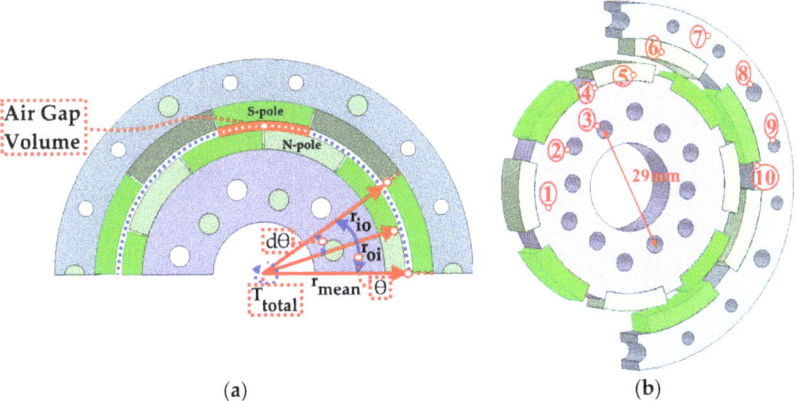

Figure 6. Model of the MCs (**a**) Parameter definitions for the VM method; (**b**) Material definitions.

However, the virtual work method [52] is a practical approach for the fast calculation of the air gap volume depending on the angular displacement between the PMs based on the energy change in the air gap and is given as follows.

$$T_{total} = \frac{dW}{d\theta} = \frac{(B_{g1})^2(V_{ag})}{2\mu_0(\theta - d\theta)} * (2p) \quad (4)$$

where dW, θ, $d\theta$, B_{g1}, V_{ag}, $2p$ are the change in stored energy in the air gap in joules, mechanical angle of a pole in radians, angular displacement between poles depending on the load, fundamental component of the flux density in the middle of the air gap in Tesla, air gap volume in m^3, and the number of poles, respectively. T_{total} is the total torque exerted on the middle of the air gap used to estimate the torque on the rotors with regard to the total number of poles. The active couplers work without any slip until the pullout torque is exceeded. The pullout torque is expressed as the maximum torque that the MC can handle.

The model of the proposed cylindrical MC comprises an inner and outer rotor (1,7), weight reduction holes (2,9), mounting holes (3,8), PM housings (4,10), and PMs (5,6), as illustrated in Figure 6b.

The torque angle (θ) is the mechanical angle between the d-axes of the inner rotor PMs and the outer rotor PMs when the MC is loaded, as shown in Figure 6a. The angle at the maximum torque is called the critical angle [3] and is calculated as in Equation (5), equivalent to 90 electrical degrees.

Mdeg represents the mechanical angle. The critical angle is 18 Mdeg and 0.314 radians for 10-poles. Preliminary sizing calculations are performed using Equation (4) and summarised in Table 2 for the 10-pole configuration at the critical angle in which ($\theta-d\theta$) is directly equal to $d\theta$. The θ is 36 Mdeg and 0.628 radians for 10-poles.

$$\theta_{critical} = \frac{360°}{(2p)*2}, \quad Mdeg \quad (5)$$

The respective air gap volume is calculated as 1289.5 mm^3. The bore diameter of the inner rotor is 29 mm due to the mounting hole diameters on the flange, as shown in Figure 4b. Similarly, the outer diameter of the inner rotor is to be a minimum of 43 mm for the model. Considering the thickness of the rotor yokes and PMs, r_{mean} is initially chosen to be 27.5 mm. With the initial assumption of an air gap length of 1.5 mm, the outer diameter of the inner rotor and the inner diameter of the outer rotor are found to be 26.75 mm and 28.25 mm, respectively. Thus, the corresponding model length is found to be 5 mm.

Table 2. Analytical design calculations by virtual work approach for preliminary sizing.

Design Outputs	10-Pole
Air gap volume, minimum required	1289.5 mm^3
Length of the model, based on r_{mean}	5 mm
(θ-dθ), dθ at critical torque angle	0.314 rad.
Critical torque angle	18 (°M)
Design Assumptions	
Middle of the air gap radius, r_{mean}	27.5 mm
Average air gap flux density	0.65 T
Air gap length	1.5 mm
Pullout torque, required	6.9 N·m

3.2. Maxwell 2D Static Analyses

The increased ability to use all processor cores and the symmetry properties tend to directly use FEM-based software in the design and optimisation of MCs [53], resulting in a tangible increase in simulation accuracy. The FEM accurately calculates the air gap flux density and transmitted torque by considering the material nonlinearities, leakage fluxes, core losses, PM magnetisation directions and temperature changes, induced eddy-current losses on PMs, and dynamic effects on the MC.

Ansys Maxwell multi-functional analyses are performed in a magnetostatic environment for static simulations and in a transient environment for dynamic simulations [54].

3.2.1. Correlation of Effective Air Gap Diameter and Model Length

The analyses are started with a 10-pole configuration to determine the effective air gap diameter and the model length for the required torque and torque density. Considering the air gap and yoke thicknesses, the analysis started with a minimum effective air gap diameter of 45 mm. The effective air gap is defined as the middle of the air gap.

Figure 7 investigates the pullout torque for the range of effective air gap diameters and model lengths. Figure 7a shows that the required pullout torque of 6.9 N·m specified in Table 1 is met with a minimum effective air gap diameter of 47 mm for a model length of 10 mm or a minimum model length of 6 mm for an effective air gap length of 67 mm.

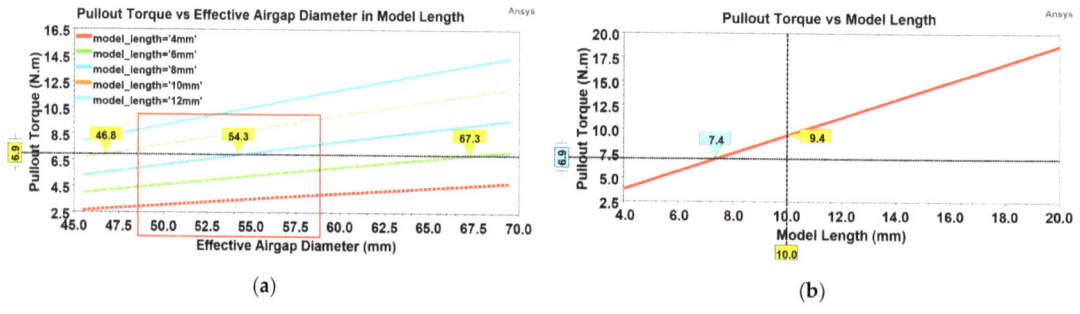

Figure 7. Pullout torque versus (**a**) Effective air gap length in model length; (**b**) Model length.

The relationship between the effective air gap diameter and the length of the MC directly affects the inertia and transferrable torque of the hybrid UAV. The choice of an effective air gap diameter as high as possible will increase the system inertia and allow modular construction for higher torque transmissions. Therefore, the effective air gap diameter is chosen to be 57.5 mm, approaching the maximum limit of the range.

On the other hand, the linear increase in the model length corresponds to an almost linear increase in the pullout torque. The required torque of 6.9 N·m is achieved with a minimum coupler length of 7.4 mm, as shown in Figure 7b. However, for easy manufacturability and higher tolerance to disturbance, the length is chosen to be 10 mm, in which case the pullout torque is 9.4 N·m, hereafter referred to as the updated torque requirement.

3.2.2. Investigation of Optimum Pole Number

In magnetic systems, such as MCs, the pole number configuration of the rotors significantly affects the transmitted torque and, thus, the torque density. The pole number of 10 for the inner and outer rotors is chosen as the optimum point because the highest pullout torque is provided, as shown in Figure 8a.

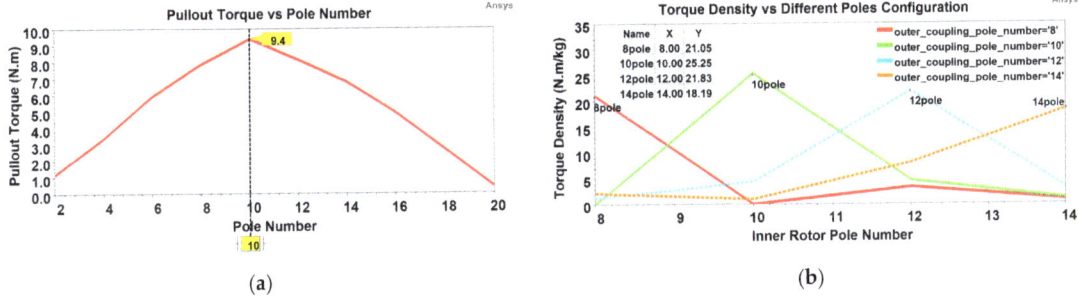

Figure 8. (**a**) Torque vs same rotor pole number; (**b**) Torque density vs different rotor pole number.

On the other hand, Figure 8b examines viable or impracticable cases of the torque density for different numbers of inner and outer rotor poles. The torque density expresses the pullout torque per unit weight in N·m/kg. The torque density decreases dramatically in the case of different inner and outer rotor pole number configurations. However, a different number of poles on both rotors is possible with the appropriate design of the modulator in the air gap. Thus, the magnetic gear concept [19] is formed.

3.2.3. Effect of Air Gap Clearance on Pullout Torque

Air gap clearance has a direct effect on the torque since it affects the total reluctance. Reducing the air gap will increase the torque, as seen in Figure 9, but it will also increase production costs and cause the rotors to rub against each other in the case of imbalance.

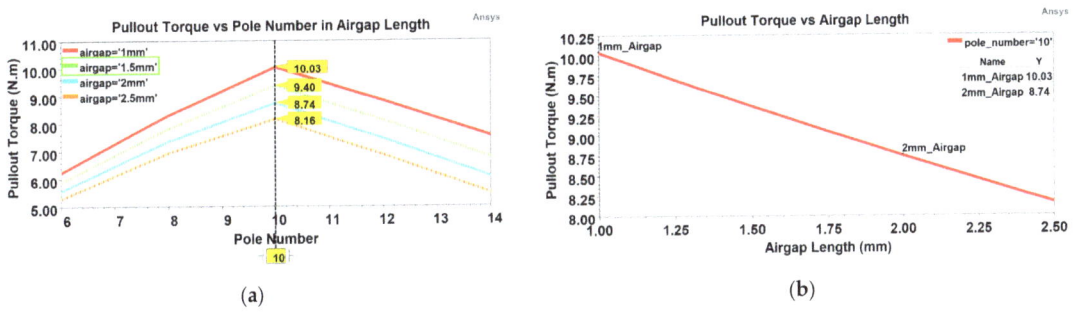

Figure 9. Pullout torque investigation versus: (**a**) Pole number in air gap length; (**b**) Air gap length.

Considering the fabrication issues and PE vibration, the air gap is set to 1.5 mm, as marked in green in Figure 9a. A twofold increase in the air gap length does not mean a twofold reduction in the torque but instead a reduction of 13%, as shown in Figure 9b.

3.2.4. Determination of PM Thickness

The PM thickness changes the average flux density in the air gap and, hence, the transmitted torque [55]. The thickness of the inner rotor PMs is set at 4 mm because the maximum increase in the torque density is met, as shown in Figure 10a. On the other hand, the thickness of the outer rotor PMs of 4 mm is chosen because the maximum weight of 375 gr is reached, as illustrated in red in Figure 10b.

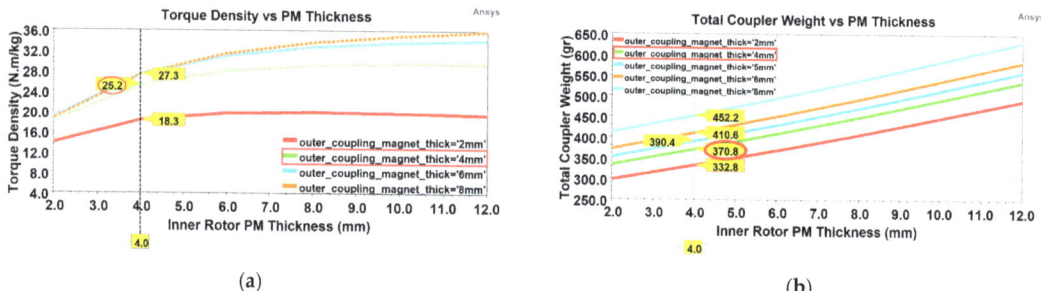

Figure 10. Inner and outer rotor PM thickness versus: (**a**) Torque density; (**b**) Coupler weight.

3.2.5. Determination of the Thickness of Rotor Yokes

The thickness of the rotor yokes should be selected carefully, as it changes the total reluctance and, hence, the air gap magnetic flux density. The minimum thickness of the rotor yokes that meet the updated torque requirement results in the minimum coupler weight.

For this purpose, the inner and outer rotor yoke thicknesses are determined to be 14 mm and 8 mm, as shown in Figures 11a and 11b, respectively. The choice of the values is evaluated based on a fraction-free approach for ease of production and in light of the minimum wall thickness necessary to eliminate material deformation during fabrication.

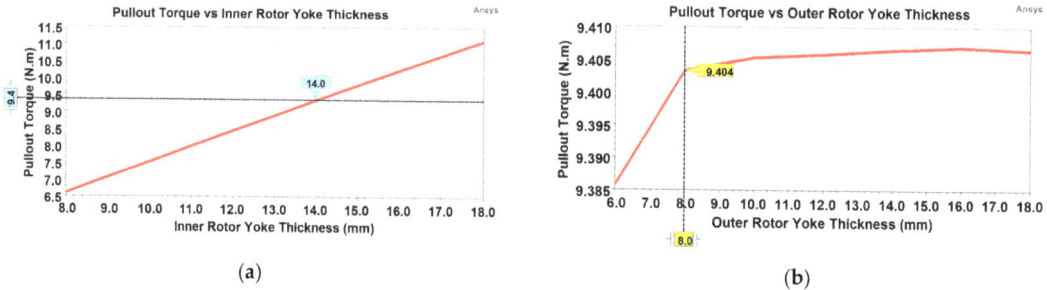

Figure 11. Pullout torque versus: (**a**) Thickness of inner yoke; (**b**) Thickness of outer yoke.

3.2.6. Investigation of the PM Embrace and Offset Effect

Pole embrace represents the ratio of the pole arc to the pole pitch. The arched PMs may not be concentric with the rotor. In the absence of a uniform air gap, the offset between the centre of the bottom and top of the PM arc is called the pole arc offset. The embrace has a more significant effect on the pullout torque, while the offset has a limited effect.

The embrace of 0.98 offers a higher torque density, as marked in red in Figure 12a. The pole arc offset may not be preferred because it negatively affects the output torque, as seen in Figure 12b, except for mandatory situations such as cogging torque. Thus, the maximum pullout torque is achieved with the maximum embrace and minimum offset.

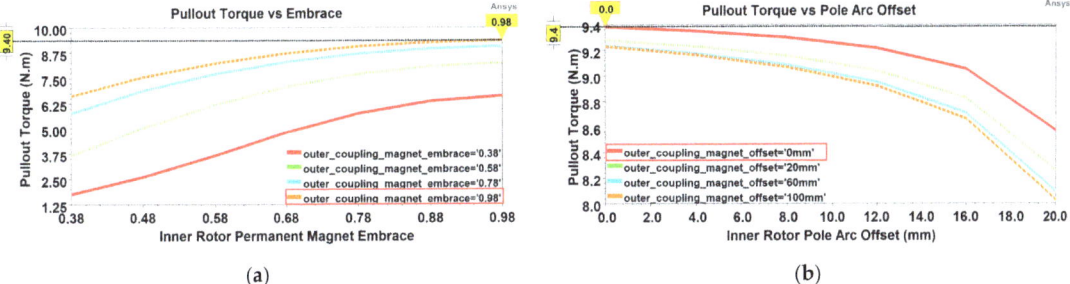

Figure 12. Pullout torque versus: (**a**) PM embrace; (**b**) PM offset.

3.3. Maxwell 2D Transient Analyses

Transient analyses allow performance outputs corresponding to the design parameters to be dynamically realised under no-load, rated-load, and overload conditions. Simulations are carried out in which the moment of inertia, the mechanical losses in terms of damping factors, and the load type acting are modelled. Thus, the moment of inertia of the inner and outer rotors is calculated as 0.42 kg-cm^2 and 1.95 kg-cm^2, respectively. Mechanical losses, i.e., wind and friction losses, ventilator losses, and bearing losses, are practically accepted at 3.5% of the output power [56]. The load type is considered such that the load varies nonlinearly with the square of the speed, such as fan load [54].

3.3.1. Comparison of Pole Types

The pole types applied to the PM machines can also be employed in the MCs. Surface-mounted PMs are more production viable than internal PMs and can be divided into three parts, as shown in Figure 13a–c. Figure 13d shows that the effect of the pole type on the torque density is minimal. However, type-1 is preferred due to its ease of installation.

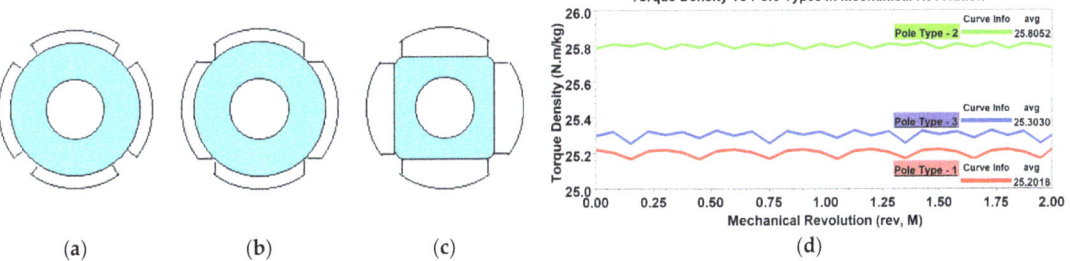

Figure 13. Pole types versus: (**a**) Type-1; (**b**) Type-2; (**c**) Type-3; (**d**) Torque density in rev,M.

3.3.2. Effect of PM Type, Grade, and Temperature on Pullout Torque

The grade, type, orientation, and operating temperature of PMs have a significant role in MC design. Figure 14 examines all implications for the 10-pole configuration at a critical angle. The pullout torque increases with the increasing PM grade, as shown in Figure 14a. On the other hand, SmCo magnets can operate at higher temperatures and in harsher conditions. Nevertheless, their energy density is lower than that of NdFeB, resulting in a lower pullout torque, as shown in Figure 14b for the same thickness of PMs.

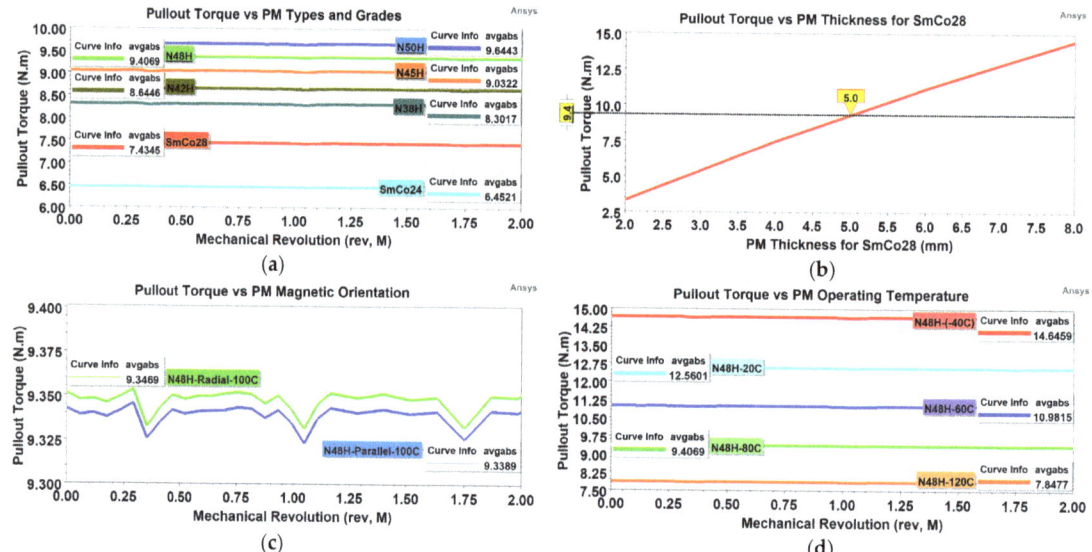

Figure 14. Effect of PM: (**a**) Grade; (**b**) Type; (**c**) Magnetic orientation; (**d**) Operating temperature.

In addition, the magnetisation direction has little effect on the pullout torque, as shown in Figure 14c. However, the operating temperature of the PMs significantly affects the torque, as shown in Figure 14d. Radially oriented N48H is preferred for accessibility.

3.3.3. Rotor Flux Density and Mesh Distribution

The flux density of the rotor yokes should be close to, but not reach, the saturation point, which is the knee point on the BH curve to achieve the maximum torque density, as shown in Figure 15a. However, the minimum wall thickness required to prevent material deformation during manufacturing and dynamic effects limits the design of the yoke thickness close to the saturation point. The yoke design is based on adjusting the yoke thicknesses as close as possible to the saturation point, as shown in Figure 15a. In this case, the outer rotor yoke thicknesses (t_{yo2} and t_{yo1}) can be a minimum of 2 mm to prevent fabrication deformation. The inner rotor inner yoke thickness (t_{yi1}) is set to a minimum of 2.5 mm to avoid reducing the mechanical strength and flywheel effect, and the inner rotor outer yoke thickness (t_{yi2}) is set to 5.24 mm to ensure the selected effective air gap diameter.

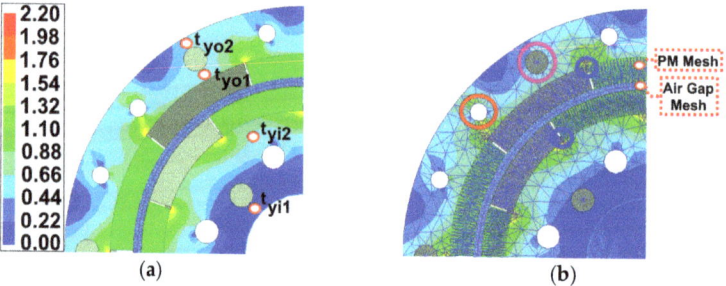

Figure 15. (**a**) Flux density distributions; (**b**) Mesh distributions.

Figure 15b exhibits the mesh distribution of the model. The critical regions, such as the air gap and PMs, are subjected to dense meshing. The total number of mesh elements is 9440. The mesh method is TAU. Thus, the solution accuracy is increased. The parts marked in pink are the mounting holes. The diameter of the weight reduction holes, marked in red, is chosen to not reduce the mechanical strength. Thus, the total weight is reduced by 4 gr.

3.3.4. Investigation of Negative Torque at Loss of Synchronisation

Loss of synchronisation (LoS) refers to the situation where the synchronisation between the rotors is disrupted by exceeding the torque limit and critical torque angle as a result of a shaft malfunction on the PE shaft. However, in the case of LoS, while the drive system is sustained by S/G, the negative torque acting from the inner rotor needs to be analysed and accounted for in the safety factor to identify the power limits of the S/G.

In the 2D simulation, the LoS torque is analysed by setting the inner rotor speed to 0 rpm and rotating the outer rotor at different speeds. In the test system, the LoS torque is measured by locking the inner rotor so that it cannot rotate and gradually rotating the outer rotor at different speeds by a speed source. As shown in Figure 16, the LoS torque at the maximum speed is −0.61 N·m for the 2D simulation and −0.6 N·m for the test results. The deviations in the results are due to the sensitivity of the sensors on the test bench and the higher moment of inertia of the test bench compared to the UAV dynamic model in the simulations.

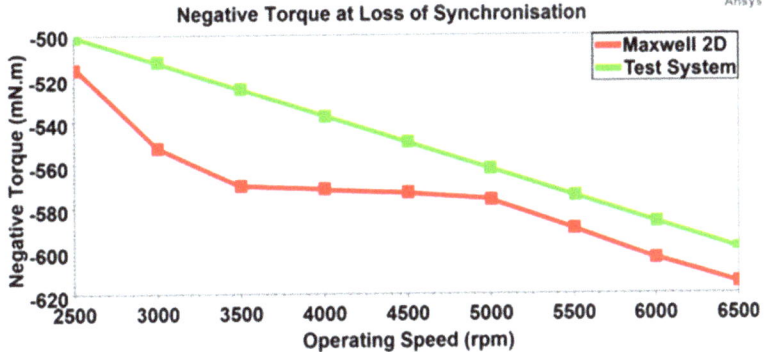

Figure 16. Negative torque at loss of synchronisation depending on operating speed.

3.4. Maxwell 3D Static and Transient Optimetric Analyses

Static simulations examine the MC behaviour in the steady state, i.e., when the inner or outer rotor shaft is locked. Therefore, model losses are not considered in the static analyses. On the other hand, transient analysis is more accurate because it considers losses, coupling effects, end effects, eddy-current losses on PMs, and material wall thickness.

3.4.1. Static Locked-Rotor Torque and Transient Torque Ripple Analyses

Locked-rotor or static torque refers to the torque capability of the coupler. It can be examined in different pole numbers depending on the torque angle, as shown in Figure 17a. The maximum static torque is provided as 9.4 N·m in the 10-pole configuration.

In the dynamic state, torque transmission in response to instantaneous load variations causes torque ripple in MCs due to the different moments of inertia of the rotors and the flywheel effect. It is 20 mN·m for the proposed model, as shown in Figure 17b. The rotors hold each other until the critical angle is exceeded, resulting in minimal torque ripple in the active MCs.

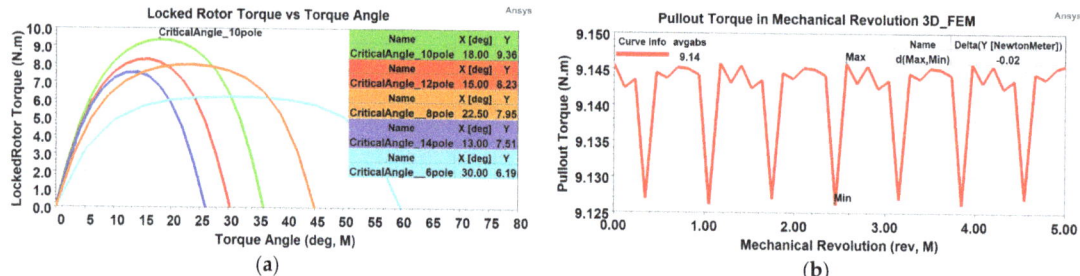

Figure 17. (**a**) Locked-rotor torque versus torque angle; (**b**) Torque ripple in mechanical revolution.

3.4.2. Investigation of Different Rotor Materials and Air Gap Flux Density

Carbon steel, electric steel, and stainless steel can be used as rotor materials. Although electrical steel has lower core losses for synchronous MCs, it does not provide an advantage in the proposed model due to its low model volume. Moreover, the design results in a higher yoke thickness due to the lower saturation point if electrical steel is preferred.

However, some exceptional cases, such as military applications, require the use of nonmagnetic materials, such as Steel-316, called yokeless design. In such cases, it is inevitable to increase the PM thickness to avoid a drastic drop in the torque of approximately 60%, as shown in Figure 18a. Steel-1020 is used as the rotor material in the production of MCs.

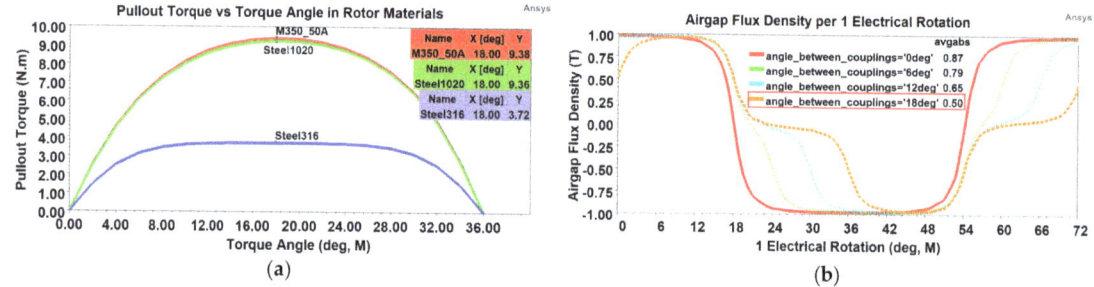

Figure 18. Investigation of: (**a**) Effect of different rotor materials on torque; (**b**) Air gap flux density.

On the other hand, the air gap flux density in the air gap is examined in Figure 18b when magnetic Steel-1020 is used as the rotor material. It is 0.5 T at the critical angle of 18 Mdeg for the proposed design, which is far from the demagnetisation point of the PMs.

3.4.3. Study of Pullout Torque Depending on Misalignment Length

As part of the worst-case scenario analysis, it is essential to examine the reduction in the pullout torque due to the misalignment length caused by the propeller pulling the system forwards until it is unable to generate thrust in the event of any extreme bearing failure.

Figure 19a examines the pullout torque depending on the misalignment. Figure 19b exhibits the misalignment. The torque decreases as the misalignment increases. Although the test results and the simulations agree with each other, the differences in the results are due to the difficulty in the precision adjustment of the misalignment length in the test system.

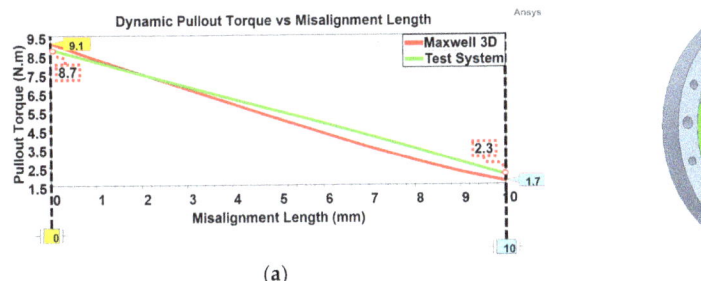 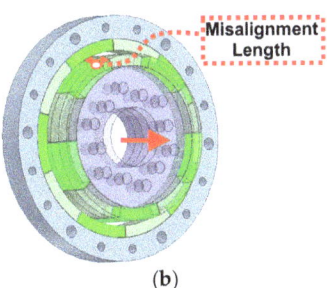

Figure 19. (a) Pullout torque depending on misalignment length; (b) Exhibition of misalignment.

3.4.4. Magnetic Coupler Efficiency and Induced Eddy-Current Losses on PMs

Magnetic coupler losses are composed of rotor core losses, induced eddy-current losses on PMs, and mechanical losses. The mechanical losses, estimated at 3.5% of the output power, are thus calculated as 0.00051 W/(rad/s)^2. On the other hand, vibration due to the natural operation of the PE, load disturbances due to different UAV operating zones, and torque fluctuations during load changes cause the torque angle to change continuously. As a result, high eddy-current losses are induced on the PMs, which increase the temperature of the PMs and reduce the transmitted torque by reducing their residual flux density, and their impact on the system should be investigated. The proposed MC design is realised in light of all these effects and the design requirements are provided.

As seen in Figure 20a, the eddy-current losses are simulated as 237 W. The efficiency of the MC is 94.3% at the maximum speed of 6500 rpm and 95% at the minimum speed of 2500 rpm, as shown in Figure 20b at the critical angle. However, to measure the efficiency on the test bench, a second transmitter is required to measure the input power in addition to the torque/speed transmitter measuring the output power. Due to the difficulty in the setup, the efficiency cannot be measured on the test bench. However, using the proposed correction coefficient, the actual efficiency can be estimated from the numerical efficiency graph.

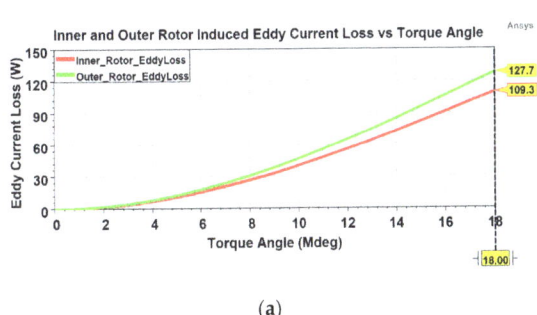 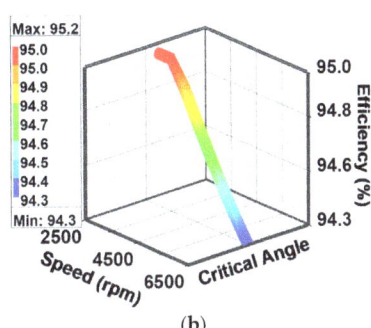

Figure 20. (a) Induced eddy-current losses on PMs; (b) Efficiency of the MC at maximum speed.

3.5. Summary List of Various MC Designs

The various MC designs obtained from the simulations are summarised in Table 3. The optimum design is realised within the system constraints, and the optimum parameters are indicated with an asterisk. The PM thickness is included in the measurement of the inner rotor outer diameter and outer rotor inner diameter. Validation of the multi-objective simulations and the tests sheds light on the safe usability of MCs in hybrid UAVs.

Table 3. Dimension and performance list of various MCs in summary.

Pole Number	8-Pole				10-Pole *	
Embrace	0.6	0.8		0.98	0.8	0.98 *
PM Thickness	3 mm	4 mm		4 mm	4 mm *	
Grade of PM	N48H	N45H	N48H	N48H	N48H *	
Outer diameter of outer rotor	79 mm	83 mm			83 mm *	
Inner diameter of outer rotor	57 mm	59 mm			59 mm *	
Outer diameter of inner rotor	54 mm	56 mm			56 mm *	
Inner diameter of inner rotor	20 mm	20 mm			20 mm *	
Air gap length	1.5 mm	1.5 mm			1.5 mm *	
Effective air gap diameter	55.5 mm	57.5 mm			57.5 mm *	
Model length	10 mm	10 mm			10 mm *	
Total weight (gr)	320	350	351	370	352	371
Pullout torque (N·m), dynamic	3.9	6.9	7.2	7.5	8.1	8.7
Torque density (N·m/kg)	12.2	19.7	20.5	20.3	23	23.45

* Optimum values.

4. Results and Discussion

This section reveals and discusses the test results and compares them with the 2D and 3D simulation data. In an optimisation process, the accuracy of the analyses is determined, and a correction coefficient is proposed based on the correlation between the results.

Various MCs with different design parameters listed in Table 3 have been produced, and some of them are illustrated in Figure 21. The upper and lower parts illustrate the outer and inner rotors of the MC, respectively.

(a)　　　(b)　　　(c)　　　(d)　　　(e)

Figure 21. Various MC productions of: (**a**) 8-pole/0.8-embrace/3 mm-PM thickness/10 mm length; (**b**) 8-pole/0.98-embrace/4 mm-PM thickness/10 mm length; (**c**) 10-pole/0.8-embrace/4 mm-PM thickness/10 mm length; (**d**) 10-pole/0.98-embrace/4 mm-PM thickness/10 mm length; (**e**) 10-pole/0.98-embrace/4 mm-PM thickness/20 mm length.

Due to the MCs being included in the group of noncontact electrical machines [57], their performance examinations are carried out in a similar way to special rotating machines, using direct or indirect test methods [58]. On the other hand, the propulsion platform, including the PE, MC, and propeller, is set up to conduct force tests on the system.

The direct test method provides more accurate results because it dynamically measures MC parameters such as the output torque and output speed with a torque/speed transmitter. Figure 22a demonstrates the installation of the MC for the direct dynamic test system, while Figure 22b shows its installation on the PE and propeller.

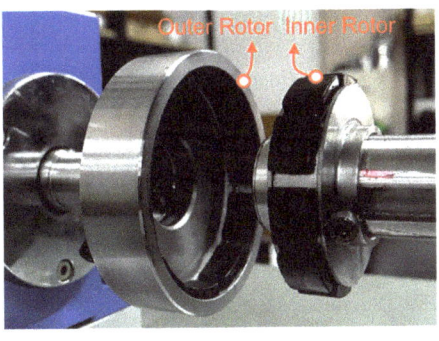 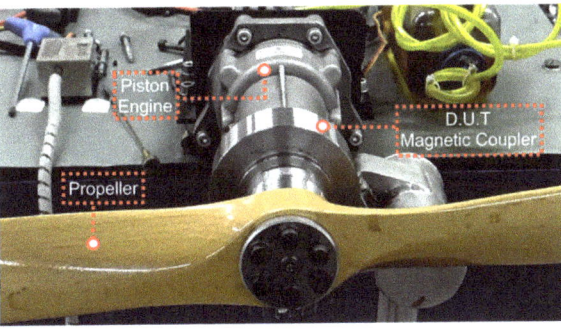

Figure 22. Assembly of (**a**) MC for dynamic tests; (**b**) MC with piston engine and propeller.

The operation principle of the direct dynamic test system shown in Figure 23a is as follows: The MC (3) device under test (D.U.T) is driven by a geared induction motor (4) with a high torque capacity. Dynamic tests are performed by gradually or directly loading the hysteresis brake for 1 h for each set of tests (1), depending on the type of test, such as no-load, rated-load, or overload. At this stage, as the load changes, the output torque and speed are measured by the torque/speed transmitter (2) and recorded by the panel.

 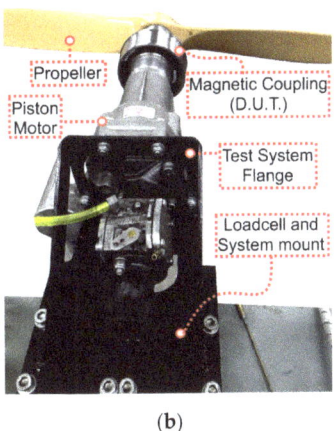

Figure 23. Test system of: (**a**) Direct dynamic method; (**b**) Thrust measurement method.

In the thrust test system shown in Figure 23b, the load cells and sensors are used to measure the force, thrust, and temperature while the PE is operated at idle speed, cruising speed, and overspeed. The sensors and transducers in the test systems are calibrated by an organisation with an international accreditation certificate [59]. In addition, the tests are performed three times in total at different times, and the average results are used.

4.1. Locked-Rotor Test Results in Summary

In the locked-rotor or static test, the shaft is heavily loaded by the hysteresis brake so that it cannot be rotated. Depending on the torque angle, torque measurement is performed by gradually adjusting the load, and the data are recorded. Figure 24a shows the results of the locked-rotor test at a critical torque angle with a 10 mm coupler length for configurations with a higher torque density, while Figure 24b exhibits the results for a lower torque density. The maximum locked-rotor torque is achieved at the configuration of 10-pole and 0.98-embrace, as marked in black in Figure 24a.

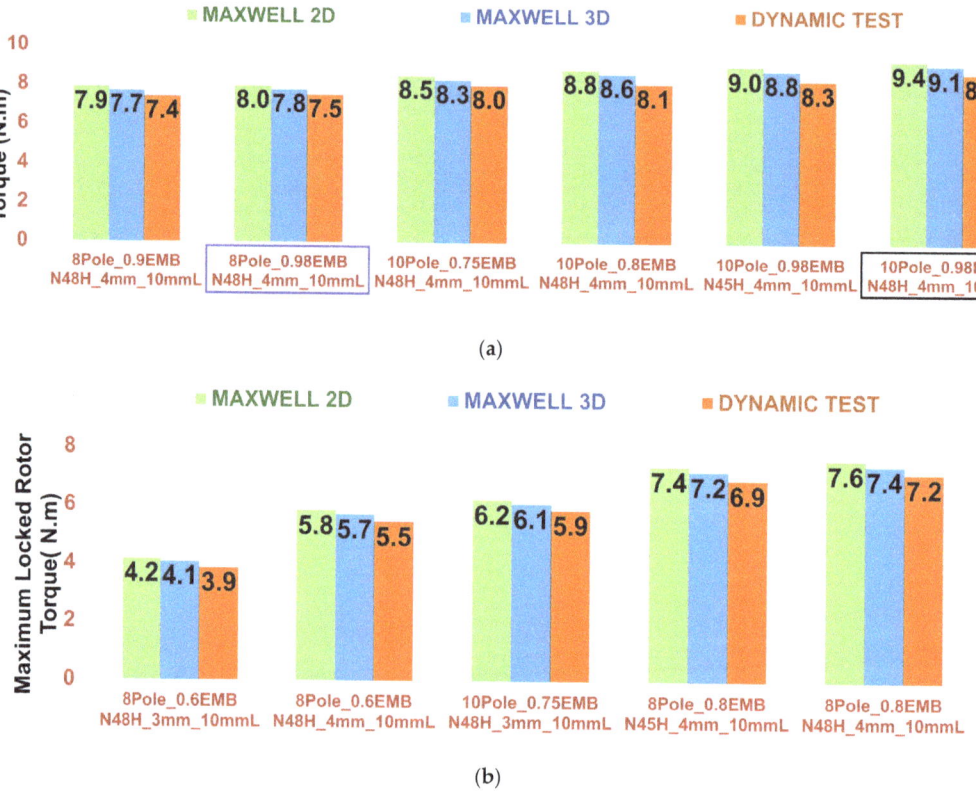

Figure 24. Locked-rotor test results of: (**a**) Higher torque density; (**b**) Lower torque density.

However, the maximum locked-rotor torque for 8 poles with the same configuration as the 10-pole structure results in a reduction of approximately 14%, as marked in blue in Figure 24a. Furthermore, if the effect of the model length is examined, a twofold increase in the length increases the locked-rotor torque by almost a factor of two.

4.2. Investigation of Pullout Torque in Transient and Static Torque versus Torque Angle

The locked-rotor and dynamic test results are consistent with each other. Therefore, the test results are plotted only for the optimum design with a configuration of 10-pole, 0.98-embrace, and 10 mm length to avoid visual pollution.

Figure 25a compares the results obtained from the Maxwell 2D and 3D simulations and dynamic tests depending on time. The average dynamic pullout torque obtained from the Maxwell 2D and 3D simulations and dynamic tests are 9.35 N·m, 9.15 N·m, and 8.7 N·m, respectively. Figure 25b plots the static torque results for the different torque angles. The maximum static torque obtained from the Maxwell 2D and 3D simulations and static tests at the critical angle of 18 Mdeg are 9.32 N·m, 9.11 N·m, and 8.72 N·m, respectively.

The Maxwell simulations are in close agreement with the test results. The numerical simulations have an acceptable margin of error compared to the test results, which is 8% for the Maxwell 2D and 4.5% for the Maxwell 3D, where the safety factor is not included. A more effective design is achieved if a relevant difference or margin of error, called a correction coefficient, is provided in the first step of the numerical design.

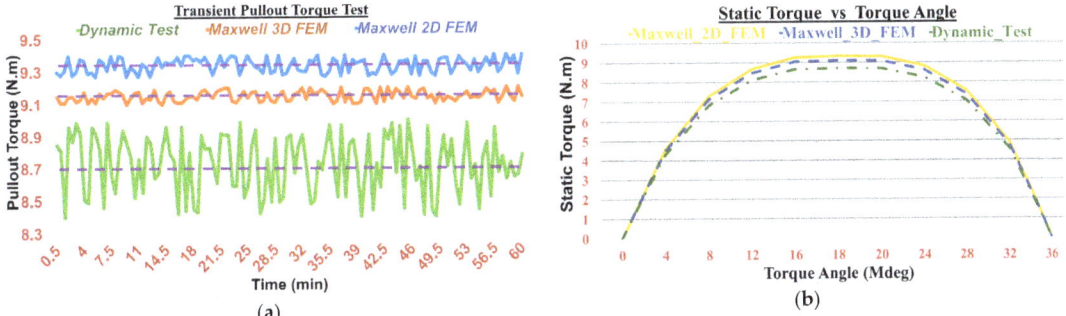

Figure 25. Test results of the optimum MC in the 10-pole, 0.98-embrace, and 10 mm length configuration: (**a**) Pullout torque in transients; (**b**) Static torque versus torque angle in magnetostatics.

In this sense, taking into account the safety factor [60], the Maxwell 2D and 3D simulations are, on average, 14% and 9% higher than the direct dynamic test results for the proposed model, respectively. As a result, a correction coefficient of 1.2 for the Maxwell 2D and 1.1 for the Maxwell 3D is proposed for the use of MCs in hybrid UAVs.

5. Conclusions

This article contributes to exploring all design parameter effects of active cylindrical MCs with multi-objective simulations based on FEMs. The magnetic design was optimised in Ansys Maxwell using optimetric and tuning tools. Increasing the number of poles results in a maximum torque density only up to a certain point. The highest pullout torque was achieved with the configuration of the 10-pole and 0.98-embrace, offering an 18% higher torque than the 8-pole configuration. Increasing the embrace provides more output torque. A 20% increase in the embrace results in a 7.5% increase in the torque density. Reducing the embrace to less than 0.6 almost halves the output torque. Increasing the offset reduces the transmitted torque by a maximum of 10%. Using PMs with a lower residual flux density reduces the pullout torque. The MCs using Nd-Fe-B PMs provide a higher torque density than the couplers using Sm-Co. A double increase in the air gap length reduces the pullout torque by 13%. The reduction in the PM thickness and yoke thickness significantly reduces the torque density. Operating the MC as close to the saturation point as possible ensures the minimum weight of the system. A direct-type dynamic test system was set up for the transient and locked-rotor tests. A thrust test system was also installed for the force tests of the MC on the PE. Exceeding the critical torque angle causes synchronisation loss between the inner and outer rotors. The torque fluctuation at load changes is approximately 0.25%. The loss of synchronisation torque at the maximum speed is -0.61 N·m. The magnetic coupler efficiency is above 94% at the maximum speed. The Maxwell 2D FEM results are higher than the 3D and dynamic tests, but the results agree with a reasonable margin of error. The test results differ by 8% with the Maxwell 2D results and 4.5% with the Maxwell 3D results. The difference is due to the density differences in the adaptive meshes, the inclusion of end-leakage effects in the 3D FEM, and temperature assignments. As a result, a correction coefficient of 1.2 for the Maxwell 2D and 1.1 for the Maxwell 3D is proposed. A comprehensive examination of the active cylindrical MCs contributes to the use of MCs for other applications such as robotics, hydraulics, automotive, medical, and pumps.

Author Contributions: Conceptualisation, S.A. and I.I.; methodology, S.A.; software, S.A.; validation, I.I. and T.S.N.; formal analysis, T.S.N.; investigation, S.A.; resources, I.I.; data curation, S.A.; writing—original draft preparation, S.A.; writing—review and editing, I.I.; visualisation, T.S.N.; supervision, I.I.; project administration, S.A.; funding acquisition, S.A. All authors have read and agreed to the published version of the manuscript.

Funding: This work in part is supported by The Scientific and Technological Research Council of TURKIYE (TUBITAK) 1501 support program with project number 3192296.

Institutional Review Board Statement: Not applicable.

Informed Consent Statement: Not applicable.

Data Availability Statement: Not applicable.

Acknowledgments: FEM is conducted with the licenced use of Ansys Electronics v2020R1. The 3D models are held with the licenced use of Solidworks software.

Conflicts of Interest: The authors declare no conflict of interest.

References

1. Arslan, S.; Iskender, I.; Navruz, T.S. Optimal Design of an In-Runner Direct-Drive Synchronous Permanent Magnet Starter/Generator for UAVs. In Proceedings of the 2022 International Symposium on Multidisciplinary Studies and Innovative Technologies (ISMSIT), Ankara, Turkey, 20–22 October 2022; pp. 543–548. [CrossRef]
2. Griffo, A.; Wrobel, R.; Mellor, P.H.; Yon, J.M. Design and Characterization of a Three-Phase Brushless Exciter for Aircraft Starter/Generator. *IEEE Trans. Ind. Appl.* **2013**, *49*, 2106–2115. [CrossRef]
3. Ose-Zala, B.; Pugachov, V. The comparison of active and reactive magnetic couplers. In Proceedings of the 2012 Electric Power Quality and Supply Reliability, Tartu, Estonia, 11–13 June 2012; pp. 1–4. [CrossRef]
4. Carpentier, A.; Galopin, N.; Chadebec, O.; Meunier, G.; Guérin, C. Application of the virtual work principle to compute magnetic forces with a volume integral method. *Int. J. Numer. Model. Electron. Netw. Devices Fields* **2013**, *27*, 418–432. [CrossRef]
5. Li, K.; Bird, J.Z.; Acharya, V.M. Ideal Radial Permanent Magnet Coupling Torque Density Analysis. *IEEE Trans. Magn.* **2017**, *53*, 1–4. [CrossRef]
6. Ravaud, R.; Lemarquand, G.; Lemarquand, V.; Depollier, C. Torque in permanent magnet couplings: Comparison of uniform and radial magnetization. *J. Appl. Phys.* **2009**, *105*, 053904. [CrossRef]
7. Akcay, Y.; Giangrande, P.; Tweedy, O.; Galea, M. Fast and Accurate 2D Analytical Subdomain Method for Coaxial Magnetic Coupling Analysis. *Energies* **2021**, *14*, 4656. [CrossRef]
8. Eker, M.; Özsoy, M. Investigation of the effect of demagnetization fault at Line Start AF-PMSM with FEM. *Acad. Platf. J. Eng. Smart Syst.* **2022**, *10*, 94–100. [CrossRef]
9. Ziolkowski, M.; Brauer, H. Fast Computation Technique of Forces Acting on Moving Permanent Magnet. *IEEE Trans. Magn.* **2010**, *46*, 2927–2930. [CrossRef]
10. Ose-Zala, B.; Onzevs, O.; Pugachov, V. Formula Synthesis of Maximal Mechanical Torque on Volume for Cylindrical Magnetic Coupler. *Electr. Control. Commun. Eng.* **2013**, *3*, 37–43. [CrossRef]
11. Kang, H.-B.; Choi, J.-Y. Parametric Analysis and Experimental Testing of Radial Flux Type Synchronous Permanent Magnet Coupling Based on Analytical Torque Calculations. *J. Electr. Eng. Technol.* **2014**, *9*, 926–931. [CrossRef]
12. Ose, B.; Pugachov, V. The Influence of Pole Pair Number and Magnets' Width on Mechanical Torque of Magnetic Coupler with Rounded Permanent Magnets. *Sci. J. Riga Tech. Univ. Power Electr. Eng.* **2011**, *28*, 63–66. [CrossRef]
13. Meng, G.Y.; Niu, Y.H. The Torque Research for Permanent Magnet Coupling Based on Ansoft Maxwell Transient Analysis. *Appl. Mech. Mater.* **2013**, *423*, 2014–2019. [CrossRef]
14. Wang, J.; Lin, H.; Fang, S.; Huang, Y. A General Analytical Model of Permanent Magnet Eddy Current Couplings. *IEEE Trans. Magn.* **2014**, *50*, 1–9. [CrossRef]
15. Kang, H.-B.; Choi, J.-Y.; Cho, H.-W.; Kim, J.-H. Comparative Study of Torque Analysis for Synchronous Permanent Magnet Coupling with Parallel and Halbach Magnetized Magnets Based on Analytical Field Calculations. *IEEE Trans. Magn.* **2014**, *50*, 1–4. [CrossRef]
16. Lubin, T.; Mezani, S.; Rezzoug, A. Simple Analytical Expressions for the Force and Torque of Axial Magnetic Couplings. *IEEE Trans. Energy Convers.* **2012**, *27*, 536–546. [CrossRef]
17. Ge, Y.-J.; Nie, C.-Y.; Xin, Q. A three dimensional analytical calculation of the air-gap magnetic field and torque of coaxial magnetic gears. *Prog. Electromagn. Res.* **2012**, *131*, 391–407. [CrossRef]
18. Niu, S.; Mao, Y. A Comparative Study of Novel Topologies of Magnetic Gears. *Energies* **2016**, *9*, 773. [CrossRef]
19. Jian, L.; Chau, K.-T. Analytical calculation of magnetic field distribution in coaxial magnetic gears. *Prog. Electromagn. Res.* **2009**, *92*, 1–16. [CrossRef]
20. Chau, K.T.; Zhang, D.; Jiang, J.Z.; Liu, C.; Zhang, Y. Design of a Magnetic-Geared Outer-Rotor Permanent-Magnet Brushless Motor for Electric Vehicles. *IEEE Trans. Magn.* **2007**, *43*, 2504–2506. [CrossRef]
21. Jian, L.; Chau, K.T. Design and analysis of a magnetic-geared electronic-continuously variable transmission system using finite element method. *Prog. Electromagn. Res.* **2010**, *107*, 47–61. [CrossRef]
22. Wang, S.; Guo, Y.; Cheng, G.; Li, D. Performance Study of Hybrid Magnetic Coupler Based on Magneto Thermal Coupled Analysis. *Energies* **2017**, *10*, 1148. [CrossRef]

23. Zheng, D.; Wang, D.; Li, S.; Shi, T.; Li, Z.; Yu, L. Eddy current loss calculation and thermal analysis of axial-flux permanent magnet couplers. *AIP Adv.* **2017**, *7*, 025117. [CrossRef]
24. Min, K.-C.; Choi, J.-Y.; Kim, J.-M.; Cho, H.-W.; Jang, S.-M. Eddy-Current Loss Analysis of Noncontact Magnetic Device with Permanent Magnets Based on Analytical Field Calculations. *IEEE Trans. Magn.* **2015**, *51*, 1–4. [CrossRef]
25. Ose-Zala, B.; Jakobsons, E.; Suskis, P. The use of Magnetic Coupler instead of Lever Actuated Friction Clutch for Wind Plant. *Elektron. Elektrotechnika* **2012**, *18*, 13–16. [CrossRef]
26. Li, Y.; Hu, Y.; Song, B.; Mao, Z.; Tian, W. Performance Analysis of Conical Permanent Magnet Couplings for Underwater Propulsion. *J. Mar. Sci. Eng.* **2019**, *7*, 187. [CrossRef]
27. Chau, K.T.; Jiang, C.; Han, W.; Lee, C.H.T. State-of-the-art electromagnetics research in electric and hybrid vehicles (invited paper). *Prog. Electromagn. Res.* **2017**, *159*, 139–157. [CrossRef]
28. Saleh, K.; Sumner, M. Sensorless Speed Control of Five-Phase PMSM Drives in Case of a Single-Phase Open-Circuit Fault. *Iran. J. Sci. Technol. Trans. Electr. Eng.* **2019**, *43*, 501–517. [CrossRef]
29. Niemenmaa, A.; Salmia, L.; Arkkio, A.; Saari, J. Modeling Motion, Stiffness, and Damping of a Permanent-Magnet Shaft Coupling. *IEEE Trans. Magn.* **2010**, *46*, 2763–2766. [CrossRef]
30. Krasilnikov, A.Y.; Krasilnikov, A.A. Calculation of losses in current-conducting screen in sealed machines and devices due to loose packing of magnets in half-clutches of magnetic clutch. *Chem. Pet. Eng.* **2011**, *47*, 392–397. [CrossRef]
31. Mei, Y.; Luo, J. Influence of Eccentric Arc Design at the Rotor on the Electromagnetic Performance of a Permanent-Magnet Synchronous Motor. *Iran. J. Sci. Technol. Trans. Electr. Eng.* **2022**. [CrossRef]
32. Fontchastagner, J.; Lefevre, Y.; Messine, F. Some Co-Axial Magnetic Couplings Designed Using an Analytical Model and an Exact Global Optimization Code. *IEEE Trans. Magn.* **2009**, *45*, 1458–1461. [CrossRef]
33. Kim, S.H.; Shin, J.W.; Ishiyama, K. Magnetic Bearings and Synchronous Magnetic Axial Coupling for the Enhancement of the Driving Performance of Magnetic Wireless Pumps. *IEEE Trans. Magn.* **2014**, *50*, 1–4. [CrossRef]
34. Cao, W.; Mecrow, B.C.; Atkinson, G.J.; Bennett, J.W.; Atkinson, D.J. Overview of Electric Motor Technologies Used for More Electric Aircraft (MEA). *IEEE Trans. Ind. Electron.* **2012**, *59*, 3523–3531. [CrossRef]
35. Zareb, M.; Nouibat, W.; Bestaoui, Y.; Ayad, R.; Bouzid, Y. Evolutionary Autopilot Design Approach for UAV Quadrotor by Using GA. *Iran. J. Sci. Technol. Trans. Electr. Eng.* **2020**, *44*, 347–375. [CrossRef]
36. Benarous, M.; Trezieres, M. Design of a cost-effective magnetic gearbox for an aerospace application. *J. Eng.* **2019**, *2019*, 4081–4084. [CrossRef]
37. Charpentier, J.; Lemarquand, G. Calculation of ironless permanent magnet couplings using semi-numerical magnetic pole theory method. *COMPEL-Int. J. Comput. Math. Electr. Electron. Eng.* **2001**, *20*, 72–89. [CrossRef]
38. Singh, R.; Lal, R.; Singari, R.; Chaudhary, R. Failure of Piston in IC Engines: A Review. *Int. J. Mod. Eng. Res.* **2014**, *4*, 1–10.
39. Moosavian, A.; Najafi, G.; Ghobadian, B.; Mirsalim, M. The effect of piston scratching fault on the vibration behavior of an IC engine. *Appl. Acoust.* **2017**, *126*, 91–100. [CrossRef]
40. 3W-Modellmotoren GmbH Home Page. Available online: https://3w-modellmotoren.de/produkt/bearcat-f8f-blau/?lang=en (accessed on 29 January 2023).
41. IEEE. IEEE Approved Draft Trial-Use Guide for Testing Permanent Magnet Machines. In *IEEE P1812/D5*; IEEE: New York, NY, USA, 2014; pp. 1–65.
42. Yang, X.; Pei, X. 15—Hybrid system for powering unmanned aerial vehicles: Demonstration and study cases. In *Hybrid Energy Systems Hybrid Technologies for Power Generation*; Elsevier: Amsterdam, The Netherlands, 2022; pp. 439–473. [CrossRef]
43. Valavanis, K.P.; Vachtsevanos, G.J. *Handbook of Unmanned Aerial Vehicles*; Springer: Berlin/Heidelberg, Germany, 2014.
44. Easy Access Rules for Unmanned Aircraft Systems (Regulations (EU) 2019/947 and 2019/945). Official Journal of the European Union. 2022. Available online: https://www.easa.europa.eu/en/downloads/110913/en (accessed on 29 January 2023).
45. Ose-Zala, B. Design Optimization of Cylindrical Magnetic Coupler Based on Calculations of Magnetic Field. Ph.D. Thesis, Department of Power and Electrical Engineering, Riga Technical University, Riga, Latvia, 2015.
46. Suti, A.; Di Rito, G.; Galatolo, R. Fault-Tolerant Control of a Dual-Stator PMSM for the Full-Electric Propulsion of a Lightweight Fixed-Wing UAV. *Aerospace* **2022**, *9*, 337. [CrossRef]
47. Quattrocchi, G.; Berri, P.C.; Vedova, M.D.L.D.; Maggiore, P. An Improved Fault Identification Method for Electromechanical Actuators. *Aerospace* **2022**, *9*, 341. [CrossRef]
48. Akcay, Y.; Giangrande, P.; Galea, M. A Novel Magnetic Coupling Configuration for Enhancing the Torque Density. In Proceedings of the 23rd International Conference on Electrical Machines and Systems (ICEMS), Hamamatsu, Japan, 24–27 November 2020; pp. 228–233. [CrossRef]
49. Ravaud, R.; Lemarquand, V.; Lemarquand, G. Analytical Design of Permanent Magnet Radial Couplings. *IEEE Trans. Magn.* **2010**, *46*, 3860–3865. [CrossRef]
50. Bossavit, A. Virtual power principle and Maxwell's tensor: Which comes first? *COMPEL Int. J. Comput. Math. Electr. Electron. Eng.* **2011**, *30*, 1804–1814. [CrossRef]
51. Yonnet, J.-P.; Hemmerlin, S.; Rulliere, E.; Lemarquand, G. Analytical calculation of permanent magnet couplings. *IEEE Trans. Magn.* **1993**, *29*, 2932–2934. [CrossRef]
52. Nagrial, M.H.; Rizk, J.; Hellany, A. Design of synchronous torque couplers. *World Acad. Sci. Eng. Technol. Int. J. Mech. Mechatron. Eng.* **2011**, *5*, 1319–1324. [CrossRef]

53. Lin, W.Y.; Kuan, L.P.; Jun, W.; Han, D. Near-Optimal Design and 3-D Finite Element Analysis of Multiple Sets of Radial Magnetic Couplings. *IEEE Trans. Magn.* **2009**, *44*, 4747–4753. [CrossRef]
54. *Maxwell Help Overview Statistical Analysis Overview, Release 20.1*; Ansys®Electronics: Canonsburg, PA, USA, 2020.
55. Ose, B.; Pugachov, V.; Orlova, S.; Vanags, J. The influence of permanent magnets' width and number on the mechanical torque of a magnetic coupler with rectangular permanent magnets. In Proceedings of the 2011 IEEE International Symposium on Industrial Electronics, Gdansk, Poland, 27–30 June 2011; pp. 761–765. [CrossRef]
56. Nachouane, A.B.; Abdelli, A.; Friedrich, G.; Vivier, S. Estimation of Windage Losses inside Very Narrow Air Gaps of High Speed Electrical Machines without an Internal Ventilation Using CFD Methods. In Proceedings of the 2016 XXII International Conference on Electrical Machines (ICEM), Lausanne, Switzerland, 4–7 September 2016. [CrossRef]
57. Suti, A.; Di Rito, G.; Galatolo, R. Novel Approach to Fault-Tolerant Control of Inter-Turn Short Circuits in Permanent Magnet Synchronous Motors for UAV Propellers. *Aerospace* **2022**, *9*, 401. [CrossRef]
58. Bojoi, R.; Cavagnino, A.; Miotto, A.; Tenconi, A.; Vaschetto, S. Radial flux and axial flux PM machines analysis for more electric engine aircraft applications. In Proceedings of the 2010 IEEE Energy Conversion Congress and Exposition, Atlanta, GA, USA, 12–16 September 2010; pp. 1672–1679. [CrossRef]
59. Ums Quality Electrical Calibration Co., Ltd. Home Page. Available online: http://www.umsankara.com.tr (accessed on 29 January 2023).
60. Zipay, J.J.; Modlin, C.T.; Larsen, C.E. The Ultimate Factor of Safety for Aircraft and Spacecraft—Its History, Applications and Misconceptions. In Proceedings of the 57th AIAA/ASCE/AHS/ASC Structures, Structural Dynamics, and Materials Conference, San Diego, CA, USA, 4–8 January 2016. [CrossRef]

Disclaimer/Publisher's Note: The statements, opinions and data contained in all publications are solely those of the individual author(s) and contributor(s) and not of MDPI and/or the editor(s). MDPI and/or the editor(s) disclaim responsibility for any injury to people or property resulting from any ideas, methods, instructions or products referred to in the content.

Article

Fault-Tolerant Control of a Dual-Stator PMSM for the Full-Electric Propulsion of a Lightweight Fixed-Wing UAV

Aleksander Suti *, Gianpietro Di Rito and Roberto Galatolo

Department of Civil and Industrial Engineering, University of Pisa, Largo Lucio Lazzarino 2, 56122 Pisa, Italy; gianpietro.di.rito@unipi.it (G.D.R.); roberto.galatolo@unipi.it (R.G.)
* Correspondence: aleksander.suti@dici.unipi.it; Tel.: +39-0502217211

Abstract: The reliability enhancement of electrical machines is one of the key enabling factors for spreading the full-electric propulsion to next-generation long-endurance UAVs. This paper deals with the fault-tolerant control design of a Full-Electric Propulsion System (FEPS) for a lightweight fixed-wing UAV, in which a dual-stator Permanent Magnet Synchronous Machine (PMSM) drives a twin-blade fixed-pitch propeller. The FEPS is designed to operate with both stators delivering power (active/active status) during climb, to maximize performances, while only one stator is used (active/stand-by status) in cruise and landing, to enhance reliability. To assess the fault-tolerant capabilities of the system, as well as to evaluate the impacts of its failure transients on the UAV performances, a detailed model of the FEPS (including three-phase electrical systems, digital regulators, drivetrain compliance and propeller loads) is integrated with the model of the UAV longitudinal dynamics, and the system response is characterized by injecting a phase-to-ground fault in the motor during different flight manoeuvres. The results show that, even after a stator failure, the fault-tolerant control permits the UAV to hold altitude and speed during cruise, to keep on climbing (even with reduced performances), and to safely manage the flight termination (requiring to stop and align the propeller blades with the UAV wing), by avoiding potentially dangerous torque ripples and structural vibrations.

Keywords: fixed-wing UAV; full-electric propulsion system; axial-flux PMSMS; fault-tolerant control; phase-to-ground short circuit; failure transient analysis

1. Introduction

The global market size of Unmanned Aerial Vehicles (UAVs) was 27.4 billion USD in 2021 and, despite the negative impact of the COVID-19 pandemic, it is expected to grow within 2026 up to 58.4 billion USD, at a Compound Annual Growth Rate (CAGR) of 16.4% [1]. Additionally, pushed by the wider objectives of the aerospace electrification, the design of next-generation long-endurance UAVs is undoubtedly moving toward the use of Full-Electric Propulsion Systems (FEPSs). Although immature nowadays in terms of reliability and energy density (e.g., lithium-ion battery packs typically range about 300 kJ/kg, which is 100 times lower than gasoline [2]), FEPSs are expected to obtain large investments in the forthcoming years, aiming to replace the conventional internal combustion motors, as well as to outclass the hybrid or hydrogen-based solutions [3]. Coherently, the global market size of electric motors is projected to grow within 2028 up to 181.9 billion USD, at a CAGR of 7.0% [4]. In particular, the segment of Permanent Magnet Synchronous Machines (PMSMs) is forecast to hold more and more significant markets, due to their advantages in terms of power density, efficiency, low torque ripple and dynamic performances. In this context, the Italian Government and the Tuscany Regional Government co-funded the project TERSA (*Tecnologie Elettriche e Radar per Sistemi aeromobili a pilotaggio remoto Autonomi*) [5], led by Sky Eye Systems (Italy) in collaboration with the University of Pisa and other Italian industries.

The TERSA project aims to develop an Unmanned Aerial System (UAS) with fixed-wing UAV, Figure 1, having the following main characteristics:

- Take-off weight: from 35 to 50 kg;
- Endurance: >6 h;
- Range: >3 km;
- Take-off system: pneumatic launcher;
- Landing system: parachute and airbags;
- Propulsion system: FEPS powering a twin-blade fixed-pitch propeller;
- Innovative sensing systems:
    ○ Synthetic aperture radar, to support surveillance missions in adverse environmental conditions;
    ○ Sense-and-avoid system, integrating a camera with a miniaturised radar, to support autonomous flight capabilities in emergency conditions.

Figure 1. Rendering of the TERSA UAV [5].

With particular reference to the activities related to the TERSA UAV propulsion system development (which this work refers to), special attention has been dedicated to the demonstration of its fault-tolerant capabilities. It is well-known that, compared with solutions based on internal combustion motors, FEPSs on UAVs would guarantee smaller CO_2 emissions, higher efficiency, lower noise, reduced thermal signature (crucial for military applications), higher service ceiling and simplified maintenance [6], but several reliability and safety issues are still open, especially for long-endurance flights in unsegregated airspaces. As relevant example, the failure rate of a simplex FEPS solution with a three-phase PMSM driven by three-leg converter typically ranges about 2.4 per thousand flight hours [7,8], which is far from the reliability and safety levels required for the airworthiness certification [9].

Provided that the weight and envelopes required by UAV applications impede the extensive use of hardware redundancy (e.g., redundant motors), the reliability enhancement of FEPSs can be achieved only through motor phase redundancy or by using unconventional converters. Different solutions are proposed in the literature, and they can be split in two categories: those applying conventional three-leg converters (using multiple phases [10,11] or multiple three-phase arrangements [12,13]), and those using four-leg converters [8,14,15]. In the latter solution, a couple of power switches are added as stand-by devices to the conventional three-leg bridge, enabling the control of the central point of the motor Y-connection. Although the four-leg solution permits to save weight, it requires an ad hoc design of the motor and its power electronics [8,16]. On the other hand, PMSMs with multiple three-phase arrangements are less compact, but they use conventional converters driven by standard techniques [13].

The failure rate of electric machines is essentially driven by faults on motor phases and converters (open-switch in a converter leg, open-phase, phase-to-ground fault, interturn short circuit, or capacitor short circuit [17]), that cover about 70% of the system fault modes [7]. Stator faults can initiate for different causes, such as dielectric breakdown, degradation of the winding insulation, thermal stress, overburden, or mechanical vibrations [18], and many research efforts have been carried out for their diagnosis and the compensation, especially for open-phase faults [8,19–21] and inter-turns short circuits [22–25], while the literature is poorer for phase-to-ground faults in electrical machines [26–28]. As discussed in [29], phase-to-ground faults fall into the short circuit faults category. Usually, a short circuit initiates as an inter-turn fault (very difficult to detect at an early stage), which typically evolves into a coil-to-coil, phase-to-phase, or into a phase-to-ground short circuit. Phase-to-ground faults are particularly dangerous, because they can cause irreversible damages to both windings and core. If the motor windings could be replaced, the core damage is irreversible and it requires the entire motor removal.

When addressing UAV applications, the basic consequence related to motor faults is the decrease or loss of thrust power, which essentially impacts on the altitude hold and/or climb capabilities of the vehicle. Together with other major UAV failures such as those affecting control actuators and sensors, the hazard mitigation requires the application of suitable fault-tolerant techniques. A comprehensive survey on methods for fault diagnosis and fault-tolerant control against UAV failures is provided by [30], and the reference highlights that the works addressing propulsion failures for fixed-wing UAVs is very limited. Most of the literature is actually focused on the effects of faults to control actuators and sensors for both single UAVs [31–34] and UAVs in formation flight [35–37], while the faults to propulsion systems are typically modelled with rough or very simplified approaches (e.g., total propulsion loss as in [38] or increase in the drivetrain friction as in [39]).

Together with reliability requirements, an FEPS for UAVs must have high compactness, high power-to-weight ratio, high torque density, and excellent efficiency. For these high-performance applications, Axial-Flux PMSMs (AFPMSMs) are preferred to conventional PMSMs with radial flux linkages [40]. In fact, although conventional PMSMs have higher technology readiness, AFPMSMs are superior in terms of weight (core material is reduced), torque-to-weight ratio (magnets are thinner), efficiency (rotor losses are minimized), and versatility (the axial air gaps are easily adjustable) [41–43]. The FEPS of the reference TERSA UAV is actually equipped with a dual-stator AFPMSM, capable of operating in both active/active and active/stand-by configurations to obtain fault-tolerant capabilities for the system [44].

This paper aims to contribute to the literature of FEPSs for fixed-wing UAVs by dealing with the fault-tolerant control design and the dynamic performance characterization of the TERSA UAV, particularly addressing the impacts of failure transients on both the motor and the vehicle performances in different flight phases (climb, cruise, flight termination/landing) if a phase-to-ground short circuit in a stator is simulated.

The basic objective of the investigation is, through a detailed fault modelling, to characterise both the fault symptoms (at both the motor level and UAV level) and the failure transients related to the application of fault-tolerant techniques. For this reason, the paper does not include the description of the health-monitoring algorithms, but it simply assumes that they exist and are capable of detecting the fault with a pre-defined latency; after that a compensation is applied (the failed stator is de-energized and the control on healthy one is activated or reconfigured).

The work is articulated as follows: the first part is dedicated to the system description and to the nonlinear FEPS model; successively, the main features of the fault-tolerant control design are presented. Finally, an excerpt of simulation results is proposed, by highlighting and discussing the effects of a phase-to-ground fault during different flight manoeuvres, and by demonstrating the effectiveness of the proposed design.

2. Materials and Methods

2.1. FEPS Description

The fault-tolerant FEPS of the TERSA UAV is basically composed of (Figure 2):

- Dual-stator AFPMSM, with surface-mounted magnets and phases in Y connection;
- Twin-blade fixed-pitch propeller (APC22 × 10E model [45]);

- Mechanical coupling joint between motor shaft and propeller;
- Two Electronic Control Units (ECUs), each one including:
  - Control/monitoring (CON/MON) module, for the implementation of the closed-loop control and health-monitoring functions;
  - Conventional three-leg converter;
  - Three Current Sensors (CSa, CSb, CSc), one per each motor phase;
  - Angular Position Sensor (APS), measuring the motor rotation;
  - Power Supply Unit (PSU), providing all ECU components with the required electrical supply;
- Two interface connectors, one for the electrical power input and the other for the data exchange with the Flight Control Computer (FCC).

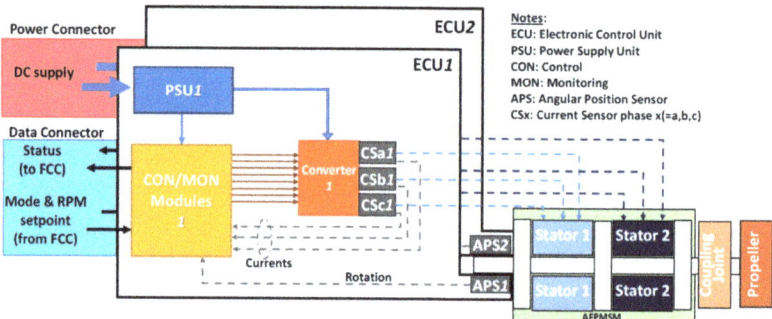

Figure 2. Schematics of the FEPS architecture.

The fault-tolerant FEPS is designed to guarantee mission accomplishment even after the failure of one of the two AFPMSM stators. Different operations of the stators are thus defined by the CON/MON modules, so that each stator can

- Be electrically supplied and controlled to apply an electrical torque on the motor shaft (*active* status);
- Be electrically supplied at the converter level to be prompt to operate, but with open/isolated phases, so that no torque is applied (*stand-by* status);
- Be de-energized at the converter level (*passive* status).

In addition, since the UAV flight termination and landing is obtained by deploying a parachute and by inflating airbags to be used as landing gears, the propeller blades must be aligned with the wing before opening the parachute to avoid interferences, and a specific control mode must be foreseen. As a consequence, four operation modes have been defined to control each stator of the AFPMSM:

(1) Flight Mission Mode (FMM), in which the stator is *active* and a speed-tracking closed-loop system is implemented, by means of two nested loops, on motor speed and currents (via Field-Oriented Control, FOC), respectively;
(2) Flight Termination Mode (FTM), in which the stator is *active* and controlled via three nested loops: the two ones of the FMM plus an outer loop on motor shaft rotation, with a predefined setpoint for the propeller alignment;
(3) Hot Stand-By (HSB), in which the stator is in *stand-by* status;
(4) Cold Stand-By (CSB), in which the stator is *passive*.

As reported in Table 1 and represented in terms of a flow chart in Figure 3, depending on the MON fault flags (generated by the health-monitoring algorithms and communicated to the FCC) and on the mission phase (received from the FCC), the CON modules can be switched to FMM, FTM, HSB or CSB modes.

Table 1. FEPS operation modes as functions of mission phases and detected faults.

Mission Phase	MON1 Fault Flag	MON2 Fault Flag	CON Mode (CON1/CON2)	FEPS Status (Stator 1/Stator 2)
Climb	off	off	FMM/FMM	Normal operation (active/active)
	off	on	FMM/CSB	Fail-operative (active/passive)
	on	off	CSB/FMM	Fail-operative (passive/active)
Cruise, Loiter, Descent	off	off	HSB/FMM	Normal operation (stand-by/active)
	off	on	FMM/CSB	Fail-operative (active/passive)
Flight termination/Landing	off	off	HSB/FTM	Normal operation (stand-by/active)
	off	on	FTM/CSB	Fail-operative (active/passive)

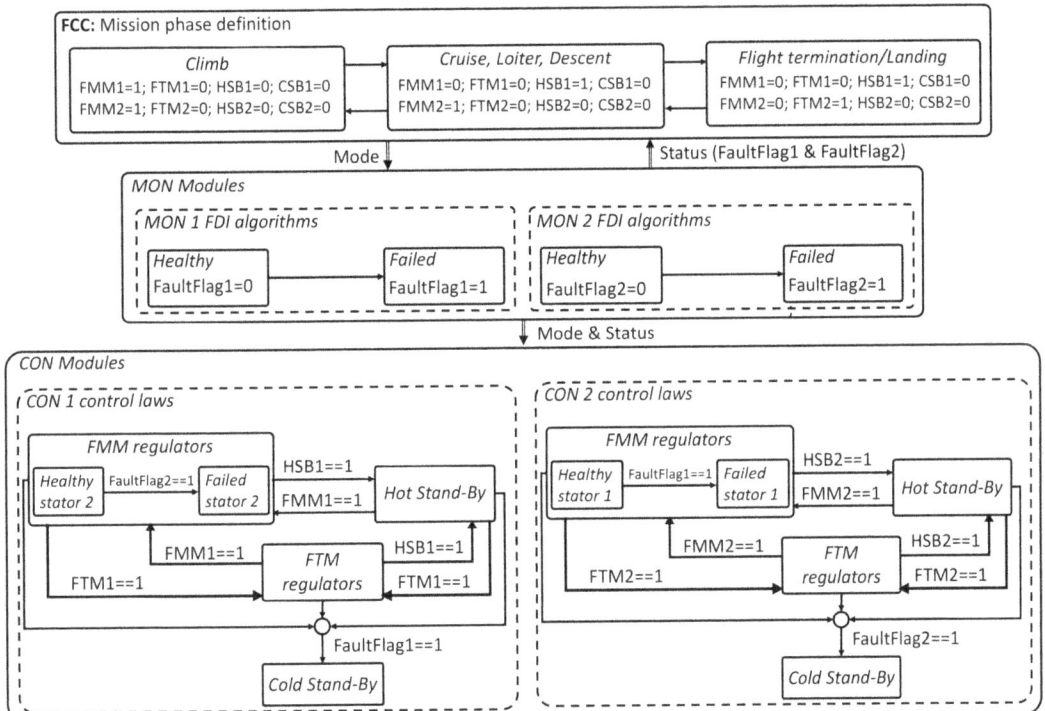

Figure 3. Flow chart defining FEPS operation modes.

2.2. Mechanical Transmission and Propeller Loads Modelling

The dynamics of the aero-mechanical section of the FEPS, providing the UAV with the thrust T_p, is schematically depicted in Figure 4a and is modelled by [8,12]:

$$\begin{cases} J_p \ddot{\theta}_p = -Q_p - C_{gb}\left(\dot{\theta}_p - \dot{\theta}_m\right) - K_{gb}\left(\theta_p - \theta_m\right) + Q_d \\ J_m \ddot{\theta}_m = Q_m + C_{gb}\left(\dot{\theta}_p - \dot{\theta}_m\right) + K_{gb}\left(\theta_p - \theta_m\right) + Q_c ' \\ Q_c = Q_{cmax} sin(n_h n_d \theta_m) \end{cases} \quad (1)$$

where J_p and θ_p, J_m and θ_m are the inertias and angles of the propeller and the motor shafts, respectively, Q_p is the propeller torque, Q_d is a gust-induced disturbance torque, Q_m is the motor torque, Q_c is the cogging torque and Q_{cmax} is its maximum amplitude, n_d is the pole pairs number, n_h is the harmonic index of the cogging disturbance, while K_{gb} and C_{gb} are the stiffness and the damping of the mechanical coupling joint.

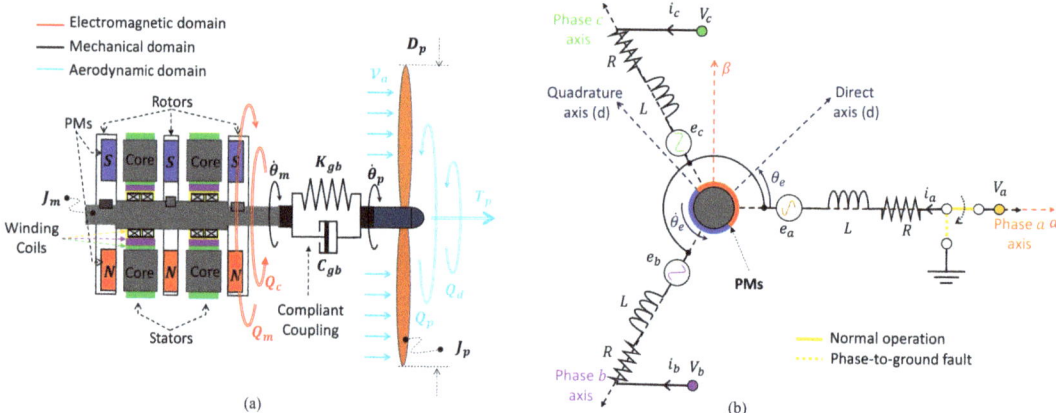

Figure 4. FEPS: (**a**) mechanical scheme; (**b**) equivalent three-phase PMSM scheme (one pole pair).

Concerning the aerodynamic torque, the FEPS is equipped with the twin-blade fixed-pitch composite propeller APC22 × 10E, which is characterized by the following thrust and torque expressions:

$$T_p = C_{T_p}\left(\dot{\theta}_p, AR\right)\rho D_p^4 \dot{\theta}_p^2, \qquad (2)$$

$$Q_p = C_{Q_p}\left(\dot{\theta}_p, AR\right)\rho D_p^5 \dot{\theta}_p^2, \qquad (3)$$

$$AR = \mathcal{V}_a / D_p \dot{\theta}_p, \qquad (4)$$

where C_{T_p} and C_{Q_p} are the nondimensional thrust and torque coefficients, AR is the propeller advance ratio, D_p is the propeller diameter, ρ is the air density, and $\mathcal{V}_a$ is the UAV forward speed.

It is worth noting that the manufacturer database provides the nondimensional coefficients C_{T_p} and C_{Q_p} only for $AR < 0.65$ [45]. This range adequately covers the FEPS operating conditions in FMM (Table 1), but it is not adequate for the FTM, where AR theoretically tends to infinite (because the propeller stops rotating), so that a loads model extension was carried out. This was carried out via Equation (5), by linearly extrapolating the coefficient trends at $AR^* = 0.65$ (Figure 5), with an approach that typically provides conservative estimates [46,47].

$$C_{X_p} = \begin{cases} C_{X_p}^{(DB)}\left(\dot{\theta}_p, AR\right) & AR \leq AR^* \\ C_{X_p}^{(DB)}\left(\dot{\theta}_p, AR^*\right) + \left.\dfrac{\partial C_{X_p}^{(DB)}}{\partial AR}\right|_{\dot{\theta}_p, AR^*} (AR - AR^*) & AR > AR^* \end{cases}, \qquad (5)$$

In Equation (5), C_{X_p} represents the thrust ($X = T$) or torque ($X = Q$) propeller coefficient, while $C_{X_p}^{(DB)}$ is the related quantity given in the manufacturer database.

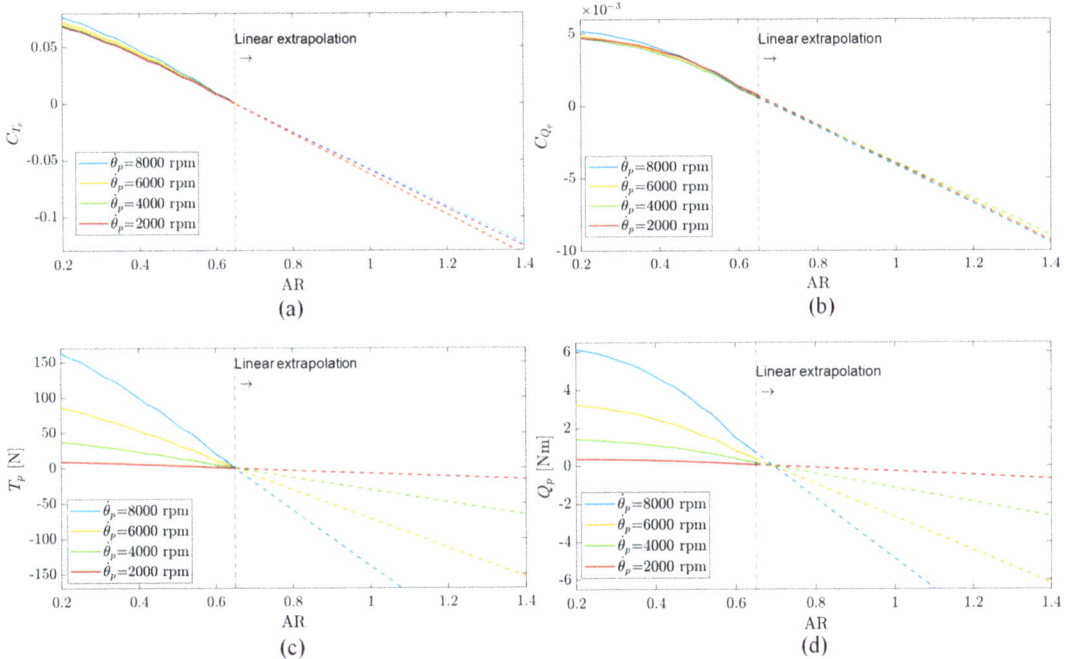

Figure 5. APC 22 × 10E propeller curves: thrust coefficient (**a**) and torque coefficient (**b**), thrust (**c**) and torque (**d**) at sea level.

2.3. Three-Phase PMSM Modelling

Apart from the architectural dissimilarity from a conventional radial-flux PMSM, the mathematical modelling of a AFPMSM is essentially identical [48,49]. With reference to the schematics in Figure 4b, and under the following assumptions [13]:

- Negligible magnetic nonlinearities of ferromagnetic parts (i.e., hysteresis, saturation);
- Each stator–rotor module is magnetically symmetric with reference to phases;
- Permanent magnets are surface-mounted, are made of rare-earth materials, and the magnet reluctance along the quadrature axis is infinite with respect to the one along the direct axis;
- Negligible magnetic coupling among phases;
- Negligible magnetic flux dispersions (i.e., secondary paths, iron losses).

The current dynamics can be described in vectorized form by [8]:

$$V_{abc} = R i_{abc} + L \frac{d}{dt} i_{abc} + e_{abc}, \tag{6}$$

$$e_{abc} = \lambda_m n_d \dot{\theta}_m \left[\sin(n_d \theta_m), \sin\left(n_d \theta_m - \frac{2}{3}\pi\right), \sin\left(n_d \theta_m + \frac{2}{3}\pi\right) \right]^T, \tag{7}$$

In Equations (5)–(6), $V_{abc} = [V_a - V_n, V_b - V_n, V_c - V_n]^T$ is the applied voltages vector, $i_{abc} = [i_a, i_b, i_c]^T$ is the stator currents vector, e_{abc} is the back-electromotive forces vector, R and L are the resistance and inductance of the phases, λ_m is the magnet flux linkage, and V_n is the neutral point voltage. The motor torque (Q_m) is thus given by:

$$Q_m = \lambda_m n_d \left[i_a \sin(n_d \theta_m) + i_b \sin\left(n_d \theta_m - \frac{2}{3}\pi\right) + i_c \sin\left(n_d \theta_m + \frac{2}{3}\pi\right) \right] \tag{8}$$

More conveniently, the motor torque can be expressed with reference to quantities in the rotating reference frame, by applying the Clarke–Park transformations [13], so that:

$$i_{\alpha\beta\gamma} = T_C i_{abc} = \sqrt{\frac{2}{3}} \begin{bmatrix} 1 & -1/2 & -1/2 \\ 0 & \sqrt{3}/2 & -\sqrt{3}/2 \\ \sqrt{2}/2 & \sqrt{2}/2 & \sqrt{2}/2 \end{bmatrix} i_{abc}, \quad (9)$$

$$i_{dqz} = T_P i_{\alpha\beta\gamma} = \begin{bmatrix} \cos(n_d\theta_m) & \sin(n_d\theta_m) & 0 \\ -\sin(n_d\theta_m) & \cos(n_d\theta_m) & 0 \\ 0 & 0 & 1 \end{bmatrix} i_{\alpha\beta\gamma}, \quad (10)$$

$$i_{dqz} = T_P T_C i_{abc} = \sqrt{\frac{2}{3}} \begin{bmatrix} \cos(n_d\theta_m) & \cos\left(n_d\theta_m - \frac{2\pi}{3}\right) & \cos\left(n_d\theta_m + \frac{2\pi}{3}\right) \\ -\sin(n_d\theta_m) & -\sin\left(n_d\theta_m - \frac{2\pi}{3}\right) & -\sin\left(n_d\theta_m + \frac{2\pi}{3}\right) \\ \sqrt{2}/2 & \sqrt{2}/2 & \sqrt{2}/2 \end{bmatrix} i_{abc} \quad (11)$$

where $i_{\alpha\beta\gamma}$ and i_{dqz} are the current vectors in the Clark and Clark–Park reference frames, respectively. By using Equation (7), we finally have:

$$Q_m = \sqrt{\frac{3}{2}} \lambda_m n_d i_q = k_t i_q, \quad (12)$$

in which k_t is the motor torque constant.

2.4. Fault-Tolerant Control System Design

The multi-mode closed-loop system of the FEPS, schematically depicted in Figure 6, has been entirely developed as a finite-state machine, by using the Matlab–Simulink–Stateflow tools, with mode switch signals that can be generated by the MON modules or overridden by the commands sent by the FCC. In FMM (Table 1), the CON modules receive the speed setpoint ($\dot{\theta}_m^\#$) from the FCC, while in FTM the angle setpoint ($\theta_m^\#$) is constant and pre-defined.

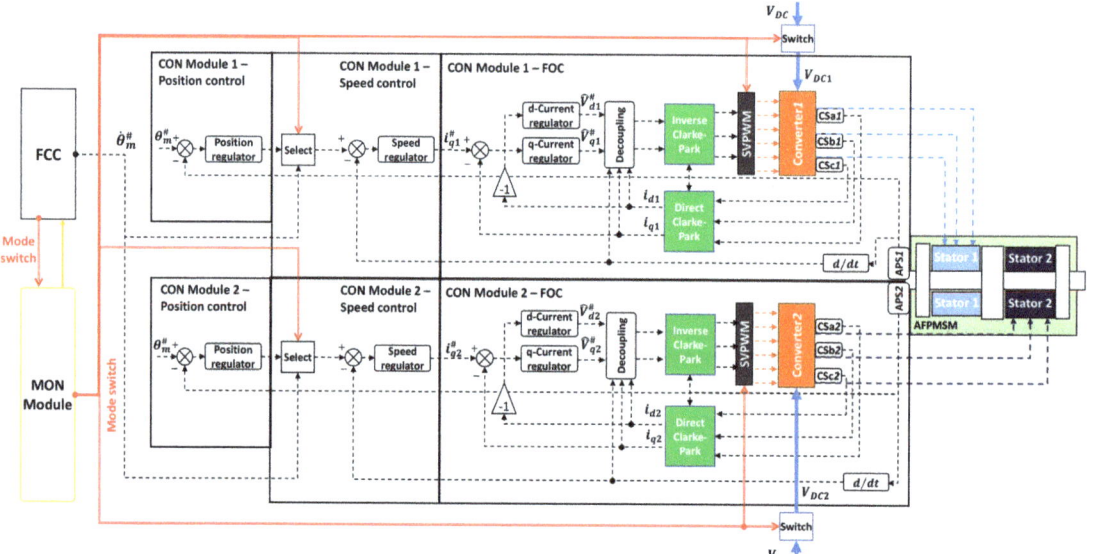

Figure 6. FEPS closed-loop architecture.

In the CON modules, all the regulators implement digital signal processing and apply proportional/integral actions on tracking error signals, plus an anti-windup function with back-calculation technique to compensate for command saturation. In particular, the generic j-th digital regulator (with $j = C, S$ and R, indicating the current, speed and rotation loops, respectively, Table A3) is governed by Equations (12)–(13):

$$y_{PI}^{(j)} = k_P^{(j)} \varepsilon^{(j)} + \frac{k_I^{(j)} T_s^{(j)}}{z-1} \left[\varepsilon^{(j)} + k_{AW}^{(j)} \left(y^{(j)} - y_{PI}^{(j)} \right) \right], \tag{13}$$

$$y^{(j)} = \begin{cases} y_{PI}^{(j)} & \left| y_{PI}^{(j)} \right| < y_{sat}^{(j)} \\ y_{sat}^{(j)} \, sgn\left(y_{PI}^{(j)} \right) & \left| y_{PI}^{(j)} \right| \geq y_{sat}^{(j)} \end{cases}, \tag{14}$$

where z is the discrete-time operator, $\varepsilon^{(j)}$ is the regulator input (tracking error), $y^{(j)}$ is the regulator output, $y_{PI}^{(j)}$ is the saturator block input (proportional–integral with reference to error, if no saturation is present), while $k_P^{(j)}$ and $k_I^{(j)}$ are the proportional and integral gains, $k_{AW}^{(j)}$ is the back-calculation anti-windup gain, $y_{sat}^{(j)}$ is the saturation limit, and $T_s^{(j)}$ is the sampling rate, Table A3.

In the MON modules, a set of monitoring algorithms are real-time executed at 10 kHz sampling rate, to detect and isolate the major FEPS faults (open-phase, shorted-phase, overheating, overcurrent, hardover, jamming, etc.), and to define the correct operation mode of the AFPMSM stators (Table 1). The maximum fault detection latency for all health-monitoring algorithms was set to 250 ms and the FEPS failure transients will be thus characterized in Section 3 with reference to this worst-case scenario.

2.5. UAV Longitudinal Dynamics Modelling

The UAV dynamics is simulated via a reduced-order model, by taking into account the longitudinal phugoid behaviour only (Figure 7). By assuming that:

- The thrust is aligned with the body frame axis (x_B);
- The aerodynamic coefficients related to the wing downwash and to the pitch rate are negligible;
- The elevator deflection (δ_e) continuously implies the pitch equilibrium;
- The angle-of-attack, the path angle and the elevator deflection are small quantities.

The UAV dynamics can be thus described by Equations (14)–(15), [50]:

$$\begin{cases} m_a \dot{V}_a = T_p \cos(\alpha - \alpha_0) - D - m_a g \sin(\gamma) \\ m_a V_a \dot{\gamma} = L - m_a g \cos(\gamma) + T_p \sin(\alpha - \alpha_0), \\ M = 0 \end{cases} \tag{15}$$

where

$$\begin{cases} L = 1/2 \rho S \mathcal{V}^2 C_L = 1/2 \rho S \mathcal{V}^2 (C_{L\alpha} \alpha + C_{L\delta_e} \delta_e) \\ D = 1/2 \rho S \mathcal{V}^2 C_D = 1/2 \rho S \mathcal{V}^2 (C_{D0} + k C_L^2) \\ M = 1/2 \rho S \mathcal{V}^2 \bar{c} (C_{m0} + C_{m\alpha} \alpha + C_{m\delta_e} \delta_e) \end{cases}, \tag{16}$$

In Equations (14)–(15), m_a, $\mathcal{V}_a$, γ, and (α_0) α are the UAV mass, forward speed, path angle, and (zero-lift) angle-of-attack; L, D, and T_p are the UAV lift, drag and thrust (Figure a and Figure); M is the total aerodynamic pitch moment; S is the wing area; $\bar{c}$ is the UAV mean aerodynamic chord; C_{m0} is the base pitch moment coefficient; $C_{m\alpha}$ and $C_{m\delta_e}$ are the pitch moment–slope coefficients; and $C_{L\alpha}$ and $C_{L\delta_e}$ are the lift-slope coefficients, while C_{D0} and k are the zero-lift drag coefficient and the induced drag factor, respectively.

To evaluate the impacts of motor failures at the vehicle level, the system simulation also includes the closed-loop control on the UAV Rate-of-Climb (RoC, Equation (16)), as described by the scheme in Figure 8.

$$RoC = \mathcal{V}_a \sin(\gamma) \tag{17}$$

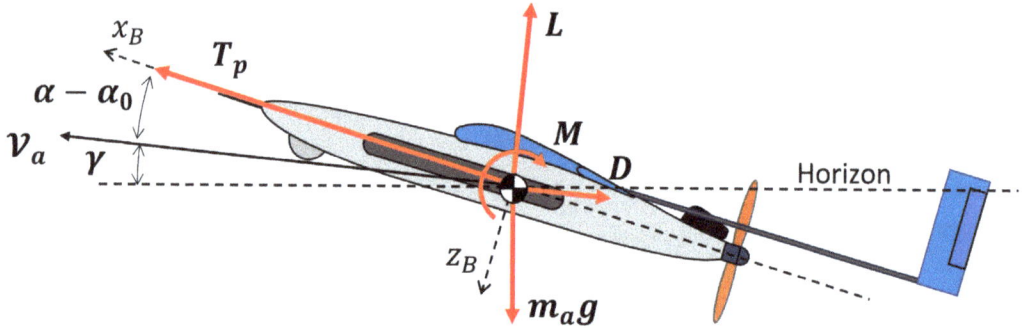

Figure 7. Reference schematics for the UAV longitudinal phugoid dynamics.

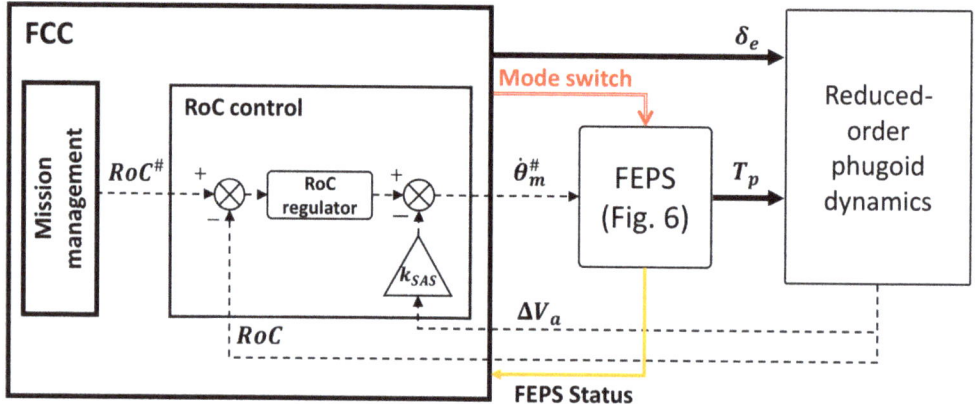

Figure 8. Rate-of-climb closed-loop system.

The RoC regulator receives the setpoint ($RoC^\#$) from the mission management functions and implements a proportional/integral action on the tracking error with back-calculation anti-windup (similarly to FEPS ones, the regulator is governed by Equations (12)–(13), in which j = RoC, Table A3). In addition, a stability augmentation system related to advance speed perturbation (loop with gain k_{SAS} in Figure 8) is also applied.

3. Results

3.1. Simulation Campaign Definition

The performances of the fault-tolerant FEPS were assessed by using a MATLAB/Simulink model of the complete system and numerically solved via the fourth-order Runge–Kutta method, using a 10^{-6} s integration step. It is worth noting that the choice of a fixed-step solver is not strictly related to the objectives of this work (in which the model is used for "off-line" simulations), but it has been selected for the next steps of the project, when the control system will be implemented in the ECU boards via automatic MATLAB compilers and executed in "real-time".

The fault-tolerant capabilities of the FEPS are tested by injecting a phase-to-ground fault in the phase A of the stator 2 (Figure 4b) in different flight manoeuvres, i.e.:

- During climb, in which the MON modules detect the fault and switch the CON modules to operate from FMM/FMM (normal operation) to FMM/CSB (fail-operative);
- During cruise, in which the MON modules detect the fault and switch the CON modules to operate from HSB/FMM (normal operation) to FMM/CSB (fail-operative);

- During flight termination/landing, in which the MON modules firstly detect the fault and switch the CON modules to operate from HSB/FMM (normal operation) to FMM/CSB (fail-operative), and then impose the transition from FMM to FTM on the active stator when the speed is adequately small. All the tests are executed by simulating the following sequence of events:
    - Start ($t = 0$ s): the FEPS works in normal operation (no faults) and drives the propeller at 5800 rpm with the UAV at 26 m/s in level flight at sea altitude;
    - FEPS command ($t = 1$ s), i.e.,
        ○ For climb, the maximum RoC of 3.5 m/s is requested by the FCC;
        ○ For cruise, the propeller speed setpoint is held;
        ○ For flight termination/landing, the propeller speed setpoint is decreased from the cruise value at a -60 rad/s^2 rate;
    - Event 1 (E1, fault injection): a phase-to-ground fault on phase a of stator 2 is imposed;
    - Event 2 (E2, fault detection and isolation): a CSB mode is set on the faulty stator;
    - Event 3 (E3, fault compensation):
        ○ For climb, the current demand for the healthy stator is doubled and the RoC setpoint is reduced to 1 m/s;
        ○ For cruise and flight termination/landing, the healthy stator is activated (250 ms delay is assumed to achieve the full electric supply) and controlled;
    - Event 4 (E4, only for flight termination/landing): the active stator is switched to operate from FMM to FTM.

To permit the evaluation of failure transient impacts on system performances, the results of two simulations will be proposed in Sections 3.2–3.4 by applying or not the system health monitoring, so that a comparison between uncompensated and compensated behaviours is documented.

3.2. Failure Transients in Climb

The simulations can be described with reference to Figure 9: firstly, the tracking performances of the closed-loop control on RoC are assessed, by requesting ($t = 1$ s) the UAV to achieve the maximum-climb rate (3.5 m/s) at 1.5g load factor; secondly, for both simulations, E1 is imposed while the UAV is performing a steady climb ($t = 12$ s). In one of the two simulations, the fault is detected and compensated (E2+E3, at $t = 12.25$ s), and the CON modules switch to operate from FMM/FMM to FMM/CSB.

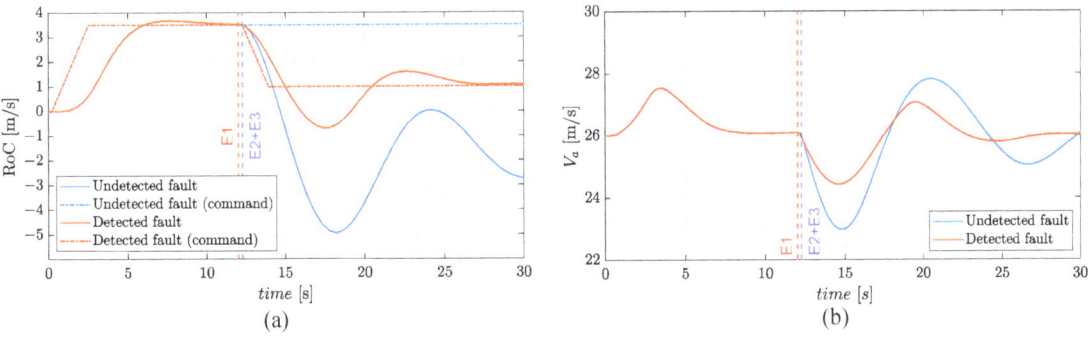

Figure 9. UAV response in climb with phase-to-ground fault on stator 2 (E1, E2 and E3 defined in Section 3.1): RoC (**a**), airspeed (**b**).

In the healthy condition, the RoC tracking is characterized by a rise time of about 5 s and negligible overshoot, Figure 9a. An undetected phase-to-ground fault drastically impacts on performances: without detection, the RoC actually goes below zero, meaning that the UAV follows a descent motion. On the other hand, the fault compensation permits to hold the UAV in climb, even if with reduced performance. If the fault is not detected, the airspeed exhibits relevant oscillations (with about a 12 s period, close to the Lanchester's phugoid prediction, i.e., $\pi\sqrt{2}\mathcal{V}_a/g$, [51]), while it rapidly recovers the cruise value if the health-monitoring is applied, Figure 9b.

The propeller speed is plotted in Figure 10a. In case of undetected fault, the faulty stator brakes down the propeller, by reducing the speed up to about 1000 rpm below the cruise value (5800 rpm), thus resulting in negative RoC, Figure 9a. If the health-monitoring is applied, immediately after the compensation, the output of the SAS block (Figure 8) increases because of the airspeed reduction, Figure 9b, as well as for the diminishing output from the RoC regulator, thus causing an initial increase in the motor demand speed (up to about $t = 16$ s). During this transient period (from 12.25 s to 16 s), the healthy stator operates in saturation condition, as confirmed by the quadrature current output in Figure 10b. It can be also observed that the phase-to-ground fault introduces relevant ripples of current (hence torque) at about 800 Hz frequency, which is twice the motor electrical frequency ($n_d \dot{\theta}_m$). The current peaks reach five times the maximum value for continuous duty cycle operations (I_{sat}), with a mean value of about $-0.5\ I_{sat}$, which produces a braking torque contribution, Figure 10b.

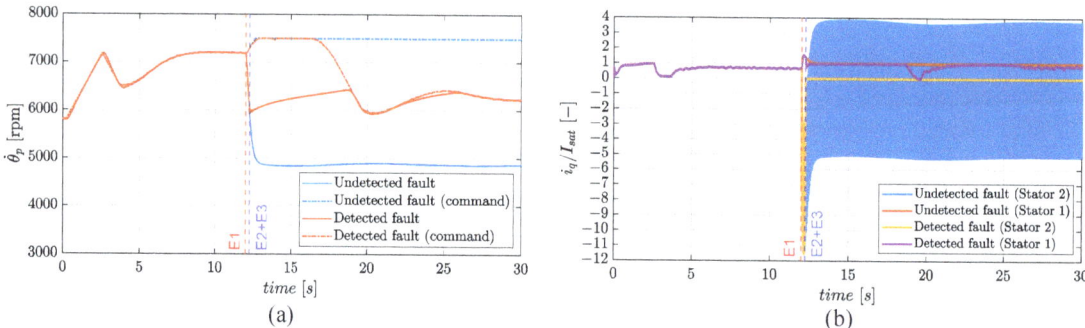

Figure 10. FEPS response in climb with phase-to-ground fault on stator 2 (E1, E2 and E3 defined in Section 3.1): propeller speed (**a**), quadrature current (**b**), with $I_{sat} = 46$ Arms.

In Figure 11, the phase currents and voltages of the two stators for the compensated case are finally shown. The results are also proposed in the time range between the fault injection and compensation to emphasize the detailed dynamic behaviours: differently from the normal operation, the phase currents in the faulty stator (although still balanced, i.e., their sum is null) are not symmetric, Figure 11a. In fact, the current in phase c roughly push-pulls with reference to the one in phase a, while in the phase b the current progressively shifts to be roughly synchronous with it. The loss of current symmetry results from the voltage grounding on pin a (Figure 11b), which implies that the phase a voltage is driven by the neutral point voltage only. As a consequence of the symmetry loss, the Clarke—Park transform on stator 2 is no longer effective, and the direct and quadrature current demands (i_{d2}, i_{q2}) become harmonic quantities, Figure 10b.

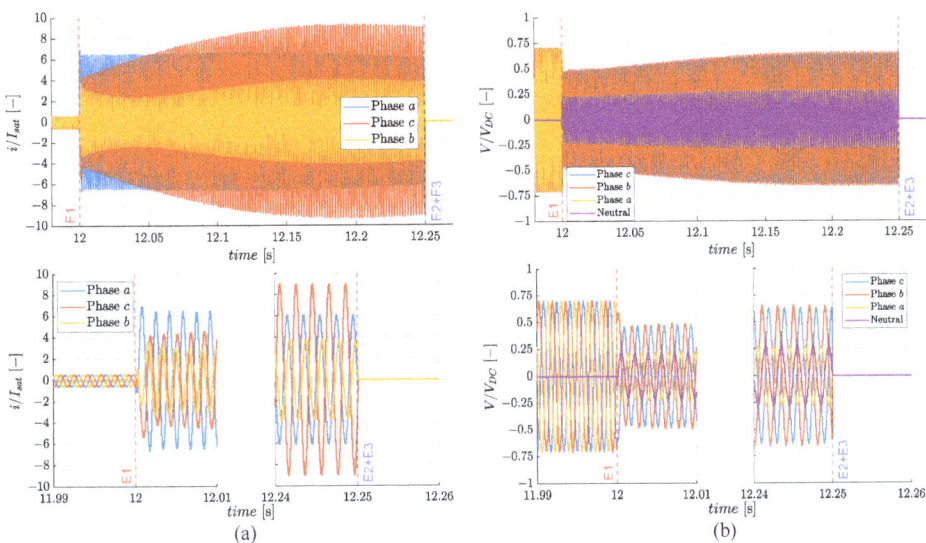

Figure 11. FEPS currents and voltages responses in climb with phase-to-ground fault on stator 2 (E1, E2 and E3 defined in Section 3.1): stator 2 currents (**a**) with $I_{sat} = 46$ Arms; stator 2 voltages (**b**) with $V_{DC} = 36$ V.

3.3. Failure Transients in Cruise

The simulations can be described with reference to Figure 12: for both simulations, E1 is imposed while the UAV is in steady cruise ($t = 1$ s); while in one of them, the fault is firstly detected (E2, at $t = 1.25$ s) and then compensated (E3, at $t = 1.5$ s), so that the CON modules are switched to operate from HSB/FMM to FMM/CSB.

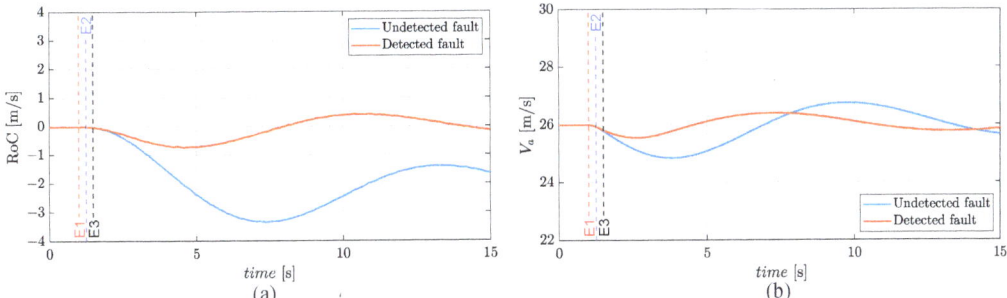

Figure 12. UAV response in cruise with phase-to-ground fault on stator 2 (E1, E2 and E3 defined in Section 3.1): RoC (**a**), airspeed (**b**).

Since in cruise the FEPS operates with one stator only, the undetected fault drastically impacts on UAV response, with the RoC that settles to about -2 m/s, Figure 12a. On the other hand, in the compensated case, the performances are fully restored. In both cases, the airspeed oscillates around the cruise value with the phugoid period, and a maximum deviation of about 1 m/s is observed, Figure 12b.

The propeller speed is plotted in Figure 13a. In the undetected case, the faulty stator brakes down the motor, reducing the speed up to about 1200 rpm below the cruise value (5800 rpm), justifying the negative RoC in Figure 12a. If the health-monitoring is applied, the delay time required for the full electric supply of the stand-by stator (from $t = 1.25$ s to $t = 1.5$ s) implies that resistive aerodynamic loads are applied to the propeller, and the speed rate diminishes, so that, immediately after the fault, the faulty stator acts as a brake.

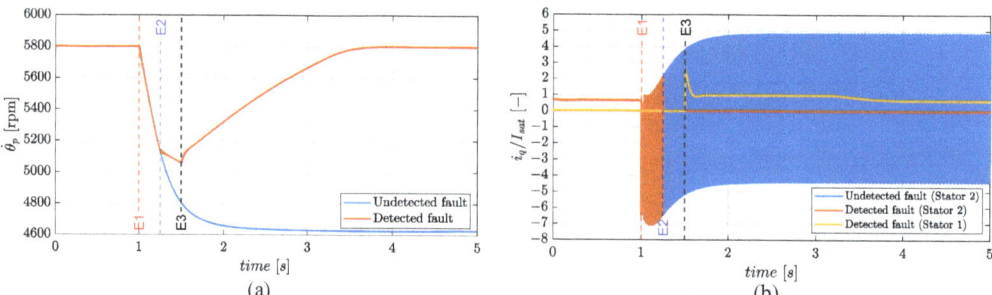

Figure 13. FEPS response in cruise with phase-to-ground fault on stator 2 (E1, E2 and E3 defined in Section 3.1): propeller speed (**a**), quadrature current (**b**), with $I_{sat} = 46$ Arms.

The above discussion is further enforced by observing the response in terms of quadrature currents, Figure 13b. As also highlighted in Section 3.2, the phase-to-ground fault introduces high-frequency ripples, and, immediately after the fault, the mean value of the quadrature current is negative. On the other hand, in steady condition, it becomes positive (0.25 I_{sat}), so that the faulty stator delivers power to the propeller.

Finally, in Figure 14, the phase currents and voltages of the two stators for the compensated case are shown. The results are also proposed in the time range between the fault injection and compensation to emphasize out the detailed dynamic behaviours. The full electric activation of the stator 1 (at $t = 1.5$ s) is characterized by relevant peaks of the phase currents (Figure 14a), up to reach about three times the maximum value in continuous duty cycle operations (I_{sat}). As discussed in Section 3.2, the phase-to-ground fault causes the loss of the currents symmetry while maintaining their balance, Figure 14b.

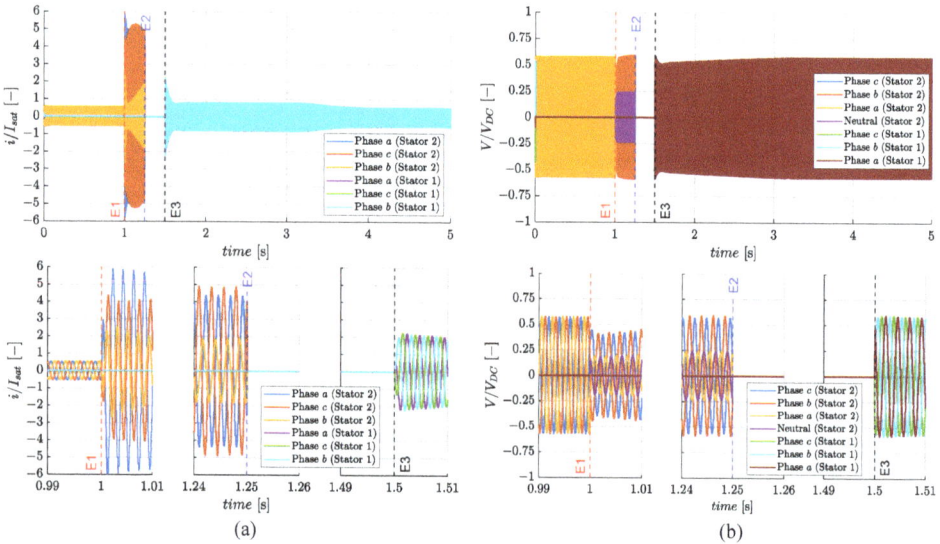

Figure 14. FEPS currents and voltages responses in cruise with phase-to-ground fault on stator 2 (E1, E2 and E3 defined in Section 3.1): stator 2 currents (**a**) with $I_{sat} = 46$ Arms; stator 2 voltages (**b**) with $V_{DC} = 36$ V.

3.4. Failure Transient and Transition from FMM to FTM in Flight Termination/Landing

The simulations can be described with reference to Figure 15: for both simulations, E1 is imposed while the UAV is performing a steady descent motion (t = 4.5 s); while in one of the two ones, the fault is firstly detected (E2, at t = 4.75 s) and then compensated (E3, at t = 5 s), so that the CON modules switch from HSB/FMM to FMM/CSB. Successively, when the speed is adequately small (<1 rad/s), the CON modules switch from HSB/FMM to HSB/FTM in the undetected case (E4a, at t = 10.4 s), and from FMM/CSB to FTM/CSB if the health-monitoring is applied (E4b, at t = 10.7 s).

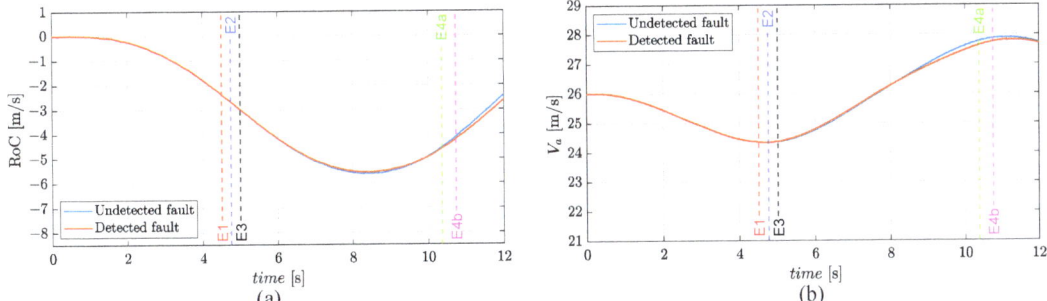

Figure 15. UAV response in flight termination/landing with phase-to-ground fault on stator 2 (E1, E2, E3 and E4 defined in Section 3.1): RoC (**a**), airspeed (**b**).

The UAV behaviour can be interpreted via Equation (12): since the elevator deflection maintains the pitch equilibrium, a reduction in the propeller speed implies a thrust reduction, which causes an oscillatory descendant trajectory, Figure 15a. On the other hand, the airspeed oscillates by following the phugoid behaviour, while keeping its mean value roughly to the one before the fault, Figure 15b.

The closed-loop tracking on propeller speed and position are reported in Figure 16. It is worth noting that the mode transition is executed when the speed is not zeroed yet, to anticipate the parachute opening, and that in both cases the position tracking to the predefined setpoint (180 deg) is correctly accomplished. The quadrature currents response (Figure 17a) also points out that, compared with the climb and cruise simulations, the failure transient for the undetected case impacts on the mechanical transmission too, because the electrical frequency sweeps down, up to equalling the drivetrain resonant frequency (located at 100 Hz), Figure 17b.

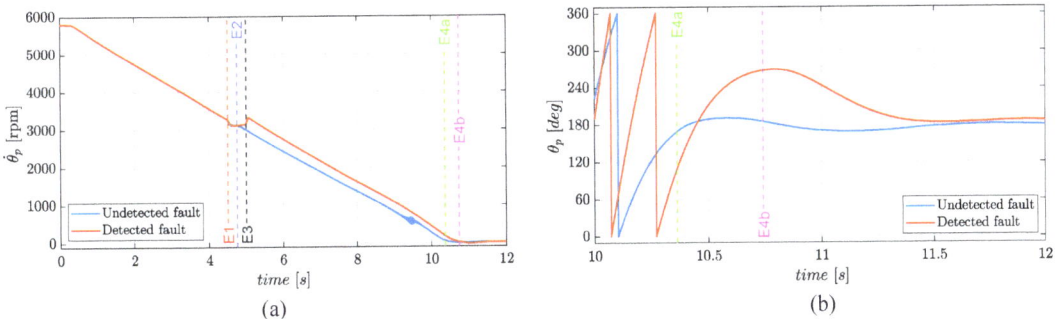

Figure 16. FEPS response in flight termination/landing with phase-to-ground fault on stator 2 (E1, E2, E3 and E4 defined in Section 3.1): propeller speed (**a**), propeller angle (**b**).

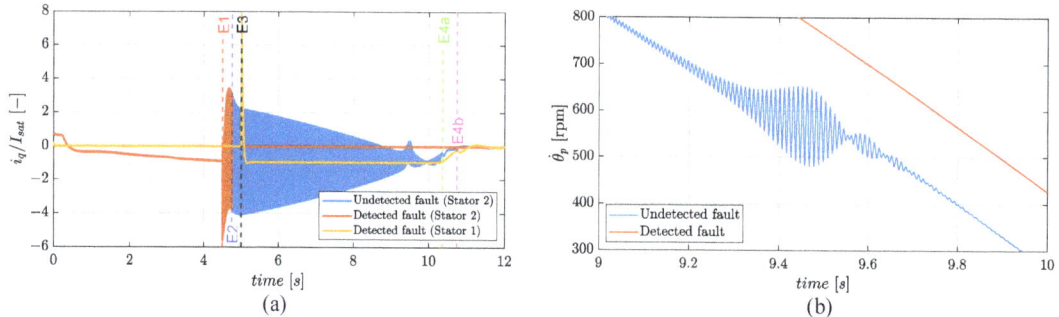

Figure 17. FEPS response in flight termination/landing with phase-to-ground fault on stator 2 (E1, E2, E3 and E4 defined in Section 3.1): quadrature currents (**a**) with $I_{sat} = 46$ Arms, propeller speed (**b**).

In Figure 18, the phase currents and voltages of the two stators for the compensated case are finally reported. The full electric activation of the stator 1 (at $t = 5$ s) is characterized by relevant peaks of the phase currents (Figure 18a), reaching up to about seven times the maximum value for continuous duty cycle (I_{sat}), while the stator 1 currents operate in saturation ($\sqrt{3}/2 I_{sat}$) until the propeller stops (at $t = 10$ s). Similarly to what was discussed in Sections 3.2 and 3.3, the phase-to-ground fault again causes the loss of the currents' symmetry while maintaining their balance (Figure 18a). Finally, it is worth noting that the reduction in voltage amplitudes is coherent with the reduction in the back-electromotive forces caused by the speed decrease (the homopolar voltage component is also represented in Figure 18b).

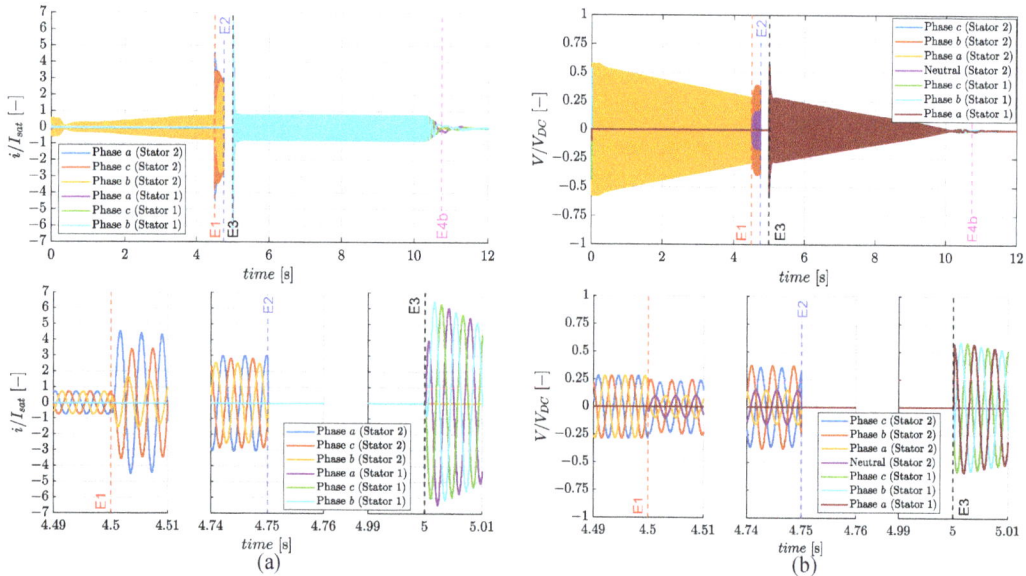

Figure 18. FEPS response in flight termination/landing with phase-to-ground fault on stator 2 (E1, E2, E3 and E4 defined in Section 3.1): stator 2 currents (**a**) with $I_{sat} = 46$ Arms; stator 2 voltages (**b**) with $V_{DC} = 36$ V.

4. Discussion

As confirmed by the results in Section 3, the phase-to-ground fault can determine damages to different parts of the electric machine. In fact, to compensate for the voltage supply lack on the faulty phase, the currents in other phases strongly increase, leading to extremely hot temperatures with consequent deterioration of the magnets. Partial or complete demagnetization may occur, since the magnet coercivity decreases with temperature. Furthermore, another contribution to demagnetization derives from the flux weakening caused by the direct current increase. The direct current in the faulty stator is characterized by high-amplitude oscillations, so the flux linkage on the quadrature axis ($\lambda_q = \lambda_m + Li_d$) can even overcome the magnet coercivity.

Furthermore, the phase-to-ground fault can also impact the structural integrity of system drivetrain (e.g., bearings). In fact, the symmetry loss of the faulty stator currents implies that the magnetic fields generated via the FOC technique are not synchronous with the rotor magnet's motion. The ripple manifests at twice the electrical frequency and potential criticalities can arise if this frequency equals the structural resonance of the drivetrain.

The failure transient's analysis proposed in Section 3 highlights that the phase-to-ground fault can strongly impact performances at the UAV level too. If the fault is not detected, the propeller torque reduces to below the one required for level flight, thus causing the UAV to fall. Furthermore, if one compares the failure transient in climb with the one in cruise, it is worth noting that the larger the speed demand for the faulty stator, the lower its torque output. During climb, the demand after the fault settles at the maximum value (7500 rpm, Figure 10a) and the mean torque to a negative value (Figure 10b); while in cruise, when the speed setpoint is kept at 5800 rpm (Figure 13a), the mean torque settles to a positive value (Figure 13b). Concerning the flight termination/landing, it is demonstrated that, even without fault detection, the FEPS is capable of stopping and aligning the propeller blades with the wing, even if the electrical frequency sweep related to the motor slowdown can generate potentially dangerous vibrations when the drivetrain resonant frequency is intercepted.

The future developments of the research will be focused on:

- System modelling validation, in particular:
  - AFPMSM model, via experimental testing with reference to normal operation (failure transient characterisation will be always simulated, but using updated parameters);
  - Propeller loads model, via CFD simulations, with special focus on the region of $AR > 0.65$ (no data from manufacturer);
  - Mechanical drivetrain model, via experimental testing;
  - UAV longitudinal dynamics, via flight data.
- System modelling enhancement, by including a thermal model of the motor that would permit assessing the effects of overcurrents caused by phase-to-ground fault, which are expected to determine severe overheating;
- Control system implementation in the ECU boards, via automatic MATLAB compilers for the "real-time" execution at a 10 kHz sampling rate.

5. Conclusions

The fault-tolerant control of the FEPS employed by the lightweight fixed-wing UAV named TERSA has been designed and verified in terms of dynamic performances by simulating major electrical faults during relevant flight manoeuvres. The reference FEPS includes a dual-stator AFPMSM operating in active/active mode during climb and in stand-by/active mode in other flight phases, and it is designed as a finite-state machine to switch the closed-loop system from speed-tracking control during the flight, to position-tracking control during flight termination/landing, when the propeller blades are aligned with the wings to avoid interference with the landing parachute opening. Thanks to a high-fidelity modelling approach, the impacts of failure transients from component levels up to UAV levels during climb, cruise and flight termination/landing manoeuvres are evaluated, by simulating a phase-to-ground fault on a phase of one of the two AFPMSM stators. The results show that if the fault is not detected, it strongly impacts on both FEPS and UAV performances, up to causing the UAV to fall, the generation of large-amplitude high-frequency torque and current oscillations, and dangerous interactions with the drivetrain structural resonance. On the other hand, if the health monitoring is applied and the fault is detected and compensated, the FEPS permits maintaining the climb capability (even if it is with reduced performance), holding the UAV altitude and speed during cruise, as well as safely managing the flight termination/landing manoeuvre.

Author Contributions: Conceptualization, methodology and investigation, A.S. and G.D.R.; software, data curation and writing—original draft preparation, A.S.; validation, formal analysis and writing—review and editing, G.D.R.; resources, supervision and visualization, G.D.R. and R.G.; project administration and funding acquisition, R.G. All authors have read and agreed to the published version of the manuscript.

Funding: This research was co-funded by the Italian Government (*Ministero Italiano dello Sviluppo Economico*, MISE) and by the Tuscany Regional Government, in the context of the R&D project "*Tecnologie Elettriche e Radar per SAPR Autonomi* (TERSA)", Grant number: F/130088/01-05/X38.

Institutional Review Board Statement: Not applicable.

Informed Consent Statement: Not applicable.

Data Availability Statement: Not applicable.

Acknowledgments: The authors wish to thank Luca Sani, from the University of Pisa (*Dipartimento di Ingegneria dell'Energia, dei Sistemi, del Territorio e delle Costruzioni*), for the support in the definition of the AFPMSM model parameters, and Ing. Francesco Schettini, from Sky Eye Systems (Italy), for the information about the aero-mechanical data and the general performances of the UAV.

Conflicts of Interest: The authors declare no conflict of interest.

Appendix A

This section contains tables reporting parameters and data related to the propulsion system model (Table A1) and the UAV model (Table A2).

Table A1. FEPS model parameters.

Definition	Symbol	Value	Unit
Stator phase resistance	R	0.025	Ω
Stator phase inductance single module	L	2×10^{-5}	H
Pole pairs number	n_d	5	-
Torque constant (single stator)	k_t	0.06	Nm/Arms
Back-electromotive force constant	k_e	0.018	V/(rad/s)
Permanent magnet flux linkage	λ_m	0.008	Wb
Maximum current (continuous duty cycle)	I_{sat}	46	Arms
Voltage supply	V_{DC}	36	V
Rotor inertia	J_{em}	8.2×10^{-3}	kg·m^2
Propeller diameter	D_p	0.5588	m
Propeller inertia	J_p	1.62×10^{-2}	kg·m^2
Joint stiffness	K_{gb}	1.598×10^3	Nm/rad
Joint damping	C_{gb}	0.2545	Nm/(rad/s)

Table A2. UAV model parameters.

Definition	Symbol	Value	Unit
UAV mass	m_a	35	kg
Air density	ρ	ISA model	kg/m^3
Reference wing area	S	1.058	m^2
Mean aerodynamic chord	$\bar{c}$	0.303	m
Lift–slope coefficient due to AoA	$C_{L\alpha}$	5.74	1/rad
Lift–slope coefficient due to elevator deflection	$C_{L\delta_e}$	0.56	1/rad
Pitch moment-slope coefficient due to AoA	$C_{m\alpha}$,	−1.1	1/rad
Pitch moment-slope coefficient due to elevator deflection	$C_{m\delta_e}$	−2.4	1/rad
Zero-lift pitch moment coefficient	C_{m0}	0.36	-
Zero-lift angle	α_0	3.5	deg
Zero-lift drag coefficient	C_{D0}	0.0491	-
Induced drag factor	k	0.0462	-

Table A3. FEPS and UAV regulators parameters.

Definition	Symbol	Value	Unit
Proportional gain of current regulator	$k_P^{(C)}$	4×10^{-4}	V/A
Integral gain of current regulator	$k_I^{(C)}$	0.8	V/A/s
Anti-windup gain of current regulator	$k_{AW}^{(C)}$	3140	A/V
Saturation limit of current regulator	$y_{sat}^{(C)}$	32	V
Proportional gain of speed regulator	$k_P^{(S)}$	5	A s/rad
Integral gain of speed regulator	$k_I^{(S)}$	15	A/rad
Anti-windup gain of speed regulator	$k_{AW}^{(S)}$	314	A rad/s
Saturation limit of speed regulator	$y_{sat}^{(S)}$	46	Arms
Proportional gain of rotation regulator	$k_P^{(R)}$	1.9	1/s
Integral gain of rotation regulator	$k_I^{(R)}$	0.19	1/s^2
Anti-windup gain of rotation regulator	$k_{AW}^{(R)}$	31.4	s
Saturation limit of rotation regulator	$y_{sat}^{(R)}$	785	rad/s
Proportional gain of RoC regulator	$k_P^{(RoC)}$	65.6	rad/m
Integral gain of RoC regulator	$k_I^{(RoC)}$	1.6	rad s/m
Anti-windup gain of RoC regulator	$k_{AW}^{(RoC)}$	10	m/rad
RoC-loop SAS gain	k_{SAS}	100	rad/m
FEPS sample time	$T_s^{(C)}, T_s^{(S)}, T_s^{(R)}$	10^{-4}	s
FCC sample time	$T_s^{(RoC)}$	10^{-3}	s

References

1. Unmanned Aerial Vehicle (UAV) Market by Point of Sale, Systems, Platform (Civil & Commercial, and Defense & Government), Function, End Use, Application, Type, Mode of Operation, MTOW, Range, and Region—Global Forecast to 2026. Research and Markets: Northbrook, IL, USA. 2021. Available online: https://www.marketsandmarkets.com/Market-Reports/unmanned-aerial-vehicles-uav-market-662.html?gclid=CjwKCAjwrqqSBhBbEiwAlQeqGiTZXREsmSFhSPuOaugcttjxTwo4L3KX7xYtWCycLbFVNWB6YuJIDhoC7ksQAvD_BwE (accessed on 4 April 2022).
2. Schlachter, F. Has the Battery Bubble Burst? *Am. Phys. Soc.* **2012**, *21*, 8. Available online: https://www.aps.org/publications/apsnews/201208/backpage.cfm (accessed on 2 April 2022).
3. Suti, A.; Di Rito, G.; Galatolo, R. Climbing performance enhancement of small fixed-wing UAVs via hybrid electric propulsion. In Proceedings of the 2021 IEEE Workshop on Electrical Machines Design, Control and Diagnosis (WEMDCD), Modena, Italy, 8–9 April 2021. [CrossRef]
4. Power, Electric Motor Market. Fortune Business Insights. Available online: https://www.fortunebusinessinsights.com/industry-reports/electric-motor-market-100752 (accessed on 4 April 2022).
5. Dipartimento di Ingegneria Civile e Industriale, Progetti istituzionali. TERSA (Tecnologie Elettriche e Radar per Sistemi aeromobili a pilotaggio remoto Autonomi). Available online: https://dici.unipi.it/ricerca/progetti-finanziati/tersa/ (accessed on 7 April 2022).
6. Mamen, A.; Supatti, U. A survey of hybrid energy storage systems applied for intermittent renewable energy systems. In Proceedings of the 2017 14th International Conference on Electrical Engineering/Electronics, Computer, Telecommunications and Information Technology (ECTI-CON), Phuket, Thailand, 27–30 June 2017. [CrossRef]
7. Cao, W.; Mecrow, B.; Atkinson, G.; Bennett, J.; Atkinson, D. Overview of Electric Motor Technologies Used for More Electric Aircraft (MEA). *IEEE Trans. Ind. Electron.* **2012**, *59*, 3523–3531. [CrossRef]
8. Suti, A.; Di Rito, G.; Galatolo, R. Fault-Tolerant Control of a Three-Phase Permanent Magnet Synchronous Motor for Lightweight UAV Propellers via Central Point Drive. *Actuators* **2021**, *10*, 253. [CrossRef]

9. NATO Standardization Agency. *STANAG 4671—Standardization Agreement—Unmanned Aerial Vehicles Systems Airworthiness Requirements (USAR)*; NATO Standardization Agency (STANAG): Brussels, Belgium, 2009.
10. Ryu, H.M.; Kim, J.W.; Sul, S.K. Synchronous-frame current control of multiphase synchronous motor under asymmetric fault condition due to open phases. *IEEE Trans. Ind. Appl.* **2006**, *42*, 1062–1070. [CrossRef]
11. Liu, G.; Lin, Z.; Zhao, W.; Chen, Q.; Xu, G. Third Harmonic Current Injection in Fault-Tolerant Five-Phase Permanent-Magnet Motor Drive. *IEEE Trans. Power Electron.* **2018**, *33*, 6970–6979. [CrossRef]
12. Bennet, J.; Mecrow, B.; Atkinson, D.; Atkinson, G. Safety-critical design of electromechanical actuation systems in commercial aircraft. *IET Electr. Power Appl.* **2011**, *5*, 37–47. [CrossRef]
13. Mazzoleni, M.; Di Rito, G.; Previdi, F. Fault Diagnosis and Condition Monitoring Approaches. In *Electro-Mechanical Actuators for the More Electric Aircraft*; Springer: Cham, Switzerland, 2021; pp. 87–117.
14. De Rossiter Correa, M.; Jacobina, C.; Da Silva, E.; Lima, A. An induction motor drive system with improved fault tolerance. *IEEE Trans. Ind. Appl.* **2001**, *37*, 873–879. [CrossRef]
15. Ribeiro, R.; Jacobina, C.; Lima, A.; da Silva, E. A strategy for improving reliability of motor drive systems using a four-leg three-phase converter. In Proceedings of the APEC 2001, Sixteenth Annual IEEE Applied Power Electronics Conference and Exposition (Cat. No. 01CH37181), Anaheim, CA, USA, 4–8 March 2001. [CrossRef]
16. Zhang, R.; Prasad, V.H.; Boroyevich, D.; Lee, F.J. Three-Dimensional Space Vector Modulation for Four-Leg Voltage-Cource Converter. *IEEE Trans. Power Electron.* **2002**, *17*, 314–326. [CrossRef]
17. Kontarcek, A.; Bajec, P.; Nemec, M.; Ambrožic, V.; Nedeljkovic, D. Cost-Effective Three-Phase PMSM Drive Tolerant to Open-Phase Fault. *IEEE Trans. Ind. Electron.* **2015**, *62*, 6708–6718. [CrossRef]
18. Bonnet, A.; Soukup, G. Cause and analysis of stator and rotor failures in three-phase squirrel-cage induction motors. *IEEE Trans. Ind. Appl.* **1992**, *28*, 921–937. [CrossRef]
19. Khalaief, A.; Boussank, M.; Gossa, M. Open phase faults detection in PMSM drives based on current signature analysis. In Proceedings of the XIX International Conference on Electrical Machines (ICEM 2010), Roma, Italy, 6–8 September 2010. [CrossRef]
20. Li, W.; Tang, H.; Luo, S.; Yan, X.; Wu, Z. Comparative analysis of the operating performance, magnetic field, and temperature rise of the three-phase permanent magnet synchronous motor with or without fault-tolerant control under single-phase open-circuit fault. *IET Electr. Power Appl.* **2021**, *15*, 861–872. [CrossRef]
21. Zhou, X.; Sun, J.; Li, H.; Song, X. High Performance Three-Phase PMSM Open-Phase Fault-Tolerant Method Based on Reference Frame Transformation. *IEEE Trans. Ind. Electron.* **2019**, *66*, 7571–7580. [CrossRef]
22. Faiz, J.; Nejadi-Koti, H.; Valipour, Z. Comprehensive review on inter-turn fault indexes in permanent magnet motors. *IET Electr. Power Appl.* **2017**, *11*, 142–156. [CrossRef]
23. Krzysztofiak, M.; Skowron, M.; Orlowska-Kowalska, T. Analysis of the Impact of Stator Inter-Turn Short Circuits on PMSM Drive with Scalar and Vector Control. *Energies* **2021**, *14*, 153. [CrossRef]
24. Arabaci, H.; Bilgin, O. The Detection of Rotor Faults By Using Short Time Fourier Transform. In Proceedings of the 2007 IEEE 15th Signal Processing and Communications Applications, Eskisehir, Turkey, 11–13 June 2007. [CrossRef]
25. Mohammed, O.A.; Liu, Z.; Liu, S.; Abed, N.Y. Internal Short Circuit Fault Diagnosis for PM Machines Using FE-Based Phase Variable Model and Wavelets Analysis. *IEEE Trans. Magn.* **2007**, *43*, 1729–1732. [CrossRef]
26. Baggu, M.M.; Chowdhury, B.H. Implementation of a Converter in Sequence Domain to Counter Voltage Imbalances. In Proceedings of the 2007 IEEE Power Engineering Society General Meeting, Tampa, FL, USA, 24–28 June 2007. [CrossRef]
27. Blánquez, F.R.; Aranda, M.; Rebollo, E.; Blázquez, F.; Platero, C.A. New Fault-Resistance Estimation Algorithm for Rotor-Winding Ground-Fault Online Location in Synchronous Machines With Static Excitation. *IEEE Trans. Ind. Electron.* **2015**, *62*, 1901–1911. [CrossRef]
28. Tan, R.H.; Ramachandaramurthy, V.K. A Comprehensive Modeling and Simulation of Power Quality Disturbances Using MATLAB/SIMULINK. In *Power Quality Issues in Distributed Generation*; Luszcz, J., Ed.; IntechOpen: London, UK, 2015. [CrossRef]
29. Pietrzak, P.; Wolkiewicz, M. On-line Detection and Classification of PMSM Stator Winding Faults Based on Stator Current Symmetrical Components Analysis and the KNN Algorithm. *Electronics* **2021**, *10*, 1786. [CrossRef]
30. Fourlas, G.K.; Karras, G.C. A Survey on Fault Diagnosis and Fault-Tolerant Control Methods for Unmanned Aerial Vehicles. *Machines* **2021**, *9*, 197. [CrossRef]
31. Freeman, P.; Pandita, R.; Srivastava, N.; Balas, G.J. Model-Based and Data-Driven Fault Detection Performance for Small UAV. *IEEE/ASME Trans. Mechatron.* **2013**, *18*, 1300–1309. [CrossRef]
32. Odendaal, H.M.; Jones, T. Actuator fault detection and isolation: An optimised parity space approach. *Control. Eng. Pract.* **2014**, *26*, 222–232. [CrossRef]
33. Cao, D.; Fu, J.; Li, Y. Fault diagnosis of actuator of Flight Control System based on analytic model (IEEE CGNCC). In Proceedings of the 2016 IEEE Chinese Guidance, Navigation and Control Conference (CGNCC), Nanjing, China, 12–14 August 2016. [CrossRef]
34. Abbaspour, A.; Yen, K.k.; Forouzannezhad, P.; Sargolzaei, A. A Neural Adaptive Approach for Active Fault-Tolerant Control Design in UAV. *IEEE Trans. Syst. Man Cybern. Syst.* **2020**, *50*, 3401–3411. [CrossRef]
35. Yin, L.; Liu, J.; Yang, P. Interval Observer-based Fault Detection for UAVs Formation with Actuator Faults. In Proceedings of the 2019 CAA Symposium on Fault Detection, Supervision and Safety for Technical Processes (SAFEPROCESS), Xiamen, China, 5–7 July 2019. [CrossRef]

36. Li, D.; Yang, P.; Liu, Z.; Liu, J. Fault Diagnosis for Distributed UAVs Formation Based on Unknown Input Observer. In Proceedings of the 2019 Chinese Control Conference (CCC), Guangzhou, China, 27–30 July 2019. [CrossRef]
37. Yu, Z.; Liu, Z.; Zhang, Y.; Qu, Y.; Su, C.Y. Distributed Finite-Time Fault-Tolerant Containment Control for Multiple Unmanned Aerial Vehicles. *IEEE Trans. Neural Netw. Learn. Syst.* **2020**, *31*, 2077–2091. [CrossRef]
38. Zogopoulos-Papaliakos, G.; Karras, G.C.; Kyriakopoulos, K.J. A Fault-Tolerant Control Scheme for Fixed-Wing UAVs with Flight Envelope Awareness. *J. Intell. Robot. Syst.* **2021**, *102*, 46. [CrossRef]
39. Haaland, O.M.; Wenz, A.W.; Gryte, K.; Hann, R.; Johansen, T.A. Detection and Isolation of Propeller Icing and Electric Propulsion System Faults in Fixed-Wing UAVs. In Proceedings of the 2021 International Conference on Unmanned Aircraft Systems (ICUAS), Athens, Greece, 15–18 June 2021. [CrossRef]
40. Huang, R.; Liu, C.; Song, Z.; Zhao, H. Design and Analysis of a Novel Axial-Radial Flux Permanent Magnet Machine with Halbach-Array Permanent Magnets. *Energies* **2021**, *14*, 3639. [CrossRef]
41. Mahmoudi, A.; Rahim, N.A.; Hew, W.P. Axial-flux permanent-magnet machine modeling, design, simulation and analysis. *Sci. Res. Essays* **2011**, *6*, 2525–2549. [CrossRef]
42. Zhao, J.; Han, Q.; Dai, Y.; Hua, M. Study on the Electromagnetic Design and Analysis of Axial Flux Permanent Magnet Synchronous Motors for Electric Vehicles. *Energies* **2019**, *12*, 3451. [CrossRef]
43. Kahourzade, S.; Mahmoudi, A.; Ping, H.W.; Mahmoudi, N.U. A Comprehensive Review of Axial-Flux Permanent-Magnet Machines. *Can. J. Electr. Comput. Eng.* **2014**, *37*, 19–33. [CrossRef]
44. Wang, Z.; Chen, J.; Cheng, M. Fault tolerant control of double-stator-winding PMSM for open phase operation based on asymmetric current injection. In Proceedings of the 17th International Conference on Electrical Machines and Systems (ICEMS), Hangzhou, China, 22–25 October 2014. [CrossRef]
45. APC Propellers TECHNICAL INFO. Available online: https://www.apcprop.com/technical-information/performance-data/ (accessed on 2 May 2021).
46. Gong, A.; Verstraete, D. Evaluation of a hybrid fuel-cell based propulsion system with a hardware-in-the-loop flight simulator. In Proceedings of the ISABE, Manchester, UK, 3–8 September 2017. Available online: https://www.researchgate.net/publication/320998946_Evaluation_of_a_hybrid_fuel-cell_based_propulsion_system_with_a_hardware-in-the-loop_flight_simulator (accessed on 28 March 2022).
47. Pivano, L.; Smogeli, Ø.N.; Fossen, T.I.; Johansen, T.A. Experimental Validation of a marine propeller thrust estimation scheme. *Nor. Soc. Autom. Control* **2007**, *28*, 105–112. [CrossRef]
48. Darba, A.; Esmalifalak, M.; Barazandeh, E.S. Implementing SVPWM technique to axial flux permanent magnet synchronous motor drive with internal model current controller. In Proceedings of the 2010 4th International Power Engineering and Optimization Conference (PEOCO), Shah Alam, Malaysia, 23–24 June 2010. [CrossRef]
49. Sabah, N.; Humod, A.T.; Hasan, F.A. Field Oriented Control of the Axial Flux PMSM Based On Multi-Objective Particle Swarm Optimization. *Technol. Rep. Kansai Univ.* **2020**, *62*, 1493. Available online: https://www.kansaiuniversityreports.com/article/field-oriented-control-of-the-axial-flux-pmsm-based-on-multi-objective-particle-swarm-optimization (accessed on 1 April 2022).
50. Rosario-Gabriel, I.; Cortés, R.H. Aircraft Longitudinal Control based on the Lanchester's Phugoid Dynamics Model. In Proceedings of the 2018 International Conference on Unmanned Aircraft Systems (ICUAS), Dallas, TX, USA, 12–15 June 2018. [CrossRef]
51. Lanchester, F.W. *Aerial Flight: Part 2, Aerodonetics*; A Constable: London, UK, 1908.

Design and Simulation Analysis of an Electromagnetic Damper for Reducing Shimmy in Electrically Actuated Nose Wheel Steering Systems

Chenfei She [1], Ming Zhang [1,*], Yibo Ge [2], Liming Tang [1], Haifeng Yin [2] and Gang Peng [2]

1. State Key Laboratory of Mechanics and Control of Mechanical Structures, College of Aerospace Engineering, Nanjing University of Aeronautics and Astronautics, Nanjing 210016, China; chfshe@nuaa.edu.cn (C.S.); tlmnuaa@nuaa.edu.cn (L.T.)
2. Nanjing Engineering Institute of Aircraft Systems, Jincheng Corporation, AVIC (NEIAS), Nanjing 211106, China; geyb@neias.cn (Y.G.); yinhf@neias.cn (H.Y.); pengg@neias.cn (G.P.)
* Correspondence: zhm6196@nuaa.edu.cn; Tel.: +86-025-8489-2384

Abstract: Based on the technical platform of electrically actuated nose wheel steering systems, a new type of damping shimmy reduction technology is developed to break through the limitations of traditional hydraulic damping shimmy reduction methods, and an electrically actuated nose wheel steering structure scheme is proposed. The mathematical model of the electromagnetic damper is established, the derivation of skin depth, damping torque and damping coefficient is completed, and the design of the shape and size of the electromagnetic damper is combined with the derivation results and the technical index of shimmy reduction. The electromagnetic field finite element simulation results show that the mathematical modeling method of the electromagnetic damper has good accuracy, and its application to the shimmy reduction module of the electrically actuated nose wheel steering system is also feasible and superior. Finally, the key factors influencing the performance of electromagnetic damper shimmy reduction are studied and analyzed, thus forming a complete electromagnetic damper shimmy reduction technology for the electrically actuated system, and laying the foundation for the design of novel all-electric aircraft and landing gear.

Keywords: electrically actuated nose wheel steering; all-electric aircraft; electromagnetic damper; electromagnetic simulation; landing gear shimmy reduction

1. Introduction

The concept of more-electric aircraft or even all-electric aircraft has been proposed and rapidly developed in order to improve energy efficiency, cut operating costs and reduce the take-off weight. In the process of progressive replacement of mechanical, hydraulic or pneumatic power sources by electromechanical actuators (EMA) [1], the validation of electromechanical nose wheel steering mechanisms has already started worldwide. In 2009, a project called Distributed and Redundant Electro-mechanical nose wheel Steering System (DRESS), jointly completed by several European aviation industry units, predicted that electromechanical integration may significantly improve the reliability and availability of nose wheel steering mechanisms [2]. In 2010, Bennett applied a fault-tolerant electromechanical actuator to the aircraft nose wheel steering system and theoretically investigated and experimentally validated it [3]. In 2010, Liao et al. studied the all-electric nose wheel steering system and introduced a design method for an electric turning control system for small aircraft landing gear based on DSP, which was experimentally verified to have the advantages of miniaturization, reliability and control flexibility [4].

Landing gear shimmy is a kind of self-excited vibration dominated by the oscillation of the wheels [5]. Understanding the shimmy phenomenon and proposing reasonable shimmy reduction measures have been a pressing problem for experts in aircraft structural dynamics, all-electric aircraft are no exception. Currently, the most effective measure to cope with the phenomenon of aircraft landing gear shimmy is the installation of a shimmy damper [6]. Dampers enable most aircraft to eliminate shimmy, or to improve shimmy problems. Most modern aircrafts use oleo dampers, except for some small aircraft and helicopters with dry friction dampers on the tail landing gear. In recent years, the rapid development of magnetorheological dampers has led to the research on landing cushioning as well as shimmy control using magnetorheological fluids. In 2021, Luong and Jo et al. mounted magnetorheological dampers on the aircraft landing gear to reduce landing impact. They also designed an intelligent controller based on supervised neural networks and verified the performance and stability of the magnetorheological dampers by crash tests [7,8]. In 2010, Chen et al. used magnetorheological dampers to suppress aircraft landing gear shimmy and designed a semi-active control strategy [9]. In 2019, Zhu et al. optimized the structure of an external magnetorheological damper on the aircraft landing gear and verified its shimmy reduction performance by means of damping characteristics tests [10]. In addition, with the development of electromechanical technology, electromagnetic dampers are proposed to be applied in the electrically actuated nose wheel steering system to realize the shimmy reduction function. Compared with the traditional friction dampers and oleo dampers, electromagnetic dampers have advantages in maintenance, environmental protection, service life, structural principle and so on.

In 2019, Jia et al. proposed an electromagnetic damper design for a soft contact robotic arm joint to cushion and unload the impact of the robotic arm in contact with the grasping target [11]. In 2013, Kou et al. used electromagnetic dampers to simulate the load force to which the motor is subjected during normal operation under laboratory conditions, and this load force could be arbitrarily adjusted within a certain range to detect some technical performance indicators of the motor [12]. In 2009, Ebrahimi designed and developed a new type of electromagnetic damper for active vehicle suspension control systems. Unlike traditional passive and semi-active control system dampers, this damper could not only adjust the damping coefficient but also convert the mechanical energy of vibration into electrical energy for reuse [13]. In 2011, Liu et al. conducted theoretical and experimental studies on electromagnetic dampers applied to auxiliary braking of large vehicles [14].

The abovementioned work studied and optimized only the nose wheel landing gear dampers of modern aircraft, which do not meet the innovative development goals of aircraft electrification, or studied and optimized electromagnetic dampers in engineering fields such as vehicle suspension systems, space docking mechanisms, and braking of high-speed trains. However, few articles have reported on the design and research of electromagnetic dampers for shimmy reduction in electrically actuated nose wheel steering systems. In view of this, this paper takes the nose landing gear of a certain jet fighter as the research object. We establish the mathematical model and 3D model of the electromagnetic damper based on the electrically actuated nose wheel steering system, and verify the feasibility and superiority of the electromagnetic damper applied to the electrically actuated nose wheel steering system by using the finite element simulation method in combination with various factors affecting the shimmy reduction performance.

2. Demand Analysis of Electromagnetic Dampers for Shimmy Reduction Performance

2.1. Operating Principle

From the theorem of electromagnetic induction, it is known that the movement of a wire cutting magnetic induction lines in a magnetic field will produce an induced current, and according to the Lenz's law, the magnetic field of the induced current will impede the movement of the wire in the magnetic field. In addition, due to the resistance of the wire itself, the induced current flowing through the wire will cause the kinetic energy of the wire to be dissipated as heat energy, thus achieving the effect of impeding the movement [15], as shown in Figure 1.

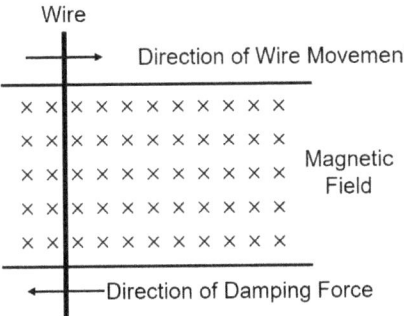

Figure 1. Basic operating principle of an electromagnetic damper.

Electromagnetic dampers are electromagnetic devices based on the abovementioned principles. According to the different excitation methods, electromagnetic dampers can be divided into three types: electrically excited electromagnetic dampers, permanent magnet electromagnetic dampers and hybrid excited electromagnetic dampers [16]. Based to the differences in structure form, electromagnetic dampers can also be divided into rotary electromagnetic dampers, single rotor disk electromagnetic dampers and double rotor disks electromagnetic dampers. Considering the working environment of the nose landing gear and the actual demand of shimmy reduction, the double rotor disks permanent magnet electromagnetic damper is finally selected to eliminate shimmy, and its structure is shown in Figure 2.

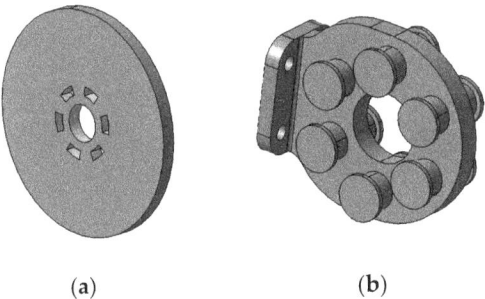

(a) (b)

Figure 2. Structural diagram of the electromagnetic damper: (**a**) Rotor disk; (**b**) Stator disk.

The double rotor disks permanent magnet electromagnetic damper mainly includes the stator disk, rotor disks and permanent magnet poles and other components. Six permanent magnets are fixed on each side of the stator disk and they are distributed along the circumference. The two adjacent poles are arranged at intervals of N and S to form three independent groups of poles. The rotor disks are commonly composed of two circular conductor disks on the top and bottom, which will leave a very small air gap when installed with the magnetic poles. This not only ensures that the relative rotation between the upper and lower rotor disks and the stator does not generate friction, but also allows the performance of the electromagnetic damper to be changed by adjusting the size of the air gap [17].

When the electromagnetic damper starts to operate, the magnetic field generated by the magnetic poles forms a circuit between the stator disk, the air gap and the upper and lower rotor disks. The polarity of the two adjacent magnets on each side is opposite, and the magnitude of the magnetic flux is related to the magnetic inductance strength of the magnetic poles. When the rotor disks rotate, the magnetic flux through the rotor disks changes, thus inducing a swirl-shaped induced current on the rotor disks, also known as induced eddy current. As shown in Figure 3, adjacent eddy currents in opposite directions are generated on the rotor disks during operation, and the corresponding induced magnetic field generates a damping torque in the opposite direction of the rotor disks' rotation, thus impeding the rotation of the rotor disks. From the perspective of energy conversion, the induced currents on the rotor disks transform the kinetic energy of the rotor disks into thermal energy, so as to achieve the effect of shimmy reduction.

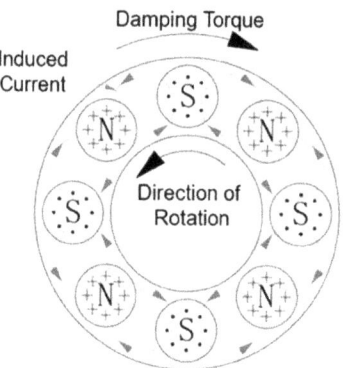

Figure 3. The damping torque and induced current of electromagnetic damper.

Compared with the traditional oleo damper, the electromagnetic damper does not need to consider the need for gas tightness during the assembly process, nor does it need regular inspection of oil quality and quantity during the working process, which greatly reduces the production costs and maintenance costs, reflecting its superiority.

2.2. Analysis of Shimmy Motion Parameters

Combining the engineering practice with the empirical data in the literature [5], the angular amplitude of shimmy is generally around 2 to 20° and the frequency is around 5 to 30 Hz. It shows the characteristics of high frequency at low amplitude and low frequency at high amplitude. Assuming that the nose wheel shimmy amplitude is A (between 2 and 20°) and the frequency is f (between 5 and 30 Hz). The angle of the nose wheel shimmy varies with time in accordance with the sine function relationship:

$$\alpha = A \sin 2\pi f t \tag{1}$$

Then the angular velocity of the nose wheel shimmy is:

$$\omega_{NW} = \frac{d\alpha}{dt} = 2\pi f A \cos 2\pi f t \tag{2}$$

From this, it can be seen that there is a phase difference of 1/4 cycle between the shimmy angle and angular velocity of the nose wheel in line with the above hypothetical law, that is, the angular velocity of the nose wheel is maximum when it passes through the neutral position where the shimmy angle is 0, and the angular velocity is 0 at the position of the maximum shimmy angle (amplitude). The maximum angular velocity is:

$$\omega_{max} = 2\pi f A \tag{3}$$

According to the assumption that 2° shimmy amplitude corresponds to 30 Hz frequency, meawhile, 20° shimmy amplitude corresponds to 5 Hz frequency and the frequency decreases linearly with the increase of shimmy amplitude, the shimmy motion parameters in the above-mentioned range of amplitude and frequency can be deduced as shown in Table 1.

Table 1. Motion parameters of shimmy.

A (°)	F (Hz)	Angular Velocity of Strut (rad/s)
2.0	30.00	6.58
4.0	27.22	11.94
6.0	24.44	16.08
8.0	21.67	19.01
10.0	18.89	20.71
11.8	16.39	21.21
12.0	16.11	21.20
14.0	13.33	20.47
16.0	10.56	18.52
18.0	7.78	15.35
20.0	5.00	10.97

2.3. Damping Requirements for Shimmy Reduction

Combined with the actual needs of the project, the design target of the damping coefficient of shimmy reduction for the research object in this paper is not less than 40 Nms/rad. The output damping coefficient of the electromagnetic damper may be amplified by the transmission mechanism and transmitted to the strut. The amplified damping coefficient should meet the design target under different shimmy motion states described in Table 1. Based on this, the design of the steering structure under the electrically actuated nose wheel steering system and the scheme of the electromagnetic damper are developed in the next sections.

3. Research on Electromagnetic Damping and Shimmy Reduction Scheme of Electrically Actuated Nose Wheel Steering Systems

3.1. Structural Scheme Design of the Electrically Actuated Nose Wheel Steering System

The overall design of the structure of the electrically actuated nose wheel steering system uses a gear-driven steering mechanism, and the electromagnetic damper is reasonably installed to better achieve the shimmy reduction function. As shown in Figure 4, the electrically actuated nose wheel steering system includes the transmission mechanism, power output source and other components. The power of the steering mechanism is provided by a servo motor, while the gear transmission mode is selected for the transmission mechanism. Other components include an electromagnetic damper, angle sensors, load sensors, etc.

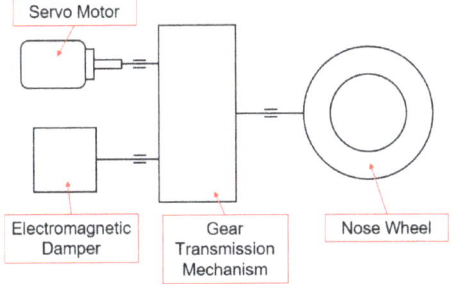

Figure 4. The design of transmission mechanism.

The schematic diagram of the gear reducer is shown in Figure 5 and the specific structural scheme design is shown in Figure 6. The nose landing gear shimmy reduction channel and the steering channel are connected in parallel in the same gear transmission mechanism. The reducer is a fixed axis gear train during the steering operation of the nose landing gear. Motor shaft gear 1 engages with gear 2, gear 2 is coaxial with 3, gear 3 engages with 4, gear 4 and 5 are duplex gears, gear 6 engages with gear 5, gear 7 is coaxial with 6 and engages with gear 8, gear 8 is coaxial with gear 9, and finally passes to the front landing gear strut gear 12. When the shimmy reduction maneuver is performed, the reducer is a compound gear train, meanwhile, gears 6–7 are planetary gears, and gears 4, 5, 8, 9 are center gears. The motor shaft is braked, the electromagnetic damper gear shaft is the input shaft, gear 13 engages with 12, gear 12 is coaxial with 11, and gear 11 engages with gear 10, driving gears 6–7 to rotate around the center gear and to be able to rotate at the same time, eventually passing to the nose landing gear strut gear 12.

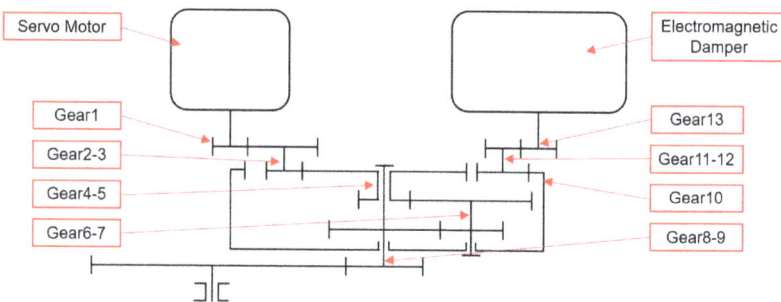

Figure 5. Schematic diagram of the gear reducer.

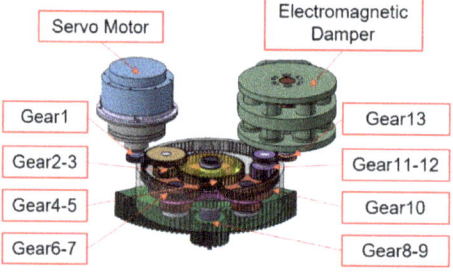

Figure 6. Specific structure of electric nose wheel steering system.

The total transmission ratio of the shimmy reduction channel of the gear reducer is 9.07, and the transmission ratio from the gear reducer to the nose wheel strut is 6, so the total transmission ratio of the shimmy reduction channel is 54.4. A 3D model of the electrically actuated steering shimmy reduction mechanism mounted on the nose landing gear of a jet fighter is shown in Figure 7.

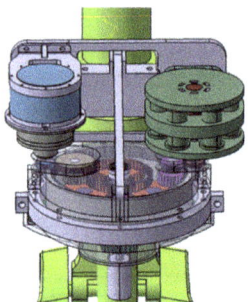

Figure 7. 3D model of the electro-mechanical nose wheel steering system.

3.2. Mathematical Modeling of the Electromagnetic Damper

3.2.1. Calculation of Eddy Current Skin Depth

When an alternating current, especially a high frequency current, flows through a conductor, it will tend to the conductor surface and flow in a thin layer near the surface, which is the skin effect in electromagnetic field theory [18]. As can be seen from Figure 3, the rotor disks generates eddy currents with alternating directions during rotating around the stator disk.

The analysis of the practical application of electromagnetism shows that the axial distribution of the eddy current density, the electric field strength and the induced magnetic field strength of the alternating eddy current of the electromagnetic damper on the rotor disk decays according to the exponential law $e^{-\sqrt{\omega\mu\gamma/2} \cdot h}$ [19], where ω is the angular frequency of the alternating eddy current, μ is the magnetic permeability of the rotor disk material, γ is the electrical conductivity (inverse of the resistivity) of the rotor disk material, and h is the vertical depth from the surface of the rotor disk. In engineering, the depth at which the amplitude of the abovementioned alternating field quantity drops to the surface value of $1/e$ is generally defined as the penetration depth of skin effect:

$$e^{-\sqrt{\omega\mu\gamma/2} \cdot \Delta} = e^{-1}$$
$$\Delta = \sqrt{2/\omega\mu\gamma} \qquad (4)$$

As can be seen from the above equation, the higher the angular frequency of the alternating eddy current or the better the conductivity and permeability of the rotor disk material, the shallower the penetration depth of the skin effect.

In addition, let the rotor disk be arranged with n magnets (for an even number to ensure that adjacent poles are reversed), then the fixed position on the rotor disk rotates for one circle and the eddy current direction changes n times, that is, it experiences $n/2$ complete alternating periods. Let the rotating angular velocity of the rotor disk be ω_n, then the rotor disk rotation period and frequency are:

$$T_{Rotor} = \frac{2\pi}{\omega_n} \qquad (5)$$

$$f_{Rotor} = \frac{1}{T_{Rotor}} = \frac{\omega_n}{2\pi} \qquad (6)$$

The eddy current alternating period, frequency and angular frequency at a fixed position on the rotor disk are:

$$T_{Ec} = \frac{T_{Rotor}}{n/2} = \frac{4\pi}{n\omega_n} \qquad (7)$$

$$f_{Ec} = \frac{1}{T_{Ec}} = \frac{n\omega_n}{4\pi} \qquad (8)$$

$$\omega = 2\pi f_{Ec} = n\omega_n/2 \qquad (9)$$

Therefore, the penetration depth of skin effect can be further expressed as:

$$\Delta = \sqrt{2/\omega\mu\gamma} = 2/\sqrt{n\omega_n\mu\gamma} \qquad (10)$$

3.2.2. Calculation of Damping Torque

Assuming that the magnetic flux density of a single magnet passing through a rotor disk is B, the rotor disk is regarded as consisting of countless small iron rods centered on the circle and having length R_2-R_1. When the rotor disk rotates, the small iron rod cuts the magnetic lines of force to excite the electromotive force, thus forming an eddy current on the surface of the rotor disk [20], as shown in Figure 8.

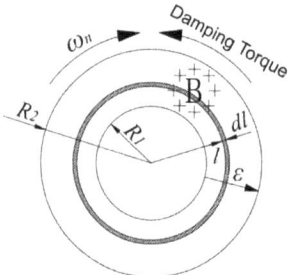

Figure 8. Analysis of the rotor disk potential and damping torque.

Let the rotor disk rotate with linear velocity V and angular velocity ω_n. Then the induced electromotive force generated on the inner and outer sides of the micro-ring with width dl is:

$$d\varepsilon = (V \times B)dl = \omega_n Bldl \tag{11}$$

The resistance value of the inner and outer sides of the micro-ring with width dl and the current value flowing along the radial direction through the inner and outer sides of this micro-ring are:

$$dR = \frac{\rho dl}{2\pi l \Delta} \tag{12}$$

$$I_{dR} = \frac{d\varepsilon}{dR} = \frac{2\pi \omega_n B l^2 \Delta}{\rho} \tag{13}$$

On the premise of uniform magnetic field, combined with ampere force formula, the damping force and damping torque of the micro-ring under the action of the magnetic pole can be calculated as follows:

$$dF = BI_{dR}dl = \frac{2\pi \omega_n B^2 l^2 \Delta}{\rho} dl \tag{14}$$

$$dT = ldF = \frac{2\pi \omega_n B^2 \Delta}{\rho} l^3 dl \tag{15}$$

For the integration of the above-mentioned micro-ring torque in the entire rotor disk (from R_1 to R_2), the damping torque of a single rotor disk under the action of magnet can be preliminarily calculated, that is:

$$T_1 = \int_{R_1}^{R_2} dT = \int_{R_1}^{R_2} \frac{2\pi \omega_n B^2 \Delta}{\rho} l^3 dl = \frac{\pi \left(R_2^4 - R_1^4\right) \omega_n B^2 \Delta}{2\rho} \tag{16}$$

It should be noted that in the derivation of the above equation, it is assumed that the entire rotor disk is located within the uniform magnetic field. However, the actual area of the magnetic poles acting on the rotor disk is approximately equal to the permanent magnet cross-sectional area (assuming that the air gap spacing is small enough). Therefore, the damping torque generated by a single rotor disk should be the above equation multiplied by the ratio of the real magnetic flux area to the rotor disk area. Setting the radius of the magnet cross section as r_{Mag}, the expression of the damping torque of the double rotor disks permanent magnet electromagnetic damper is:

$$T_n = 2T_1 \cdot \frac{n\pi r_{Mag}^2}{\pi \left(R_2^2 - R_1^2\right)} = \frac{n\pi r_{Mag}^2 \left(R_2^2 + R_1^2\right) \omega_n B^2 \Delta}{\rho} \tag{17}$$

Based on the derivation of the penetration depth Δ of the skin effect in Equation (10), it can be further obtained that:

$$T_n = \frac{n\pi r_{Mag}^2 \left(R_2^2 + R_1^2\right)\omega_n B^2}{\rho} \cdot 2\sqrt{\frac{\rho}{n\omega_n \mu}} = 2\pi r_{Mag}^2 \left(R_2^2 + R_1^2\right) B^2 \sqrt{\frac{n\omega_n}{\mu\rho}} \quad (18)$$

From the above derivation results, it can be seen that if we want to increase the damping torque, we need to increase the magnetic flux density through the rotor disk or choose rotor disk materials with lower permeability and lower resistivity and larger size permanent magnets, provided that the rotor disk shape size is determined. At the same time, the greater the angular speed of the rotor disk, the larger the output damping torque. In the structural scheme of the electrically actuated nose wheel steering system proposed in Section 3.1, the nose wheel strut is connected to the electromagnetic damper through a transmission device with transmission ratio $i = 54.4$. The transmission efficiency of commonly used 8-grade cylindrical spur gears is 0.97, i.e., $\eta_{Gear} = 0.97$. The transmission efficiency of the bearings is 0.99, i.e., $\eta_{Bearing} = 0.99$, so the total transmission efficiency is 0.81, i.e., $\eta = \eta_{Gear}^5 \cdot \eta_{Bearing}^5 = 0.81$. The angular velocity and damping torque of the nose wheel shimmy are related to the angular velocity and damping torque of the electromagnetic damper itself as:

$$\omega_n = i\omega_{Strut} \quad (19)$$

$$T_{Strut} = i\eta T_n \quad (20)$$

According to Equation (18), the damping torque of strut shimmy is:

$$T_{Struct} = i\eta T_n = 2i^{\frac{3}{2}}\eta \pi r_{Mag}^2 \left(R_2^2 + R_1^2\right) B^2 \sqrt{\frac{n\omega_{Strut}}{\mu\rho}} \quad (21)$$

The relationship between the nose landing gear damping torque and the design parameters of the electromagnetic damper and the nose wheel shimmy speed is thus established.

3.2.3. Calculation of Damping Coefficient

Whether the nose wheel dampers can meet the damping requirements of aircraft taxiing on the ground is judged mainly by how many damping coefficients they can provide. The damping coefficient of dampers or struts is usually defined in engineering applications by the ratio of damping torque and angular velocity, and then the damping coefficients of the dampers and struts are:

$$h_n = \frac{T_n}{\omega_n} = 2\pi r_{Mag}^2 \left(R_2^2 + R_1^2\right) B^2 \sqrt{\frac{n}{\mu\rho\omega_n}} \quad (22)$$

$$h_{Strut} = \frac{T_{Strut}}{\omega_{Strut}} = i^2\eta h_n = 2\pi i^2 \eta r_{Mag}^2 \left(R_2^2 + R_1^2\right) B^2 \sqrt{\frac{n}{\mu\rho\omega_n}} \quad (23)$$

From Equations (22) and (23), it can be seen that the damping coefficient decreases if the rotor disk angular velocity increases. The next step is to design the specific scheme of the electromagnetic damper based on the demand of the damping coefficient of the shimmy reduction, the structural scheme of the electrically actuated nose wheel steering system and the derivation of the theoretical calculation.

3.3. Design of the Electromagnetic Damper

Based on the operating principle of the electromagnetic damper, the materials selected for each component are shown in Table 2.

Table 2. Material selection of electromagnetic damper components.

Component	Material
Permanent magnet	NdFe35
Stator disk	Steel stainless
Rotor disk	Steel 1010

The resistivity of steel 1010 is 5×10^{-7} Ωm and the permeability is 1.8×10^{-3} H/m. This scheme uses a permanent magnet with the magnetic flux density of 0.86 T. Assuming that the air gap between the magnetic pole of the permanent magnet and the rotor disk is small enough. In order to meet the target of the damping coefficient of not less than 40 Nms/rad, according to Equations (18) and (21), dimensions of electromagnetic damper components can be obtained which are shown in Table 3.

Table 3. Dimensions of electromagnetic damper components.

Component	Diameter (mm)		Height (mm)
	Outer Diameter	Inner Diameter	
Permanent magnet		35	18
Stator disk	140	52	18
Rotor disk	140	35	10

The damping torque and damping coefficient of the electromagnetic damper and the strut under the shimmy motion distribution for this set of design parameters are shown in Table 4.

Table 4. Distribution of damping torque and damping coefficient with different shimmy state.

ω_n (rad/s)	ω_{Strut} (rad/s)	T_n (Nm)	T_{Strut} (Nm)	h_{Strut} (Nms/rad)
357.95	6.58	11.45	504.34	76.65
649.54	11.94	15.42	679.39	56.90
874.75	16.08	17.89	788.41	49.03
1034.14	19.01	19.45	857.25	45.09
1126.62	20.71	20.31	894.76	43.20
1153.82	21.21	20.55	905.49	42.69
1153.28	21.20	20.54	905.28	42.70
1113.57	20.47	20.19	889.56	43.46
1007.49	18.52	19.20	846.13	45.69
835.04	15.35	17.48	770.32	50.18
596.77	10.97	14.78	654.21	59.36

From the data in Table 4, it can be seen that the theoretical calculated values of the damping coefficient of the electromagnetic damper amplified to the strut by the transmission mechanism are all satisfactory according to the design index. In the actual installation, there is an air gap between the rotor disk and the magnetic poles, which will cause the magnetic flux density of the poles to be attenuated on the rotor disks. In order to verify the correctness of the mathematical model and theoretical calculations, and to study in depth the shimmy reduction characteristics of the double rotor disks permanent magnet electromagnetic damper applied to the electrically actuated nose wheel steering system, finite element simulation in the electromagnetic field is required.

4. Electromagnetic Field Simulation of the Electromagnetic Damper

4.1. Static Simulation Results and Analysis of the Electromagnetic Damper

The finite element simulation software generally used for electromagnetic devices such as electromagnetic dampers is Maxwell. Maxwell is an interactive software package that uses finite element analysis (FEA) to solve 3D electrostatic, magnetostatic, eddy current, and transient problems. Using Maxwell we can compute:

- Static electric fields, forces, torques, and capacitances caused by voltage distributions and charges;
- Static magnetic fields, forces, torques, and inductances caused by DC currents, static external magnetic fields, and permanent magnets;
- Time-varying magnetic fields, forces, torques, and impedances caused by AC currents and oscillating external magnetic fields;
- Transient magnetic fields caused by electrical sources and permanent magnets.

The magnetostatic solver in Maxwell software is selected for the static magnetic field simulation of the electromagnetic damper. In a magnetostatic solution, the magnetic field is produced by DC currents flowing in conductors/coils and by permanent magnets. The electric field is restricted to the objects modeled as real (non-ideal) conductors. The electric field existing inside the conductors as a consequence of the DC current flow is totally decoupled from the magnetic field. Thus, as far as magnetic material properties are concerned, the distribution of the magnetic field is influenced by the spatial distribution of the permeability. There are no time variation effects included in a magnetostatic solution, and objects are considered to be stationary. The energy transformation occurring in connection with a magnetostatic solution is only due to the ohmic losses associated with the currents flowing in real conductors.

The magnetostatic field solution verifies the following two Maxwell's equations:

$$\nabla \times \vec{H} = \vec{J} \tag{24}$$

$$\nabla \cdot \vec{B} = 0 \tag{25}$$

with the following constitutive (material) relationship being also applicable:

$$\vec{B} = \mu_0 \left(\vec{H} + \vec{M} \right) = \mu_0 \cdot \vec{H} + \mu_0 \cdot \mu_r \cdot \vec{M_p} \tag{26}$$

where $\vec{H}(x,y,z)$ is the magnetic field strength; $\vec{B}(x,y,z)$ is the magnetic flux density; $\vec{J}(x,y,z)$ is the conduction current density; $\vec{M_p}(x,y,z)$ is the permanent magnetization; $\mu_0 = 4 \cdot \pi \cdot 10^{-7}$ H/m is the permeability of vacuum; μ_r is the relative permeability.

The 3D numerical model of the electromagnetic damper is established according to the form dimensions in Table 3. To simplify the calculation, the central cross section of the stator disk is selected as the even symmetry surface, as shown in Figure 9, and the distribution of magnetic flux density on the rotor disk when it is stationary is analyzed.

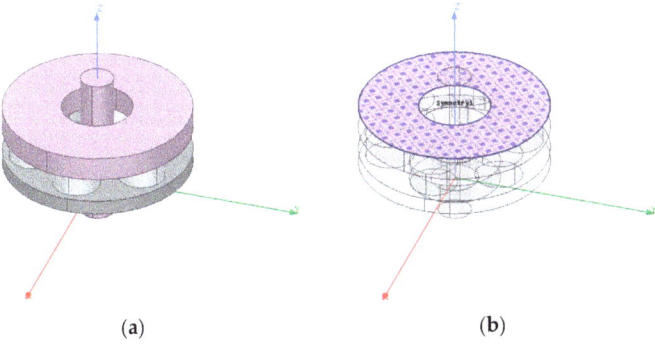

Figure 9. Finite element simulation model of the electromagnetic damper: (**a**) Simplified model; (**b**) Central cross section of the stator disk.

Firstly, the parameters of relative permeability, conductivity and coercivity of the permanent magnet poles are corrected, and the simulation results of the magnetic flux density of magnetic poles are shown in Figure 10. The magnetic field strength directions of the adjacent magnetic poles are opposite, and the magnetic flux density provided by each pole is calculated between 0.852–0.865 T by the fields calculator, which is basically consistent with the properties of the permanent magnet actually selected for this scheme.

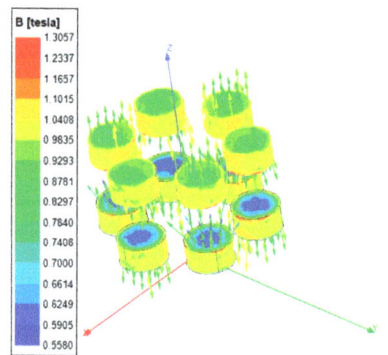

Figure 10. Magnetic flux density distribution of permanent magnet poles(static).

The distribution of magnetic flux density on the rotor disks under the action of magnetic poles close to the real conditions is shown in Figure 11.

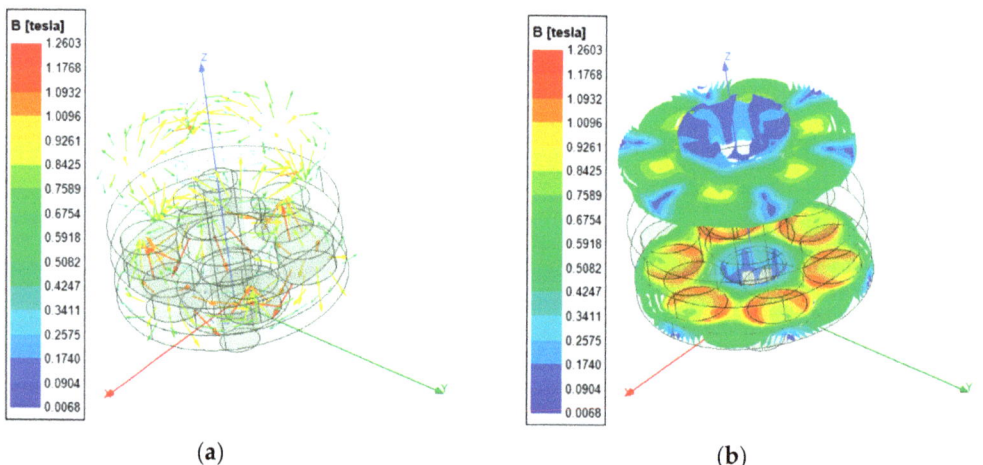

Figure 11. Static magnetic flux density distribution of the rotor disks: (**a**) Vector distribution; (**b**) Field distribution.

It can be seen from the static magnetic flux density distribution on the rotor disks:

(1) In the two areas of the rotor disk below the adjacent poles, the magnetic flux densities are in opposite directions but they are equal in values.
(2) The magnetic field superposition at the circumferential edge of the area corresponding to the magnetic poles leads to the maximum magnetic flux density, which can reach 1.26T, and gradually decays in all directions.

4.2. Dynamic Simulation Results and Analysis of the Electromagnetic Damper

The dynamic simulation of the electromagnetic damper is also completed in Maxwell finite element software, and the difference with the static simulation is that the solver is changed to transient. In the 3D transient (time domain), the solver uses the $\vec{T} - \Omega$ formulation. Motion (translational or cylindrical/non-cylindrical rotation) is allowed, excitations-currents and/or voltages-can assume arbitrary shapes as functions of time, nonlinear BH material dependencies are also allowed. For a simpler formulation of problems where motion is involved, Maxwell uses a particular convention and uses the fixed coordinate system for the Maxwell's equations in the moving and the stationary part of the model. Thus the motion term is completely eliminated for the translational type of motion while for the rotational type of motion a simpler formulation is obtained by using a cylindrical coordinate system with the z axis aligned with the actual rotation axis.

The formulation used by the Maxwell transient module supports Master-Slave boundary conditions and motion induced eddy currents everywhere in the model, in the stationary as well as in the moving parts of the model. Mechanical equations attached to the rigid-body moving parts allows a complex formulation with the electric circuits being strongly coupled with the finite element part and also coupled with the mechanical elements whenever transient mechanical effects are included by users in the solution. In this case the electromagnetic force/torque is calculated using the virtual work approach. For problems involving rotational type of motion a "sliding band" type of approach is followed and thus no re-meshing is done during the simulation.

The following two Maxwell's equations are relevant for transient (low frequency) applications:

$$\nabla \times H = \sigma(E) \quad (27)$$

$$\nabla \times E = -\frac{\partial B}{\partial t} \quad (28)$$

The following equation directly results from the above two equations:

$$\nabla \times \frac{1}{\sigma} \nabla \times H + \frac{\partial B}{\partial t} = 0 \quad (29)$$

The final result is a formulation where vector fields are represented by first order edge elements and scalar fields are represented by second order nodal unknowns.

The rotor disks' output damping torque is taken as the target parameter for the study. Because there are many operating conditions involved in the distribution of the shimmy motion, the set of motion parameters ω_n = 1034.14 rad/s in Table 4 is used as an example to analyze the dynamic magnetic field simulation results of the electromagnetic damper. The output damping torque of the double rotor disks under this operating condition is shown in Figure 12.

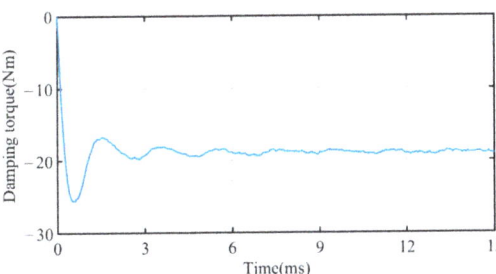

Figure 12. Output damping torque of the electromagnetic damper.

As can be seen from Figure 12, when the rotor disks start to move suddenly in the magnetic field generated by the poles of the permanent magnet, vibration is generated and the value of the output damping torque temporarily fluctuates. After 6 ms, the motion of the rotor disk stabilizes and the output damping torque remains constant around 18.8 Nm, which is close to the theoretical calculated value of the mathematical model under this condition with small deviation. It also shows that in engineering practice, electromagnetic dampers can quickly respond to and suppress the instantaneous shimmy of nose wheels, which demonstrates its superiority compared with traditional oleo dampers.

The following is an in-depth analysis of the simulation results of the rotor disks magnetic field strength distribution, magnetic flux density and current density during stable operation.

As shown in Figure 13, the magnetic field strength on the surface of the rotor disks during the stable operation of the electromagnetic damper is in the range of 8517–127690 A/m. The magnetic field strength distribution is more concentrated in the area close to the permanent magnets, and the magnetic poles of the two adjacent concentrated areas are opposite.

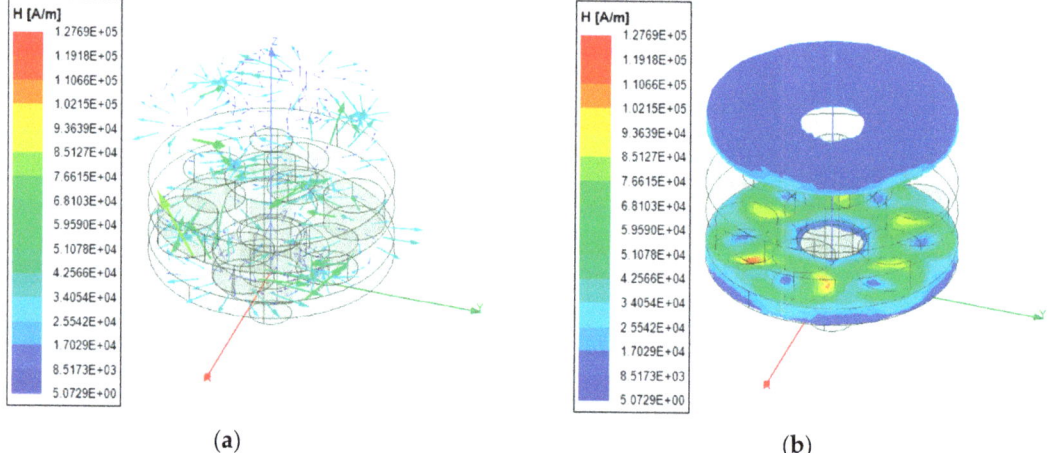

Figure 13. Transient magnetic field strength distribution of the rotor disks: (**a**) Vector distribution; (**b**) Field distribution.

As shown in Figure 14, the distribution of magnetic flux density on the rotor disks is also concentrated in the area close to the permanent magnets. When the electromagnetic damper is operating, the induced magnetic field on the rotor disks will be generated, and after superimposing the magnetic field of the permanent magnet poles through the rotor disks, the magnetic flux density is about 0.86–2.15 T, which is significantly larger than the static magnetic flux density of the permanent magnet pole in Figure 10.

As shown in Figure 15, the induced eddy current density of the rotor disks is in the range of 1.25×10^7–4.69×10^7 A/m^2. Observing the vector distribution, it can be found that the adjacent eddy currents on the surface of the rotor disks are in opposite directions, which proves the correctness of the electromagnetic damper using the induced eddy currents to convert kinetic energy into thermal energy. Meanwhile, observing the field distribution, it can be found that the induced eddy currents on the rotor disks are mainly concentrated on the surface near the pole side, which also proves the existence of skin effect and is consistent with the mathematical model.

The above simulation results are based on the set of motion parameters ω_n = 1034.14 rad/s. In order to study the shimmy reduction performance of the electromagnetic damper in the electrically actuated nose wheel steering system, all shimmy motion states listed in Table 4 need to be considered, and the output damping torque of the electromagnetic damper corresponding to each operating condition is shown in Figure 16.

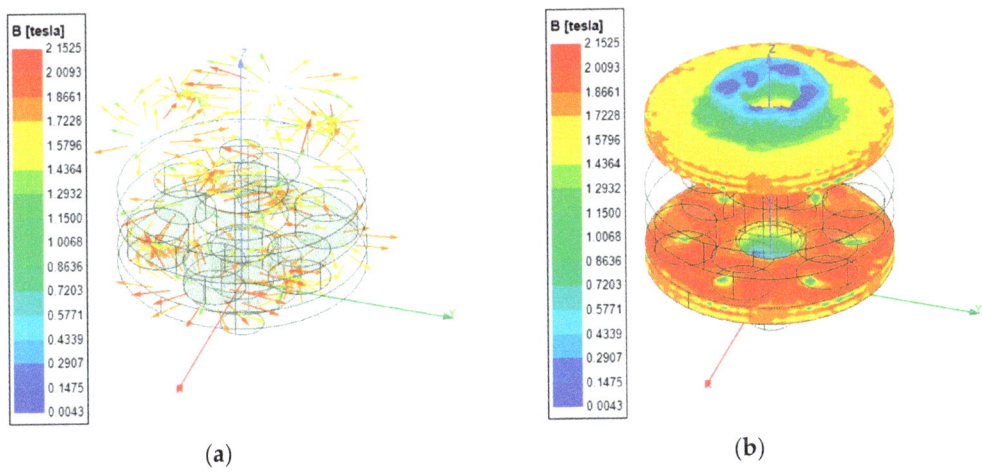

Figure 14. Transient magnetic flux density distribution of the rotor disks: (**a**) Vector distribution; (**b**) Field distribution.

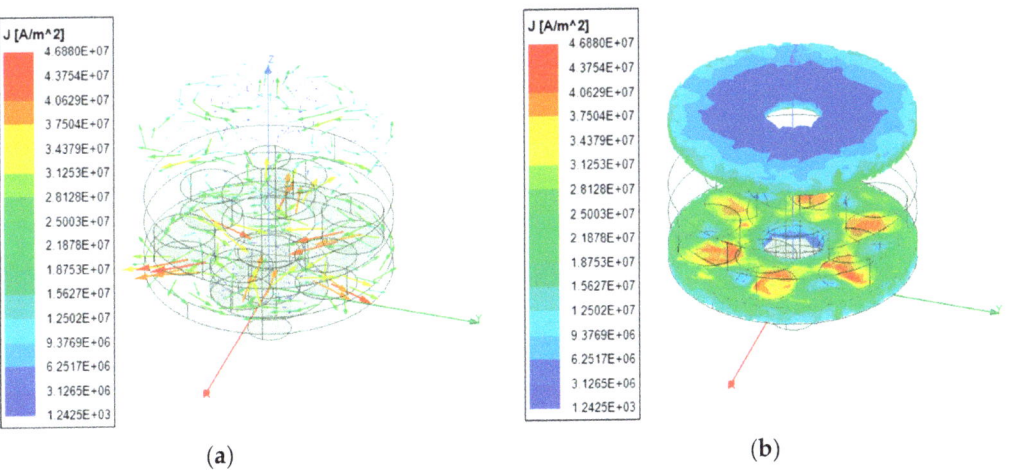

Figure 15. Transient eddy current density distribution of the rotor disks: (**a**) Vector distribution; (**b**) Field distribution.

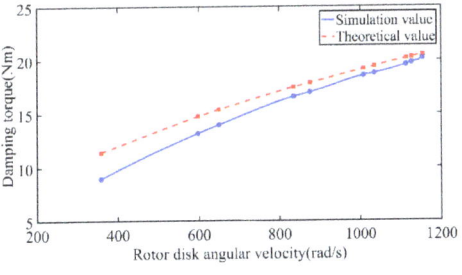

Figure 16. Distribution of damping torque with angular velocity of the rotor disks.

By comparing the simulation curve and the theoretical calculation curve of the output damping torques of the electromagnetic damper under various operating conditions, it can be found that the simulation value and the theoretical value are closer, which shows that:

(1) The mathematical model of the electromagnetic damper is reasonable;
(2) The application of electromagnetic dampers to the shimmy reduction function of electrically actuated nose wheel steering systems is also feasible, but there are some differences in the values at some data points and the slopes of the two curves are not identical, because:

- When calculating the output damping torques of the electromagnetic damper using Equation (18), the magnetic flux density of the permanent magnet source is assumed to be the magnetic flux density on the surface of the rotor disks, and the magnetic losses caused by the air gap and resistance and the induced magnetic field generated by the rotor disks themselves during rotation are ignored;
- The magnetic permeability of the permanent magnet material is simply taken as a constant value in the numerical calculation. However, from the hysteresis curve of the permanent magnet material in the finite element simulation, the magnetic field strength is not linearly related to the magnetic flux density, and the relative permeability is also not linear;
- The demagnetization effect of the permanent magnet is not taken into account in the whole mathematical modeling process of the electromagnetic damper, while the experimental results in the literature [21] show that the demagnetization effect is actually real. The magnetic flux density distribution of the permanent magnet magnetic pole at the relative angular velocity of 1034.14 rad/s is shown in Figure 17, and comparing the static magnetic flux density distribution of the pole in Figure 10, it can be found that the finite element simulation takes this demagnetization effect into account;

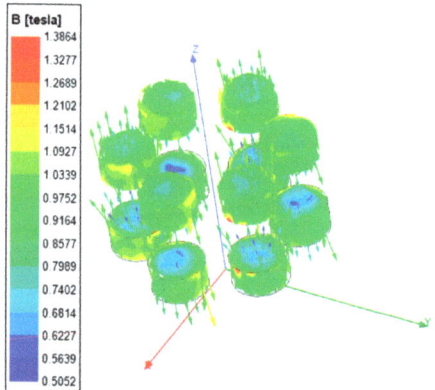

Figure 17. Magnetic flux density distribution of permanent magnet poles (at 1034.14 rad/s).

By observing Figure 16, it can be found that when the rotor disk rotates at a low speed, the theoretical value of its output damping torque is greater than the simulation value, because in the process of theoretical derivation, the mechanical power of the rotor disk is considered to be fully converted into electromagnetic power, while the electromagnetic loss is ignored. In addition, the offsetting effect of opposite poles when multiple permanent magnets work simultaneously is not taken into account in the theoretical derivation. When the rotor disk rotates at high speed, the simulation value of the output damping torque is closer to the theoretical value because the induced magnetic field superimposed on the surface of the rotor disk is large enough, which can not only ignore the demagnetization effect of the magnetic poles at high speed, but also make up for the power loss in the theoretical calculation.

We can draw the curves between the damping coefficient of the electromagnetic damper amplified by the transmission mechanism and its own angular velocity, as shown in Figure 18.

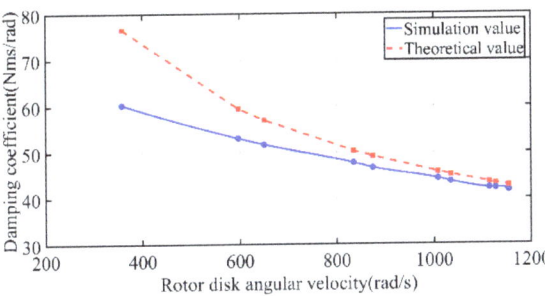

Figure 18. Distribution of damping coefficient with angular velocity of rotor disk.

It can be seen from Figure 18 that the damping coefficient of shimmy reduction provided by the electromagnetic damper varies with the shimmy frequency when the transmission mode is determined. Figures 16 and 18 show that the damping torque increases while the damping coefficient decreases as the angular velocity of the rotor disk increases, which is consistent with the derivation of Equations (18) and (23). Therefore, when evaluating the damping performance of electromagnetic dampers, we should pay more attention to the damping coefficient under high frequency shimmy.

4.3. Effect of Various Factors on the Performance of Electromagnetic Dampers

According to the operating principle and magnetic field simulation of electromagnetic dampers, there are many factors that affect the performance of electromagnetic dampers to reduce shimmy, here the two key factors, the dimensions R_2 and R_1 of the rotor disks and air gap width δ, are studied in detail, the former has direct reference value for the structural design of electromagnetic dampers under the premise of known shimmy reduction index requirements, the latter is the most important way to adjust the output damping torque after the structural scheme of the electromagnetic damper is determined.

As discussed in Section 4.2, because there are more operating conditions involved in the shimmy motion distribution, and the shimmy reduction design index usually requires the minimum shimmy reduction damping coefficient of the damper, the set of motion parameters $\omega_n = 1153.82$ rad/s in Table 4 is used as an example to analyze the influence of the rotor disks' outer diameter R_2 and air gap width δ on the shimmy reduction performance of the electromagnetic damper, respectively.

4.3.1. Study of the Effect of Rotor Disks' Dimensions R_2 and R_1 on Electromagnetic Damping

The rotor disks with $R_2 = 80$ mm and $R_1 = 20$ mm are used as the starting point and scaled by 1.25, 1.5, 1.75, 2, 2.25 and 2.5, respectively. The results of the damping coefficients after being enlarged by the transmission mechanism are shown in Figure 19. The outer dimensions of the rotor disks enlarged by 1.75 are the rotor disks' outer diameter $R_2 = 140$ mm and the inner diameter $R_1 = 35$ mm selected for this scheme.

From Figure 18, it can be seen that by changing the dimensions of the electromagnetic damper, the damping coefficient amplified by the transmission mechanism ranges from 2.83 to 195.29 Nms/rad, and there is an approximate quadratic power relationship with the dimensions of the rotor disks of the electromagnetic damper, which once again verifies the correctness of the derivation of the damping coefficient in the mathematical model.

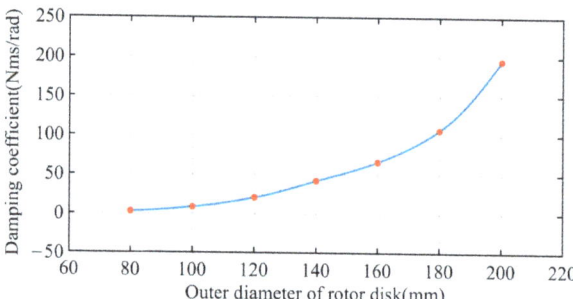

Figure 19. Distribution of damping coefficient with the outer diameter of rotor disk.

4.3.2. Study of the Effect of Air Gap Width δ on Electromagnetic Damping

The width of the air gap between the rotor disks and the permanent magnet poles is generally obtained within the range of 0.5–3 mm. Too small will lead to the thermal expansion and collision of the rotor disks and the permanent magnet poles during operating, while too large will lead to a sharp decline in the performance of the electromagnetic damper shimmy reduction. Since the air gap is composed of air, whose relative permeability is 1, and other parts such as the rotor disks and the stator disk are composed of high permeability materials, so the magnetic resistance of other structures can be neglected relative to the air gap. Taking the air gap width δ as 0.2, 0.3, 0.5, 1, 1.5, 2, 2.5, 3 and 3.5 mm, respectively, the corresponding output torque of the electromagnetic damper is obtained as shown in Figure 20.

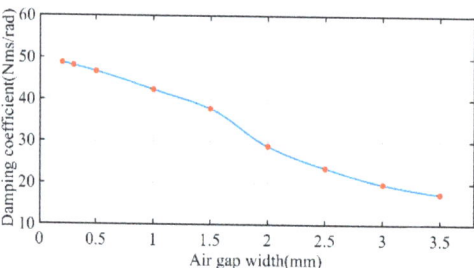

Figure 20. Distribution of damping torque with the air gap width of the rotor disks.

For permanent magnet electromagnetic dampers, the magnetoresistance increases as the air gap width increases, while the flux is the ratio of the magneto motive force to the magnetoresistance. The flux decreases with increasing magnetoresistance and constant magneto motive force, producing a flux density in the magnetic saturation region, which decreases more slowly, and then in the non-saturation region, which decreases rapidly. From Figure 20, it can be seen that during the reduction of the air gap width from 3.5 mm to 0.2 mm, the damping coefficient of shimmy reduction gradually increases from 17.55 Nms/rad to 49.79 Nms/rad, while the change of the damping torque of shimmy reduction slows down during the continued reduction from 1.5 mm, indicating that the induced magnetic field on the rotor disks has started to enter the saturation state. Therefore, within a certain range, the magnetic flux density can be increased by reducing the air gap to increase the output damping torque of the electromagnetic damper, but the air gap should not be too small considering the influence of mechanical processing precision and material thermal expansion and contraction.

5. Conclusions

In this paper, an electromagnetic damping method for an electrically actuated nose wheel steering system is studied, a structural scheme of the electrically actuated nose wheel steering system is proposed, and a new electromagnetic damping shimmy reduction device integrated in this system is designed. In order to verify the shimmy reduction performance of the proposed electromagnetic damper, the corresponding mathematical model of the electromagnetic damper is established, and then the specific scheme design of the electromagnetic damper is carried out and parametric modeling and electromagnetic field simulation are conducted according to the design index of shimmy reduction damping coefficient, and the following conclusions are obtained:

(1) After correcting the assumption that the rotor disk is located in the uniform magnetic field, which was commonly used in the previous derivation process, and combining the calculation of the skin depth, the derived equation for the output damping torque of the electromagnetic damper is closer to the simulated value;

(2) The electromagnetic damper designed in this paper can provide a damping coefficient of not less than 40 Nms/rad under the conditions of shimmy amplitude between 2–20° and frequency between 5–30 Hz, which not only meets the requirements of the index, but also overcomes the disadvantages of relying on the hydraulic power source compared with the traditional oleo dampers. It also has the advantages of higher reliability, lower maintenance cost and faster response;

(3) There is an approximate quadratic relationship between the dimensions R_2 and R_1 of the rotor disks and the damping coefficient. The preliminary structural design of electromagnetic dampers' dimensions can be based on this relationship when the shimmy reduction index requirements are given;

(4) The air gap width δ of the electromagnetic damper designed in this paper can be adjusted to obtain a range of 17.55–49.79 Nms/rad for the damping coefficient. In the actual engineering application, the required damping coefficient can be obtained by adjusting the air gap width δ of the electromagnetic damper for different shimmy conditions.

(5) The application of electromagnetic damping technology to strut damping is the first comprehensive and systematic study of this technology at the theoretical and simulation levels, which can be applied not only to the design of electrically actuated nose wheel steering systems for various types of aircraft, but also to other impact cushioning and vibration energy recovery fields.

Author Contributions: Conceptualization: C.S. and M.Z.; Investigation: M.Z., Y.G., H.Y. and G.P.; Methodology: C.S., M.Z. and Y.G.; Project administration: M.Z.; Software: C.S. and L.T.; Supervision: M.Z. and Y.G.; Visualization: C.S., H.Y. and G.P.; Writing—original draft: C.S. and L.T.; Writing—review & editing, M.Z. All authors have read and agreed to the published version of the manuscript.

Funding: This research was funded by the Aeronautical Science Foundation of China under Grant No. 20182852021.

Institutional Review Board Statement: Not applicable.

Informed Consent Statement: Not applicable.

Data Availability Statement: Not applicable.

Acknowledgments: The author would like to thank the reviewers and the editors for their valuable comments and constructive suggestions that helped to improve the paper significantly.

Conflicts of Interest: The authors declare no conflict of interest.

Nomenclature

The following nomenclatures are used in this manuscript:

A	Nose wheel shimmy amplitude
f	Nose wheel shimmy frequency
α	Nose wheel shimmy angle
ω_{NW}	Angular velocity of nose wheel shimmy
ω_{max}	Max angular velocity of the nose wheel shimmy
ω	Angular frequency of the eddy current
μ	Magnetic permeability of the rotor disk material
γ	Electrical conductivity of the rotor disk material
h	Vertical depth from the surface of the rotor disk
Δ	Penetration depth of skin effect
ω_n	Rotation angular velocity of the rotor disk
T_{Rotor}	Rotation period of the rotor disk
f_{Rotor}	Rotation frequency of the rotor disk
T_{Ec}	Alternating period of the eddy current
f_{Ec}	Frequency of the eddy current
B	Magnetic flux density of a single magnet passing through a rotor disk
V	Rotation linear velocity of the rotor disk
$d\varepsilon$	Induced electromotive force of the micro-ring
dl	Width of the micro-ring
dR	Resistance of the micro-ring
I_{dR}	Current value flowing along the radial direction of the micro-ring
ρ	Resistivity of the rotor disk material
dF	Damping force of the micro-ring
dT	Damping torque of the micro-ring
R_2	Outer radius of the rotor disk
R_1	Inner radius of the rotor disk
T_1	Damping torque of a single rotor disk
r_{Mag}	Radius of the magnet cross section
T_n	Damping torque of the electromagnetic damper
i	Transmission ratio
η_{Gear}	Transmission efficiency of the 8-grade cylindrical spur gear
$\eta_{Bearing}$	Transmission efficiency of the bearing
η	Total transmission efficiency
ω_{Strut}	Angular velocity of the nose landing gear strut
T_{Strut}	Damping torque of the nose landing gear strut
h_n	Damping coefficient of the electromagnetic damper
h_{Strut}	Damping coefficient of the nose landing gear strut
$\vec{H}$	Magnetic field strength
$\vec{B}$	Magnetic flux density
$\vec{J}$	Conduction current density
$\vec{Mp}$	Permanent magnetization
μ_0	Permeability of vacuum
μ_r	Relative permeability

References

1. Cláudio, A.L.; James, C.; Kamran, E.S. Rotor Position Synchronization in Central-Converter Multi-Motor Electric Actuation Systems. *Energies* **2021**, *14*, 7485. [CrossRef]
2. George, I.; Stephane, D. Dress: Distributed and Redundant Electro-Mechanical Nose Wheel Steering System. *Sae Int. J. Aerosp.* **2009**, *2*, 46–53.
3. Bennett, C.; William, J. Fault Tolerant Electromechanical Actuators for Aircraft. Ph.D. Thesis, Newcastle University, Newcastle, UK, 2010.
4. Liao, P.; Zhang, X.; Li, W. Design of Electrical Swerve Control System for Small Aircraft Undercarriages. *Small Spec. Electr. Mach.* **2010**, *38*, 4.

5. Feng, F.; Nie, H.; Zhang, M.; Peng, Y. Effect of Torsional Damping on Aircraft Nose Landing-Gear Shimmy. *J. Aircr.* **2014**, *52*, 561–568. [CrossRef]
6. Zhu, P. *Shimmy Theory and Anti-Shimmy Measures*, 1st ed.; National Defense Industry Press: Beijing, China, 1984; pp. 9–23.
7. Luong, Q.-V.; Jo, B.-H.; Hwang, J.-H.; Jang, D.-S. A Supervised Neural Network Control for Magnetorheological Damper in an Aircraft Landing Gear. *Appl. Sci.* **2022**, *12*, 400. [CrossRef]
8. Jo, B.-H.; Jang, D.-S.; Hwang, J.-H.; Choi, Y.-H. Experimental Validation for the Performance of MR Damper Aircraft Landing Gear. *Aerospace* **2021**, *8*, 272. [CrossRef]
9. Chen, D.; Gu, H.; Wu, D. Semi-active Control of Landing Gear Shimmy Based on Magneto-rheological (MR) Damper. *China Mech. Eng.* **2010**, *21*, 1401–1405.
10. Zhu, S.; Wang, G. Design and Experimental Research of External Coil Type Magnetorheological Shimmy Damper. *Mach. Tool Hydraul.* **2019**, *47*, 12–17.
11. Jia, C.; Xia, Y.; Chu, M.; Zhang, X. Post-capture Angular Momentum Management of Space Robot with Controllable Damping Joints. In Proceedings of the 2019 IEEE 2nd International Conference on Automation, Electronics and Electrical Engineering (AUTEEE), Shenyang, China, 22–24 November 2019; pp. 638–642.
12. Kou, B.; Li, L.; Jin, Y.; Pan, D. System for Testing Linear Motor Characteristics. CN Patent 102096042B, 13 February 2013.
13. Ebrahimi, B. Development of Hybrid Electromagnetic Dampers for Vehicle Suspension Systems. Ph.D. Thesis, University of Waterloo, Waterloo, ON, Canada, 2009.
14. Liu, C.; Jiang, K.; Zhang, Y. Design and Use of an Eddy Current Retarder in an Automobile. *Int. J. Automot. Technol.* **2011**, *12*, 611–616. [CrossRef]
15. Heald, D.; Mark, A. Magnetic Braking: Improved Theory. *Am. J. Phys.* **1988**, *56*, 521–522. [CrossRef]
16. Kou, B.; Jin, Y.; Zhang, H.; Zhang, L.; Zhang, H. Development and Application Prospects of the Electromagnetic Damper. *Proc. Chin. Soc. Electr. Eng.* **2015**, *35*, 3132–3143.
17. Gay, S.E.; Ehsani, M. Parametric Analysis of Eddy-Current Brake Performance by 3-D Finite-Element Analysis. *IEEE Trans. Magn.* **2006**, *42*, 319–328. [CrossRef]
18. Fujita, T.; Kitade, K.; Yokoyama, T. Development of Original End Point Detection System Utilizing Eddy Current Variation Due to Skin Effect in Chemical Mechanical Polishing. *Jpn. J. Appl. Phys.* **2011**, *50*, 05–09. [CrossRef]
19. Jow, H.-M.; Ghovanloo, M. Design and Optimization of Printed Spiral Coils for Efficient Transcutaneous Inductive Power Transmission. *IEEE Trans. Biomed. Circuits Syst.* **2007**, *1*, 193. [CrossRef] [PubMed]
20. He, R.; Yi, F.; He, J. A Computation Method for Braking Torque of Eddy Current R19etarder. *Automot. Eng.* **2004**, *26*, 197–200.
21. Baranski, M.; Szelag, W.; Lyskawinski, W. Experimental and simulation studies of partial demagnetization process of permanent magnets in electric motors. *IEEE Trans. Energy Convers.* **2021**, *99*, 1. [CrossRef]

Article

Optimization of the Wire Diameter Based on the Analytical Model of the Mean Magnetic Field for a Magnetically Driven Actuator

Zhangbin Wu, Hongbai Bai, Guangming Xue * and Zhiying Ren

School of Mechanical Engineering and Automation, Fuzhou University, Fuzhou 350116, China
* Correspondence: yy0youxia@163.com; Tel.: +86-150-0514-1625

Abstract: A magnetic field induced by an electromagnetic coil is the key variable that determines the performance of a magnetically driven actuator. The applicability of the empirical models of the coil turns, static resistance, and inductance were discussed. Then, the model of the mean magnetic field induced by the coil was established analytically. Based on the proposed model, the sinusoidal response and square-wave response were calculated with the wire diameter as the decision variable. The amplitude and phase lag of the sinusoidal response, the time-domain response, steady-state value, and the response time of the square-wave response were discussed under different wire diameters. From the experimental and computational results, the model was verified as the relative errors were acceptably low in computing various responses and characteristic variables. Additionally, the optimization on the wire diameter was carried out for the optimal amplitude and response time. The proposed model will be helpful for the analytical analysis of the mean magnetic field, and the optimization result of the wire diameter under limited space can be employed to improve the performance of a magnetically driven actuator.

Keywords: mean magnetic field; wire diameter; coil; sinusoidal response; square-wave response

Citation: Wu, Z.; Bai, H.; Xue, G.; Ren, Z. Optimization of the Wire Diameter Based on the Analytical Model of the Mean Magnetic Field for a Magnetically Driven Actuator. *Aerospace* 2023, 10, 270. https://doi.org/10.3390/aerospace10030270

Academic Editor: Gianpietro Di Rito

Received: 26 December 2022
Revised: 7 March 2023
Accepted: 8 March 2023
Published: 10 March 2023

Copyright: © 2023 by the authors. Licensee MDPI, Basel, Switzerland. This article is an open access article distributed under the terms and conditions of the Creative Commons Attribution (CC BY) license (https://creativecommons.org/licenses/by/4.0/).

1. Introduction

The magnetically driven actuator has been widely used in plenty of engineering fields, including vibration reduction or control, ultra-precision machining, acting fluidic valves, etc. [1–4]. Magnetically driven actuators have also been introduced to quite commonly actuate an aerospace device [5–15], including the electro-hydraulic servo valve.

Optimization of the actuator is quite important to improve the actuator's performance. A magnetic field was generally chosen as the optimization objective function as it influences the output performance of the actuator directly, and is the simplest variable to optimize the actuator, compared to the magnetization/magnetic induction intensity or the displacement. Taking the giant magnetostrictive actuator which employs the giant magnetostrictive material (GMM) as its actuation core as an example, Figure 1 summarizes the generally used optimization methods. From the point of view of magnetic fields, the optimization of actuator performance was generally converted to the promotion of the mean magnetic field in the GMM area, which is equivalent to the maximization of the magneto motive force (MMF) distributed on GMM. Additionally, two methods were used to promote the MMF on GMM, respectively, improving the MMF ratio occupied by GMM and increasing the total MMF.

The first optimization method was accomplished based on some magnetic field models from a "field" or "circuit" method [16–19]. Liang Yan et al. [20] and HyoYoung Kim et al. [21] proposed a mathematic model based on the Biot–Savart law and the finite element model to formulate the three-dimensional magnetic field distribution in a spherical actuator. Abdul Ghani Olabi et al. [22] also established the finite element model of a magnetostrictive actuator for analyzing the magnetic field in the actuator. The proposed models supplied the mean values and distribution characters of magnetic devices, which were quite helpful for the magnetic circuit optimization. Due to the complex magnetic circuit of the hybrid excitation generator used in an energy conversion system, Huihui Geng et al. [23] proposed an analytical method of the main magnetic field, where the Carter coefficient and rotor magnetomotive force were taken as the objective variables. Compared with traditional methods, the proposed method can improve the accuracy of the outputted magnetic field. Jaewook Lee et al. [24] adopted a simplified finite element model to execute structural topology optimization for the high magnetic force of a linear actuator, and they found that the use of a periodic ladder structure was best for magnetic field manipulation. Kim Tien, Xulei Yang et al. [14,25] utilized the finite element model to analyze the distribution of the magnetic field in a giant magnetostrictive actuator separately. By adjusting the permeability of the parts appropriately, the uniformity and mean intensity of the magnetic field within the material could be improved. Some other modeling and optimizing methods for the magnetic field within specified structures can also supply effective references [17,26–29]. On the whole, the circuit model was always used to form a magnetic field model for an analytical analysis. The finite element model [19,30,31] was commonly used to promote magnetic field uniformity. For the mean magnetic field applied to the giant magnetostrictive material, the positively proportional model vs. the coil current [1–3,16,19,28,32–35] was quite commonly used. Then, the closed circuit was verified to be helpful for higher magnetic field intensity [15] as it improved the proportional factor.

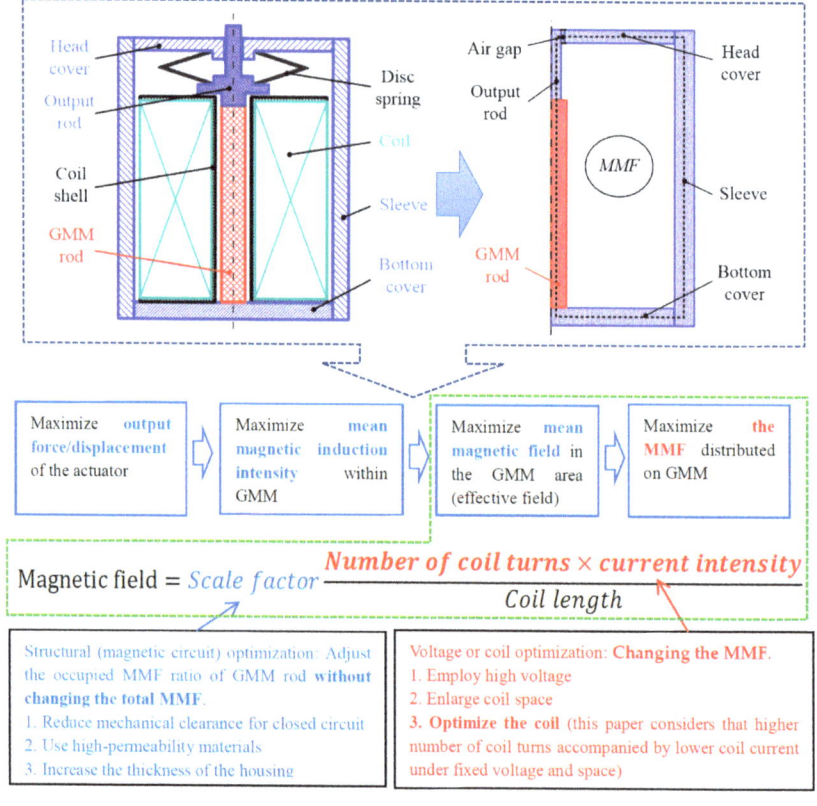

Figure 1. Generally used optimization methods for the giant magnetostrictive actuator: Refs. [16–31] based on the first method and Refs. [36–44] based on the second one.

For the second optimization method, an appropriate voltage waveform or the winging method was considered to directly promote the total MMF and the magnetic field induced by a coil [2]. A high threshold with a low-holding voltage has been widely used in an electromagnetic injector. C.B. Britht et al. [36] and G. Xue et al. [37,38] introduced this type of voltage to stimulate a giant magnetostrictive device. Additionally, it was comprehensively verified that the introduced voltage promoted the response time of the coil current, and the magnetic field, quite efficiently. Manh Cuong Hoang et al. [39] proposed an optimization method of the magnetic field for an electromagnetic actuation system. The maximum magnetic and gradient fields were significantly enhanced by the proposed algorithm compared to the conventional independent control. Haoying Pang et al. [40] proposed a novel spherical coil for the atomic sensor, where the magnetic field uniformity was improved along the axis. Yiwei Lu et al. [18] introduced the magnetically shielded room to enhance the coil magnetic field and reduce power loss for a multi-coil system. Cooperated with the non-dominated sorting genetic algorithm, the design reached prominent reductions in total current and power loss. Yundong Tang et al. [41] introduced two correcting coils to improve the uniformity of the magnetic field for a solenoid coil, while it was not so convenient when the coil space was limited as the correcting coils should have occupied some axial spaces. Some other optimization methods for the coil or contactor [42–44] can also provide useful references for optimizing the magnetic field induced by an electromagnetic coil.

Based on the second optimization method, this paper focuses on coil optimization when the volume of the magnetically driven actuator suitable for an electro-hydraulic servo valve is limited. In this paper, the dynamic magnetic field was tested based on the linear relationship between the magnetic field and coil current. Then, the dimension parameter, static resistance, and static induction were modeled based on empirical equations or mathematical fitting. The mean magnetic field within the coil was modeled, especially its functional relationship with respect to the wire diameter. Then, the sinusoidal and square-wave responses were calculated, and the important characteristic parameters of these responses were extracted. From the calculated and tested results, the influence of the wire diameter on the mean magnetic field was discussed comprehensively for an optimal selection of the wire diameter. During analysis, the relative errors in computing various variables were also given to verify the precision of the proposed model and effectiveness of the optimization. For the magnetically driven actuator, optimized results can be employed to promote the amplitude and response speed of the mean magnetic field, and then to improve the actuator performance.

2. Experimental Methods
2.1. Test Principle

The dynamic magnetic field intensity or magnetic flux density was always measured "indirectly" based on Ampere's circuital theorem or Faraday's law of induction. Based on the former principle, the induced magnetic field and coil current have an ideal positive proportion relationship, which has been a commonly used analytical model of the mean magnetic field in some magnetically driven actuators, especially the giant magnetostrictive actuator. Based on this measuring principle, as long as the coil current is measured, the accurate mean of the magnetic field in a dynamic type can be obtained.

The model was easily given by adding a proportional coefficient to the magnetic field model in an infinitely long solenoid [3,5,14,16,19,33–35]

$$H = C_{HI}\frac{NI}{L} \qquad (1)$$

where H is the magnetic field intensity and I is the current intensity within the coil, C_{HI} is the proportional coefficient of the mean magnetic field intensity; its value belongs to (0,1), N is the number of the coil turn, and L is the coil length.

The following optimization was based on Equation (1)—the optimization is effective as long as the mean magnetic field in the magnetically driven actuator is in direct proportion to the product of the number of coil turns and current intensity. For a hollow coil, Equation (1) was not only capable of computing the mean magnetic field within homogeneous medium, but was also suitable to the local mean magnetic field as long as the whole magnetic circuit was filled locally uniformly and did not have too many reluctance numbers. Considering Equation (1) is suitable for most giant magnetostrictive actuators and some micro-displacement electromagnetic actuators; the optimization proposed in this paper is suitable to these types of actuators.

2.2. Experiment Setup and Parameters

The experimental system was shown in Figure 2. As illustrated in Figure 2a,c, the computer controlled PS3403D digital oscilloscope (with an embedded signal generator) to generate the required waveform signals. The generated signals were then amplified by an ATA304 power amplifier and inputted into the two ends of the coil. The input voltage at both ends of the coil was differentially collected and the coil current was measured by a TA189A current clamp. The measured voltage and current data were delivered into the digital oscilloscope and then into the computer for processing.

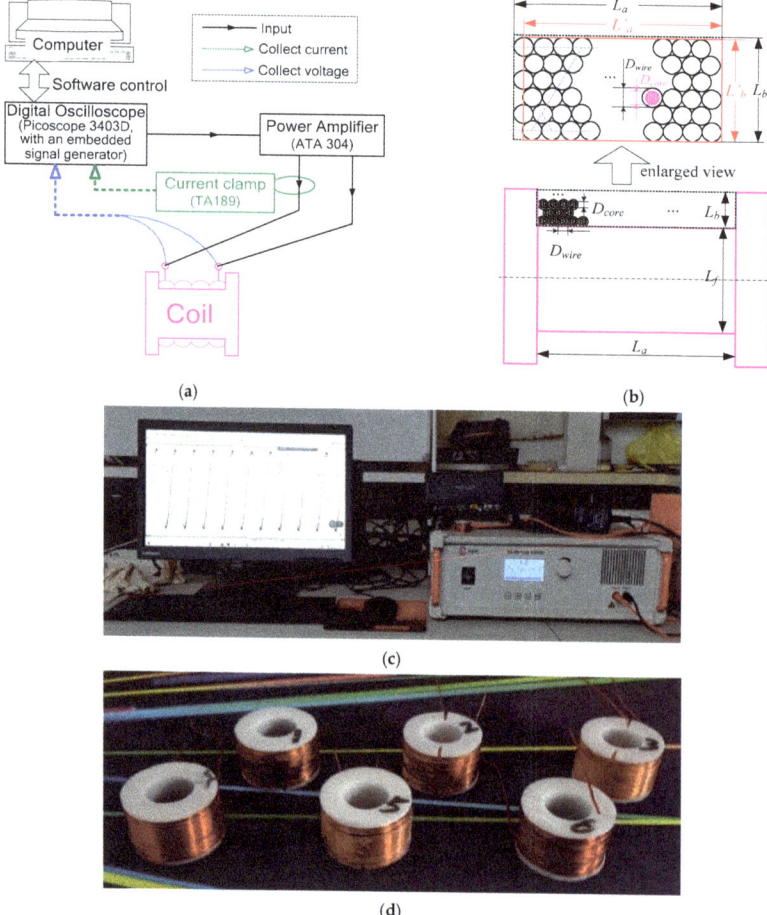

Figure 2. Experimental system and wound coils: (**a**) block diagram of the experimental system; (**b**) dimensioned sectional drawing of the coil, (**c**) photograph of the experimental system; (**d**) photograph of the coils.

Figure 2b,d supplied the sectional drawing and photograph of the coils, where L_a, L_b, L_f represented the coil length, coil thickness, and diameter of the skeleton shaft, respectively, and D_{wire} and D_{core} were the enameled wire diameter and copper core diameter, respectively. The coils were tightly wound by the use of standard enameled wires. Since this article focuses on the optimization of the coil itself, it is not necessary to consider the influence of the iron core or other parts in an actuator. The parameters of the coils are given in Table 1, and some necessary parameters of the skeleton and material are supplied in Table 2. Considering the value of C_{HI} does not affect the increasing or decreasing relationship between the variables; C_{HI} will have no effect on the optimization results. C_{HI} is specified as 0.8 here.

Table 1. The main parameters of the coils.

Coil Label	External Diameter (D_{wire}) [mm]	Core Diameter (D_{wire}) [mm]	Number of Coil Turns (N) [Null]	Resistance (R) [Ω]	Inductance (L) [mH]
Coil 1	0.31	0.27	837	18.325	10.933
Coil 2	0.39	0.35	537	7.472	4.487
Coil 3	0.49	0.44	342	2.994	1.789
Coil 4	0.60	0.55	229	1.342	0.801
Coil 5	0.69	0.64	175	0.767	0.459
Coil 6	0.80	0.74	124	0.410	0.243

Table 2. The main parameters of the skeleton and material.

Parameter (Variable) [Unit]	Value
Coil length (L_a) [mm]	16.5
Coil thickness (L_b) [mm]	6.8
Diameter of skeleton shaft (L_f) [mm]	18.2
Resistivity of copper (ρ) [Ω·m]	1.71×10^{-8}
Proportional coefficient (C_{HI}) [null]	0.8

3. Data Processing and Analysis

3.1. Inherent Characteristic Parameters of Coils

3.1.1. Dimension Parameters

Standard enameled wire has a nominal diameter of the external wire or the copper core. Then, a certain functional relationship can be supplied between the enameled wire diameter D_{wire} and the copper core diameter D_{core}. Figure 3 shows the actual values of D_{wire} and D_{core} and the fitted results using linear functions. It can be seen from Figure 3 that the diameter of copper core is approximately linear vs. the external diameter of enameled wire. With and without an intercept, the fitted linear equations were determined as $D_{core} = 0.9687 D_{wire} - 0.03214$ and $D_{core} = 0.9394 D_{wire}$, respectively. The linear function with an intercept was quite accurate as the relative error was lower than 1.52% when D_{wire} was higher than 0.3 mm and lower than 2.55 mm. In contrast, the linear function without an intercept was not so accurate since the relative error was higher than 5% under some conditions, especially when D_{wire} was quite low.

Though the positively proportional relationship was not suitable to a wide range of dimensions, it may be feasible when the D_{wire} changed within a relatively narrow interval. The coils used in this paper were wound by the wires with diameters of 0.3~0.8 mm. Executing a simple linear fitting, Table 3 supplies the results and relative errors of the two line equations. From computation, the linear equation with intercept was $D_{core} = 0.962 D_{wire} - 0.0277$ and had a relative error lower than 0.84%. In comparison, the linear equation without an intercept $D_{core} = 0.898 D_{wire}$ also had high precision as the relative error was lower than 3.2%. Thus, it is acceptable to use a positively proportional function to describe the relationship between D_{core} and D_{wire} when D_{wire} changes within a narrow interval, which is quite convenient for the following optimization.

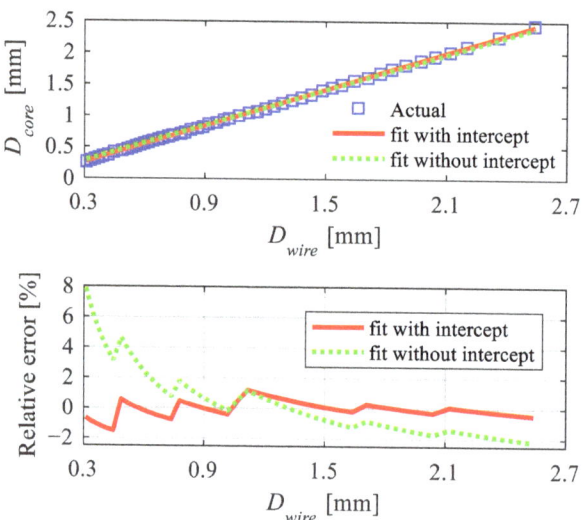

Figure 3. Actual and fitted values of D_{core} simultaneously supplied the relative errors of the fitting lines with and without an intercept.

Table 3. Linear fitting between D_{core} and D_{wire} when $D_{wire} \in [0.3, 0.8]$.

D_{wire}	Tested D_{core}	D_{core} from $0.962 D_{wire} - 0.0277$	Relative Error of $0.962 D_{wire} - 0.0277$ (%)	D_{core} from $0.898 D_{wire}$	Relative Error of $0.898 D_{wire}$ (%)
0.31	0.27	0.2705	0.1926	0.2784	3.1037
0.39	0.35	0.3475	−0.7200	0.3502	0.0629
0.49	0.44	0.4437	0.8364	0.4400	0.0045
0.6	0.55	0.5495	−0.0909	0.5388	−2.0364
0.69	0.64	0.6361	−0.6125	0.6196	−3.1844
0.8	0.74	0.7419	0.2568	0.7184	−2.9189

From the sectional drawing shown in Figure 2b, it can be observed that winding a coil was equivalent to arranging the cross-sectional area of the wire in the rectangular area supplied by the coil skeleton. The coil turns must be an integer; while L_a or L_b was not exactly the integral multiple of D_{wire}, the effective length L_a' and thickness L_b' were a little lower than L_a and L_b, respectively. Coil length or thickness was not fully utilized, and the available area was $L_a' \times L_b'$, which was slightly less than the actual area.

From Figure 2b, the turn number per layer was $\lfloor L_a/D_{wire} \rfloor$ and the number of layers was $\lfloor \frac{L_b - D_{wire}}{\sqrt{3}D_{wire}/2} \rfloor + 1$, so that the accurate value of coil turns was

$$\begin{aligned} N &= C'_f \left\lfloor \frac{L_a}{D_{wire}} \right\rfloor \left(\left\lfloor \frac{L_b - D_{wire}}{\sqrt{3}D_{wire}/2} \right\rfloor + 1 \right) \\ &\leq C'_f \left(\frac{L_a L_b}{\sqrt{3}D_{wire}^2/2} - 0.155 \frac{L_a}{D_{wire}} \right) \\ &\approx C_f \frac{L_a L_b}{\pi D_{wire}^2/4} \end{aligned} \qquad (2)$$

where C'_f was introduced to describing the winding effect, C_f was the filling factor of the enameled wire. $L_a L_b$ was the axis-sectional area of the coil and $\pi D_{wire}^2/4$ was the cross-sectional area of single enameled wire.

From Equation (2), the assumption that N was positively proportional to the ratio of the cross-sectional area of the coil skeleton to D_{wire} was conditional. That was, with the effectiveness of Equation (2), determined by the weight of $C_f' 0.155\, L_a/D_{wire}$ in the total coil turns. Additionally, the relative error of the positively proportional function was $0.155/(1.155\, L_b/D_{wire} - 0.155) \times 100\%$, which was determined by L_b/D_{wire}.

L_b/D_{wire} determined the number of layers and the relative error, which are displayed in Figure 4. From the calculation results, the relative error of $C_f L_a L_b/(\pi D_{wire}^2/4)$ computing N decreased with L_b/D_{wire} increasing. To guarantee that the relative error of Equation (2) is lower than 5.0% in computing N, it should be met that $L_b > 2.8\, D_{wire}$. That is, the coil should be wound with three layers at least. When $L_b < 2.8\, D_{wire}$, one should use $L_b' = \left\lfloor \frac{L_b - D_{wire}}{\sqrt{3} D_{wire}/2} \right\rfloor + 1$ instead of L_b for computations. For the coils in this paper, the values of L_b/D_{wire} under different D_{wire} were higher than $6.8/0.8 = 8.5$ so that the approximate expression in Equation (2) has enough precision.

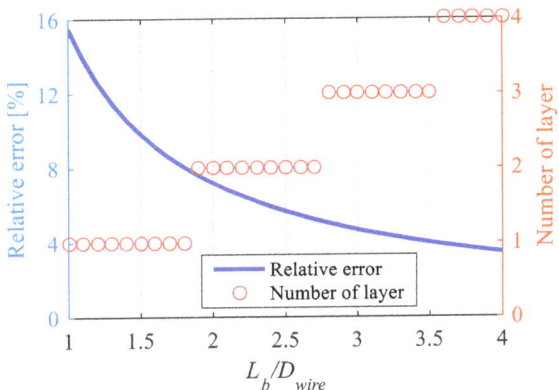

Figure 4. Relative errors of the positively proportional function of the available area of coil skeleton vs. the cross-sectional area of enameled wire.

For convenience, the value of C_f was determined by the mean values of N and D_{wire} so that $C_f = 0.57$. The effect of Equation (2) computing N is shown in Table 4. From the calculation results, the model of coil turn can predict the practical coil turn effectively as the relative error was lower than 2.8%.

Table 4. Coil turns from the test and model.

Coil Label	Coil Turns from Test	Coil Turns from Model [1]	Relative Error (%)
1	837	847.33	1.23
2	537	535.36	−0.30
3	342	339.15	−0.83
4	229	226.19	−1.23
5	175	171.03	−2.27
6	124	127.23	2.61

[1] Cannot be an integer.

From above analysis, the empirical equations in describing the relationships between D_{core}, D_{wire} and N were written as

$$\begin{cases} \hat{D}_{core} = 0.962 D_{wire} - 0.0277 \text{ or } 0.898 D_{wire} \\ N = C_f \dfrac{L_a L_b}{\pi D_{wire}^2/4} \end{cases} \quad (3)$$

3.1.2. Static Resistance and Static Inductance

Static resistance and inductance are the key parameters to determine the current response of a coil with an unobvious skin effect. Based on the empirical expression of inductance L and the basic equation of resistance R, the model can be easily established as

$$\begin{cases} L = 4\pi C_{L0} N^2 = \dfrac{C_L C_f^2}{D_{wire}^4} \\ R = \rho \dfrac{N(L_f+L_b)}{D_{core}^2/4} = 16\rho C_f \dfrac{L_a L_b (L_f+L_b)}{\pi D_{core}^2 D_{wire}^2} \end{cases} \quad (4)$$

where C_{L0} and C_L were two parameters dependent on L_a, L_b, L_f while independent of other variables and met $C_L = 64\, C_{L0}(L_a L_b)^2/\pi$; ρ was the resistivity of copper.

Figure 5 displays the relationships between L, R, and D_{wire} from the experiment and computation. From the results, it was easily reached that both L and R were monotonically decreasing functions vs. D_{wire}. More specifically, as concluded from the expression of N in Equation (3) and D_{core} = 0.898 D_{wire}, both R and L were inversely proportional functions vs. D_{wire}^4 (also N^2). The model was in good agreement with the experiment as the relative errors of the model in computing R and L were lower than 3.1% and 2.8%, respectively.

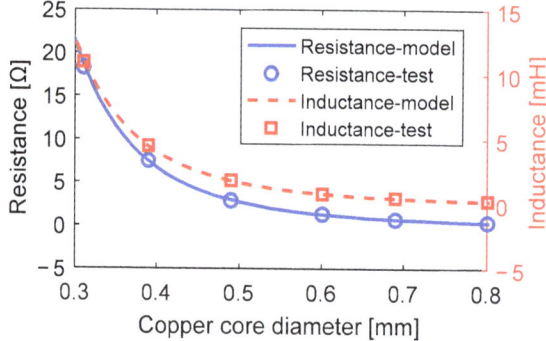

Figure 5. Curves of the static resistance and static inductance vs. the enameled wire diameter.

3.2. Sinusoidal Response

Equivalent to the series connection of an inductor and a resistor, the electromagnetic coil was generally modeled as a first-order linear time-invariant system model. Additionally, the amplitude-frequency and phase-frequency characteristics are the most important characteristics of the sinusoidal response of the coil.

Stimulated by a sinusoidal voltage $U(t) = U_{amp}\sin(\omega t)$, the current response within the coil can be calculated by $I(t) = I_{amp}\sin(\omega t - \varphi_I)$, where ω is the angular frequency of the input and φ_I is the phase lag of the coil current compared to the voltage. From the theory of the linear time-invariant system, the amplitude ratio function is $A_I = I_{amp}/U_{amp} = 1/(R^2 + \omega^2 L^2)^{1/2}$, and $\tan\varphi_I = \omega L/R$. Substituting Equation (4) into these expressions, one obtains

$$\begin{cases} A_I = \dfrac{\pi D_{core}^2 D_{wire}^2}{C_f L_a \sqrt{\left[\dfrac{16\rho L_b \cdot}{(D_f+L_b)}\right]^2 + \left(\dfrac{\omega C_L C_f D_{core}^2}{D_{wire}^2}\right)^2}} \approx \dfrac{0.806\pi D_{wire}^4}{C_f L_a \sqrt{\left[\dfrac{16\rho L_b \cdot}{(D_f+L_b)}\right]^2 + (0.806\omega C_L C_f)^2}} \\ \varphi_I = \arctan\dfrac{\omega C_L C_f}{16\rho L_b(D_f+L_b)} \cdot \dfrac{D_{core}^2}{D_{wire}^2} \approx \arctan\dfrac{0.0504\omega C_L C_f}{\rho L_b(D_f+L_b)} \end{cases} \quad (5)$$

From the empirical equation of the mean magnetic field given in Equation (1), the amplitude radio to inputted voltage of the magnetic field A_H and the lagging phase of the magnetic field φ_H can be easily reached as $C_{HI}NA_I/L_a$ and $\varphi_H = \varphi_I$. By substituting Equation (5) into these two equations, one obtains

$$\begin{cases} A_H = \dfrac{4C_{HI}L_bD_{core}^2}{L_a\sqrt{\left[\dfrac{16\rho L_b \cdot}{(D_f+L_b)}\right]^2 + \left(\dfrac{\omega C_L C_f D_{core}^2}{D_{wire}^2}\right)^2}} \approx \dfrac{3.226C_{HI}L_bD_{wire}^2}{L_a\sqrt{\left[\dfrac{16\rho L_b \cdot}{(D_f+L_b)}\right]^2 + (0.806\omega C_L C_f)^2}} \\ \varphi_H = \arctan\dfrac{\omega C_L C_f}{16\rho L_b(D_f+L_b)} \cdot \dfrac{D_{core}^2}{D_{wire}^2} \approx \arctan\dfrac{0.0504\omega C_L C_f}{\rho L_b(D_f+L_b)} \end{cases} \quad (6)$$

Changing the frequency from 10 Hz to 1000 Hz, Figure 6 shows the tested and calculated amplitude ratios and phase lags of the magnetic field with respect to the inputted voltage. To demonstrate the influence of the wire diameter more clearly, the wire diameter was plotted on the horizontal axis. From the tested and calculated results, a wider wire diameter is quite helpful for a higher magnetic field amplitude as the amplitude ratio increased faster with an increase in wire diameter. On the contrary, the wire diameter has little influence on the phase lag of the magnetic field, which represents the response time of the magnetic field from 0 to some required proportion of a steady-state value.

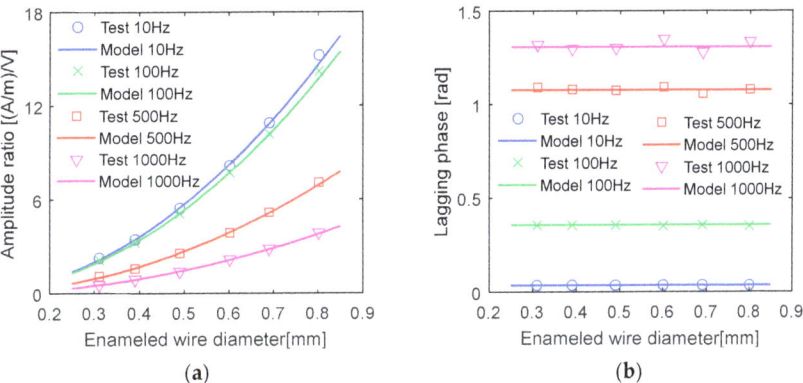

Figure 6. Amplitude ratio and phase lag of the magnetic field under different wire diameters: (**a**) curves of the amplitude ratio vs. enameled wire diameter; (**b**) curves of the phase lag vs. enameled wire diameter.

Figure 7 shows the relative errors of the model under various frequencies. For predicting the amplitude ratio, the calculation error was lower than 2.0% when the wire diameter was between 0.39 mm and 0.69 mm. The model accuracy was a little lower when the wire diameter was wider than 0.8 mm or narrower than 0.31 mm, as the relative errors at these points were higher than 5%; this was acceptable as the errors were still lower than 6.4%. For computing the lagging phase, the relative errors under different parameters, including various frequencies and wire diameters, were lower than 3.2%, which showed high precision of the model in predicting the lagging phase of the magnetic field. A low calculation accuracy regarding the computing amplitude ratio was mainly caused by poor winding when the coil wire was quite thin or thick. On the whole, the proposed models for the magnetic field amplitude and lagging phase were verified by the low relative errors under most conditions.

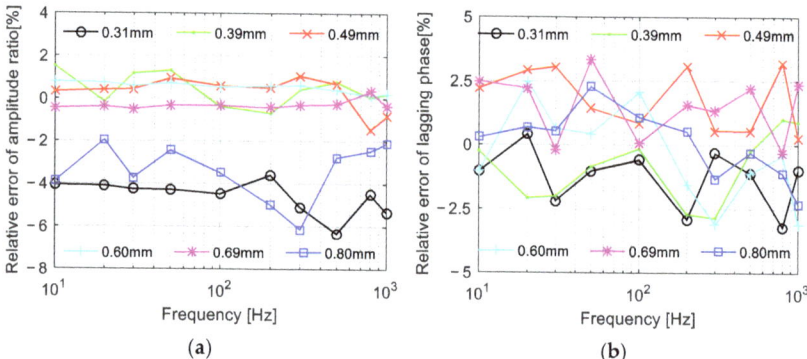

Figure 7. The relative errors of the model in computing the coil current under a harmonic voltage: (**a**) the relative errors of computing the amplitude ratio; (**b**) the relative errors of computing the phase lag.

Figure 8 shows the relationships between A_H and D_{wire}^2. From Figure 8, the linear relationship between the amplitude ratio of the magnetic field and the square of the wire diameter was verified as the tested points under a certain frequency were roughly plotted in a line passing through the origin.

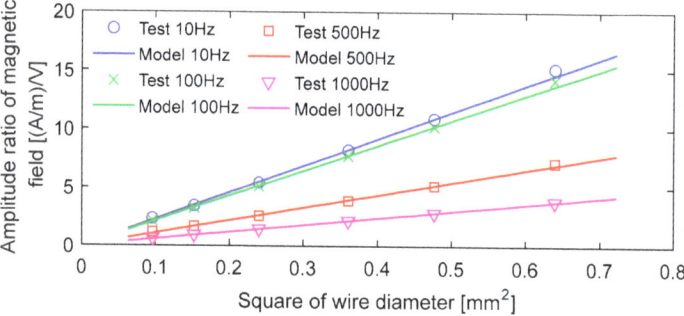

Figure 8. Linear curves of A_H-D_{wire}^2 from the model and test under different frequencies.

3.3. Square-Wave Response

3.3.1. Time-Domain Response

In addition to the sinusoidal voltage, the direct current (DC) square-wave voltage is frequently used, especially to drive an on–off-type actuator.

For the square-wave response, more attention should be paid to the transient-state process. Additionally, based on the first-order linear time-invariant system, the transient-state current within the coil is

$$I(t) = \frac{U_{st}}{R} + (I_0 - \frac{U_{st}}{R})e^{-\frac{R}{L}t} \tag{7}$$

where I_0 is the initial value of the coil current, U_{st} is the steady-state amplitude of the voltage. Equation (7) was suitable to both the charging and discharging process of the coil. For charging, $I_0 = 0$. For discharging, $U_{st} = 0$.

From Equation (4), the reciprocal of the time-constant used in Equation (7) was

$$\frac{R}{L} = \frac{\rho(L_f + L_b)D_{wire}^2}{4C_{L0}C_f L_a L_b D_{core}^2} \tag{8}$$

By substituting Equations (7) and (8) into $H(t) = C_{HI}NI(t)/L_a$, one obtains the transient-state response of the magnetic field

$$\begin{aligned}
H(t) &= \frac{C_{HI}C_f L_b}{\pi D_{wire}^2/4}\left[\frac{U_{st}}{R} + (I_0 - \frac{U_{st}}{R})e^{-\frac{R}{L}t}\right] \\
&= \frac{4C_{HI}C_f L_b}{\pi D_{wire}^2}\left[\frac{\pi D_{core}^2 D_{wire}^2 U_{st}}{16\rho C_f L_a L_b(L_f+L_b)} + \left(I_0 - \frac{\pi D_{core}^2 D_{wire}^2 U_{st}}{16\rho C_f L_a L_b(L_f+L_b)}\right)e^{-\frac{\rho(L_f+L_b)D_{wire}^2}{4C_{L0}C_f L_a L_b D_{core}^2}t}\right] \\
&= \frac{C_{HI}D_{core}^2}{4\rho L_a(L_f+L_b)}U_{st} + \frac{4C_{HI}C_f L_b}{\pi D_{wire}^2}I_0 e^{-\frac{\rho(L_f+L_b)D_{wire}^2}{4C_{L0}C_f L_a L_b D_{core}^2}t} - \frac{C_{HI}D_{core}^2}{4\rho L_a(L_f+L_b)}U_{st}e^{-\frac{\rho(L_f+L_b)D_{wire}^2}{4C_{L0}C_f L_a L_b D_{core}^2}t}
\end{aligned} \quad (9)$$

The inputted voltage was generated with an amplitude of 2 V and a high-voltage duration of 20 ms to guarantee the coil current reaching the steady state. The time-domain magnetic fields are shown in Figure 9. From the test and model, the proposed model precisely calculated the amplitudes and effectively described the curve shapes under different wire diameters as the transient-state results were also quite close.

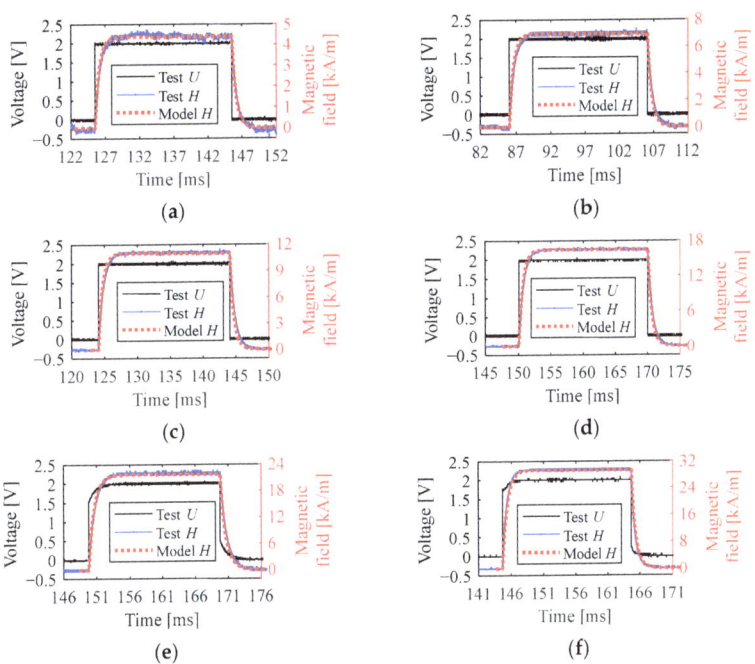

Figure 9. The dynamic magnetic field under square-wave input: (**a**) D_{wire} = 0.31 mm; (**b**) D_{wire} = 0.39 mm; (**c**) D_{wire} = 0.49 mm; (**d**) D_{wire} = 0.60 mm; (**e**) D_{wire} = 0.69 mm; (**f**) D_{wire} = 0.80 mm.

3.3.2. Steady-State Value and Response Time

The steady-state value and response time are the most important characteristic parameters of the on–off-type actuator. When the square-wave voltage maintains a high level for a long enough duration, the steady-state response of the magnetic field H_{st} can be easily acquired from Equation (9), as

$$H_{st} = \frac{C_{HI}D_{core}^2}{4\rho L_a(L_f+L_b)}U_{st} \approx \frac{0.2016 C_{HI}D_{wire}^2}{\rho L_a(L_f+L_b)}U_{st} \quad (10)$$

By exacting the mean value of the magnetic field in the steady-state stage, Figure 10 shows the curves of H_{st} vs. D_{wire} from the test and model. From the tested and calculated results, a higher D_{wire} is helpful for a higher H_{st}. More specifically, H_{st} was positively proportional to D_{wire}^2, as explained in Equation (10). Thus, the change law of the H_{st} under the square wave was the same as one of the magnetic field amplitudes under the sinusoidal voltage. It was easily illustrated that both the functions of $1/(R^2 + \omega^2 L^2)^{1/2}$ and $1/R$ can be expressed by the quartic function vs. the wire diameter approximately. In addition, the relatively errors under different wire diameters were less than 1.2% thus, the model can predict the steady-state magnetic field quite effectively.

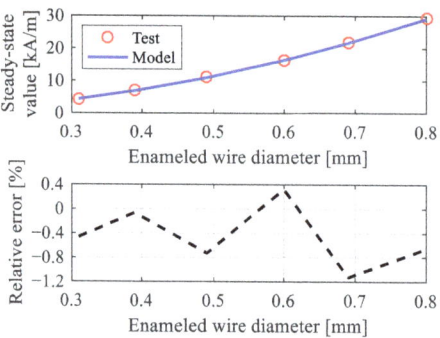

Figure 10. Steady-state magnetic field of the square-wave response from the test and model.

There are two commonly used response times—the time from 0 to the required intensity and the time from 0 to the required proportion of the steady-state value. The former was especially concerned when the high-opening voltage was employed and the latter was generally concerned when the standard square-wave voltage was employed (or the duty cycle was adjusted but not the amplitude in the voltage wave).

By imposing I_{req} the required intensity of the coil current and substituting I_{req} into Equation (7), the response time t_{Iv} can be reached, as

$$t_{Iv} = \frac{L}{R} \ln\left(1 + \frac{I_{req}}{U_{st}/R - I_{req}}\right)$$

$$= \frac{C_L C_f}{16\rho L_b (D_f + L_b)} \frac{D_{core}^2}{D_{wire}^2} \ln\left(1 + \frac{I_{req}}{\pi D_{core}^2 D_{wire}^2 U_{st} / \left[16\rho C_f L_a L_b (D_f + L_b)\right] - I_{req}}\right) \quad (11)$$

Similarly, by substituting the required magnetic field H_{req} into Equation (9), one obtains the response time to the specified magnetic field intensity t_{Hv}, as

$$t_{Hv} = \frac{L}{R} \ln\left(1 + \frac{L_a H_{req}}{C_{HI} N U_{st}/R - L_a H_{req}}\right)$$

$$= \frac{C_L C_f}{16\rho L_b (D_f + L_b)} \frac{D_{core}^2}{D_{wire}^2} \ln\left(1 + \frac{L_a H_{req}}{C_{HI} D_{core}^2 U_{st} / \left[4\rho (D_f + L_b)\right] - L_a H_{req}}\right) \quad (12)$$

$$\approx \frac{0.0504 \omega C_L C_f}{\rho L_b (D_f + L_b)} \ln\left(1 + \frac{L_a H_{req}}{0.2016 C_{HI} D_{wire}^2 U_{st} / \left[\rho (D_f + L_b)\right] - L_a H_{req}}\right)$$

From the calculated result, it can be observed that a thicker wire is better for reducing both the response time of the coil current and one of the magnetic fields, while the change degree is different. The enameled wire diameter has more influence on the response time of the coil current than that of the magnetic field as t_{Iv} is in the function form of $a \ln[1 + b/(cx^2 - b)]$ vs. D_{wire} while t_{Hv} is expressed by $a \ln[1 + b/(cx^4 - b)]$ vs. D_{wire}.

The response time to a specified proportion of the steady-state value can be easily deduced from Equations (10) and (12). For a given proportion p, the corresponding intensities of the coil current and magnetic field are, respectively, $I_{req} = p(U_{st}/R)$ or $H_{req} = p[C_{HI}N(U_{st}/R)/L_a]$. By substituting the two expressions into Equations (10) and (11), one obtains the response time to a specified proportion, as

$$\begin{aligned} t_{Hp} = t_{Ip} &= \frac{L}{R}\ln\left(\frac{1}{1-p}\right) \\ &= \frac{C_L C_f}{16\rho L_b (D_f + L_b)} \frac{D_{core}^2}{D_{wire}^2} \ln\left(\frac{1}{1-p}\right) \\ &\approx \frac{0.0504\omega C_L C_f}{\rho L_b (D_f + L_b)} \ln\left(\frac{1}{1-p}\right) \end{aligned} \quad (13)$$

where p is a constant belonging to (0, 1).

Compared to t_{Hv}, the factors influencing t_{Hp} were almost independent of D_{wire}. More specifically, t_{Hp} was just determined by the ratio of L/R. The value of L/R was only slightly influenced by D_{wire}; optimizing the wire diameter would be helpless to promote this type of response speed.

Figure 11 shows the two types of response times from the tested and calculated results; the specified intensities H_{req} were 3 kA/m, 3.5 kA/m, and 4 kA/m, and the specified proportions p were 0.7, 0.8, and 0.9. Just as predicted by the model, H_{req} was effectively reduced by increasing D_{wire}. Furthermore, H_{req} declined fast first and then slowly with D_{wire} increasing. For the value of t_{Hp}, it changed slightly with D_{wire} increasing. The model was verified as the calculated results were consistent with the experimental data.

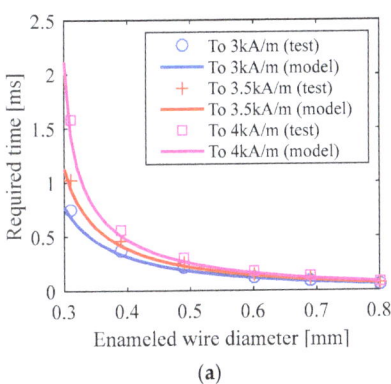 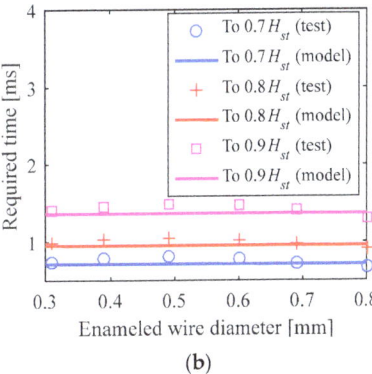

Figure 11. The response time of the square-wave response from the test and model: (**a**) the time from 0 to specified intensities, respectively, of 3 kA/m, 3.5 kA/m, and 4 kA/m; (**b**) the time from 0 to the specified proportions of the steady-state response, respectively, of 0.7, 0.8, and 0.9.

On the whole, increasing the wire diameter is quite helpful for reducing the response time from 0 to a specified value of the coil current or magnetic field, while failing to improve the response speed from 0 to the steady-state or any other proportional value. Therefore, for an electromagnetic actuator stimulated by a high-open-low-hold-type voltage, a coil with a wider wire diameter will be stimulated more quickly to save the response time of the whole actuator, while when a traditional square-wave voltage is introduced, an adjustment in the wire diameter is helpless.

4. Conclusions

An analytical model of the mean magnetic field for the hollow cylindrical coil used in a magnetically driven actuator was proposed in this paper. Additionally, the selection of the enameled wire diameter was optimized for a high-amplitude and fast-response magnetic field based on the model.

(1) The resistance and inductance are inversely proportional functions vs. the quartic of the enameled wire diameter. Under the sinusoidal voltage, a wider wire diameter is quite helpful for a higher magnetic field amplitude while it has little influence on the phase lag of the magnetic field. Under the square-wave voltage, the steady-state magnetic field was positively proportional to the square of the wire diameter, as a wider wire diameter is helpful for a higher steady-state magnetic field. Regarding the response speed, increasing the wire's diameter is helpful for reducing the response time from 0 to the specified intensity, while it is helpless to improve the response speed from 0 to the steady-state or any other proportional value.

(2) The proposed model was verified as the calculated results from the model were in good agreement with the experimental results. Specifically, the relative errors of the model in computing the resistance and the inductance were lower than 3.1% and 2.8%, respectively. For predicting the sinusoidal response, the errors were lower than 6.4% (lower than 2.0% under most conditions) in computing the amplitude and lower than 3.2% in computing the lagging phase. For predicting the square-wave response, the model calculated the amplitudes with errors lower than 1.2% and described the curve shape effectively.

This paper was devoted to the promotion of the output performance of the whole magnetically driven actuator without considering the coil quality factor or power loss. Further work can focus on reducing the power loss of the coil.

Author Contributions: Conceptualization. G.X.; Data curation, Z.W.; Formal analysis, Z.W.; Investigation: Z.W.; Software and visualization, Z.W. and G.X.; Methodology, G.X.; Validation, G.X. and H.B.; Funding acquisition, H.B. and G.X.; Writing—review and editing, Z.W. and G.X.; Writing—review and editing, H.B., G.X. and Z.R.; Supervision: Z.R. All authors have read and agreed to the published version of the manuscript.

Funding: This work was supported by the Young and Middle-aged Teachers Education and Research Project (Science and Technology) of Fujian Province (No. JAT220016) and First Batch of Yin Ling Fund (No. ZL3H39).

Data Availability Statement: The data presented in this study are available on request from the corresponding author.

Acknowledgments: The authors thank Zhaoshu Yang (in China Astronaut Research and Training Center) and Tuo Li (in Officers College of PAP) for their improvement in the English of this article.

Conflicts of Interest: The authors declare no conflict of interest.

References

1. Zhao, T.; Yuan, H.; Pan, H.; Li, B. Study on the rare-earth giant magnetostrictive actuator based on experimental and theoretical analysis. *J. Magn. Magn. Mater.* **2018**, *460*, 509–524. [CrossRef]
2. Xue, G.; Zhang, P.; Li, X.; He, Z.; Wang, H.; Li, Y.; Ce, R.; Zeng, W.; Li, B. A review of giant magnetostrictive injector (GMI). *Sensor. Actuat. A-Phys.* **2018**, *273*, 159–181. [CrossRef]
3. Braghin, F.; Cinquemani, S.; Resta, F. A model of magnetostrictive actuators for active vibration control. *Sensor. Actuat. A-Phys.* **2011**, *165*, 342–350. [CrossRef]
4. Sobczyk, M.; Wiesenhütter, S.; Noennig, J.R.; Wallmersperger, T. Smart materials in architecture for actuator and sensor applications: A review. *J. Intell. Mater. Syst. Struct.* **2022**, *33*, 379–399. [CrossRef]
5. Zhang, X.; Zhang, Y.; Guo, Y.; Li, L. Engineering Vibration Control Based on Rotating Actuator. *J. Nanjing Univ. Aeronaut. Astronaut.* **2018**, *50*, 601–610.
6. McIvor, B.; Chahl, J. Energy Efficiency of Linear Electromagnetic Actuators for Flapping Wing Micro Aerial Vehicles. *Energies* **2020**, *13*, 1075. [CrossRef]
7. Li, Y.; Wu, J.; Hu, J.; Wang, Q.; Liu, C. Analysis of Factors Influencing Peak Torques and Peak Losses of Rotary Voice Coil Actuators Used in Aerospace. *IEEE Access* **2021**, *9*, 57120–57126.
8. Li, W.; Shi, G.; Asme. Review of Aerospace Actuator Technology. In Proceedings of the 33rd Bath/ASME International Symposium on Fluid Power and Motion Control (FPMC), online, 9–11 September 2020.

9. Kang, J.G.; Kwon, J.Y.; Lee, M.S. A Dynamic Power Consumption Estimation Method of Electro-mechanical Actuator for UAV Modeling and Simulation. *Int. J. Aeronaut. Space Sci.* **2022**, *23*, 233–239. [CrossRef]
10. Ashpis, D.E.; Thurman, D.R. Dielectric Barrier Discharge (DBD) Plasma Actuators for Flow Control in Turbine Engines: Simulation of Flight Conditions in the Laboratory by Density Matching. *Int. J. Turbo Jet-Engines* **2019**, *36*, 157–173. [CrossRef]
11. Roussel, J.; Budinger, M.; Ruet, L. Preliminary Sizing of the Electrical Motor and Housing of Electromechanical Actuators Applied on the Primary Flight Control System of Unmanned Helicopters. *Aerospace* **2022**, *9*, 473. [CrossRef]
12. Misra, R.; Wisniewski, R.; Zuyev, A. Attitude Stabilization of a Satellite Having Only Electromagnetic Actuation Using Oscillating Controls. *Aerospace* **2022**, *9*, 444. [CrossRef]
13. Yang, X.; Zhu, Y.; Ji, L.; Fei, S.; Guo, Y. Experimental investigation and characteristic analysis of a giant magnetostrictive materials-based electro-hydrostatic actuator. *AcAAS* **2016**, *37*, 2839–2850.
14. Yang, X.; Zhu, Y.; Fei, S.; Ji, L.; Guo, Y. Magnetic field analysis and optimization of giant magnetostrictive electro-hydrostatic actuator. *JAerP* **2016**, *31*, 2210–2217.
15. Fang, S. *Research on Robotic Motion Planning and Compliance Control for Ultrasonic Strengthening of Aviation Blade Surface*; Beijing Jiaotong University: Beijing, China, 2021.
16. Grunwald, A.; Olabi, A.G. Design of a magnetostrictive (MS) actuator. *Sensor. Actuat. A-Phys.* **2008**, *144*, 161–175. [CrossRef]
17. Shi, X.; Ren, C.; Li, Y.; Ding, G. Design, Modeling, and Optimization of a Bistable Electromagnetic Actuator with Large Deflection. *IEEE Magn. Lett.* **2021**, *12*, 2102505. [CrossRef]
18. Lu, Y.W.; Yang, Y.; Zhang, M.; Wang, R.M.; Zhu, B.L.; Jiang, L. Magnetic Enhancement-Based Multi-Objective Optimization Design of the Large-Scale High-Intensity Homogeneous Magnetic Field Coil System. *IEEE Trans. Magn.* **2022**, *58*, 9. [CrossRef]
19. Yan, H.; Gao, H.; Hao, H.; Zhang, Z.; Zhuang, F. Design and Simulation of Exciting Coil in Rare Earth Giant Magnetostrictive Actuator. *Mech. Sci. Technol. Aerosp. Eng.* **2019**, *38*, 1569–1575.
20. Yan, L.; Wu, Z.; Jiao, Z.; Chen, C.-Y.; Chen, I.M. Equivalent energized coil model for magnetic field of permanent-magnet spherical actuators. *Sensor. Actuat. A-Phys.* **2015**, *229*, 68–76. [CrossRef]
21. Kim, H.; Kim, H.; Gweon, D. Magnetic field analysis of a VCM spherical actuator. *Sensor. Actuat. A-Phys.* **2013**, *195*, 38–49. [CrossRef]
22. Olabi, A.G.; Grunwald, A. Computation of magnetic field in an actuator. *Simul. Model. Pract. Theory* **2008**, *16*, 1728–1736. [CrossRef]
23. Geng, H.; Zhang, X.; Yan, S.; Tong, L.; Ma, Q.; Xu, M.; Zhang, Y.; Han, Y. Magnetic field analysis and performance optimization of hybrid excitation generators for vehicles. *Sustain. Energy Technol. Assess.* **2022**, *52*, 102200. [CrossRef]
24. Lee, J.; Dede, E.M.; Banerjee, D.; Iizuka, H. Magnetic force enhancement in a linear actuator by air-gap magnetic field distribution optimization and design. *Finite Elem. Anal. Des.* **2012**, *58*, 44–52. [CrossRef]
25. Yang, X.; Zhu, Y.; Zhu, Y. Characteristic investigations on magnetic field and fluid field of a giant magnetostrictive material-based electro-hydrostatic actuator. *Proc. Inst. Mech. Eng. Part G J. Aerosp. Eng.* **2018**, *232*, 847–860. [CrossRef]
26. Golda, D.; Culpepper, M.L. Modeling 3D magnetic fields for precision magnetic actuators that use non-periodic magnet arrays. *Precis. Eng.* **2008**, *32*, 134–142. [CrossRef]
27. Chen, Z.; Yang, X.; Li, S.; Zhang, Z.; Chen, Y. Dynamic modeling of stack giant magnetostrictive actuator with magnetic equivalent network considering eddy current effect. *J. Appl. Phys.* **2022**, *131*, 224503. [CrossRef]
28. Chen, L.; Zhu, Y.; Ling, J.; Zhang, M. Development and Characteristic Investigation of a Multidimensional Discrete Magnetostrictive Actuator. *IEEE-ASME T. Mech.* **2022**, *27*, 2071–2079. [CrossRef]
29. Ramirez-Laboreo, E.; Roes, M.G.L.; Sagues, C. Hybrid Dynamical Model for Reluctance Actuators Including Saturation, Hysteresis, and Eddy Currents. *IEEE-ASME T. Mech.* **2019**, *24*, 1396–1406. [CrossRef]
30. Lyu, Z.; Zhang, J.; Wang, S.; Zhao, X.; Li, T. Optimal design of multi-coil system for generating uniform magnetic field based on intelligent optimization algorithm and finite element method. *J. Beijing Univ. Aeronaut.* **2019**, *45*, 980–988.
31. Sun, J.; Ren, J.; Le, Y.; Wang, H. Analysis of Air-Gap Magnetic Field and Structure Optimization Design of Hollow-Cup Motor. *Aerospace* **2022**, *9*, 549. [CrossRef]
32. Niu, M.; Yang, B.; Yang, Y.; Meng, G. Modelling and parameter design of a 3-DOF compliant platform driven by magnetostrictive actuators. *Precis. Eng.* **2020**, *66*, 255–268. [CrossRef]
33. Zhou, J.; He, Z.; Shi, Z.; Song, J.; Li, Q. Design and experimental performance of an inertial giant magnetostrictive linear actuator. *Sensor. Actuat. A-Phys.* **2020**, *301*, 111771. [CrossRef]
34. Liu, H.; Jia, Z.; Wang, F.; Zong, F. Research on the constant output force control system for giant magnetostrictive actuator disturbed by external force. *Mechatronics* **2012**, *22*, 911–922. [CrossRef]
35. Xue, G.; Zhang, P.; He, Z.; Li, D.; Yang, Z.; Zhao, Z. Displacement model and driving voltage optimization for a giant magnetostrictive actuator used on a high-pressure common-rail injector. *Mater. Des.* **2016**, *95*, 501–509. [CrossRef]
36. Bright, C.B.; Garza, J.C. *Possible Very High Speed Rate Shaping Fuel Injector*; SAE Technical Paper 2007-01-4113; SAE: Warrendale, PA, USA, 2007.
37. Xue, G.; Zhang, P.; He, Z.; Li, D.; Huang, Y.; Xie, W. Design and experimental study of a novel giant magnetostrictive actuator. *J. Magn. Magn. Mater.* **2016**, *420*, 185–191. [CrossRef]
38. Xue, G.; Ge, J.; Ning, P.; Zhou, J.; Wang, K.; Cheng, Z.; Pei, G. Simulation studies on the boot shape injection of a giant magnetostrictive injector. *Sci. Rep.* **2021**, *11*, 22999. [CrossRef]

39. Hoang, M.C.; Kim, J.; Park, J.-O.; Kim, C.-S. Optimized magnetic field control of an electromagnetic actuation system for enhanced microrobot manipulation. *Mechatronics* **2022**, *85*, 102830. [CrossRef]
40. Pang, H.Y.; Duan, L.H.; Quan, W.; Wang, J.; Wu, W.F.; Fan, W.F.; Liu, F. Design of Highly Uniform Three Dimensional Spherical Magnetic Field Coils for Atomic Sensors. *IEEE Sens. J.* **2020**, *20*, 11229–11236. [CrossRef]
41. Tang, Y.D.; Jin, T.; Flesch, R.C.C.; Gao, Y.M. Improvement of solenoid magnetic field and its influence on therapeutic effect during magnetic hyperthermia. *J. Phys. D-Appl. Phys.* **2020**, *53*, 6. [CrossRef]
42. Rong, C.; He, Z.B.; Xue, G.M.; Zhou, J.T.; Zhao, Z.L. Physics-based modeling and multi-objective parameter optimization of excitation coil for giant magnetostrictive actuator used on fuel injector. *Meas. Control* **2022**, *55*, 421–436. [CrossRef]
43. Fang, S.; Chen, Y.; Yang, Y. Optimization design and energy-saving control strategy of high power dc contactor. *Int. J. Electr. Power Energy Syst.* **2020**, *117*, 105633. [CrossRef]
44. Bright, C.B.; Faidley, L.; Witthauer, A.; Rickels, E.; Donlin, T. *Programmable Diesel Injector Transducer Test Results*; SAE Technical Paper 2011-01-0381; SAE: Warrendale, PA, USA, 2011.

Disclaimer/Publisher's Note: The statements, opinions and data contained in all publications are solely those of the individual author(s) and contributor(s) and not of MDPI and/or the editor(s). MDPI and/or the editor(s) disclaim responsibility for any injury to people or property resulting from any ideas, methods, instructions or products referred to in the content.

Article

A Preliminary Top-Down Parametric Design of Electromechanical Actuator Position Control

Jean-Charles Maré

INSA-Institut Clément Ader (CNRS UMR 5312), 31400 Toulouse, France; jean-charles.mare@insa-toulouse.fr

Abstract: A top-down process is proposed and virtually validated for the position control of electromechanical actuators (EMA) that use conventional cascade controllers. It aims at facilitating the early design phases of a project by providing a straightforward mean that requires simple algebraic calculations only, from the specified performance and the top-level EMA design parameters. This makes it possible to include realistic control considerations in the preliminary sizing and optimisation phase. The position, speed and current controllers are addressed in sequence. This top-down process is based on the generation and use of charts that define the optimal position gain, speed loop second-order damping factor and natural frequency with respect to the specified performance of the position loop. For each loop, the control design formally specifies the required dynamics and the digital implementation of the following inner loop. A noncausal flow chart summarises the equations used and the interdependencies between data. This potentially allows changing which ones are used as inputs. The process is virtually validated using the example of a flight control actuator. This is achieved with resort to the simulation of a realistic lumped-parameter model, which includes any significant functional and parasitic effects. The virtual tests are run following a bottom–up approach to highlight the pursuit and rejection performance. Using low-, medium- and high-excitation magnitudes, they show the robustness of the controllers against nonlinearities. Finally, the simulation results confirm the soundness of the proposed process.

Keywords: actuator; aerospace; electromechanical; flight control; friction; modelling; position control; preliminary design; simulation; validation

Citation: Maré, J.-C. A Preliminary Top-Down Parametric Design of Electromechanical Actuator Position Control. *Aerospace* 2022, 9, 314. https://doi.org/10.3390/aerospace9060314

Academic Editor: Gianpietro Di Rito

Received: 28 March 2022
Accepted: 6 June 2022
Published: 9 June 2022

Publisher's Note: MDPI stays neutral with regard to jurisdictional claims in published maps and institutional affiliations.

Copyright: © 2022 by the author. Licensee MDPI, Basel, Switzerland. This article is an open access article distributed under the terms and conditions of the Creative Commons Attribution (CC BY) license (https://creativecommons.org/licenses/by/4.0/).

1. Introduction

The last decade has seen significant progress in electromechanical technology for actuation. In the range of some kilowatts or some tens of kilonewtons, they provide attractive solutions compared with the servohydraulic (or so-called conventional) technology [1]. This evolution is particularly observed in aerospace, which is looking for greener actuation for flight controls, landing gears and engines.

For many applications, electromechanical actuators (EMAs) have already reached the highest technology readiness level, TRL9, which enables them to be put into service. However, it appears that EMAs for aerospace cannot be standardised easily, as opposed to those devoted to industrial applications. This mainly comes from the specificity of requirements and constraints that concern the geometrical integration, the reliability, the mission profiles (including four-quadrant operation with numerous and rapid changes between quadrants) and the certifiability and development assurance level (DAL). The EMA control design itself is driven by these considerations.

Although commercially off-the-shelf drives for industrial applications include efficient self-tuning features [2], each aerospace actuation project requires a specific activity for control design, which must suit the application constraints and development timing in a systems-engineering (SE) frame [3]. There are potentially many candidate types of controllers that today offer extended possibilities: for example, R-S-T digital polynomial controllers (combining parallel R, series S and feedforward T corrections), state feedback controllers with an estimator, nonlinear controllers or adaptive controllers, for example [4–7]. On their side, EMAs have numerous technology imperfections (e.g., friction and backlash) that are highly sensitive to the operating point. This generally greatly penalises the applicability, robustness and certifiability of these advanced controllers for safety-critical applications such as flight controls. This is why production EMAs involve quite conventional control strategies, which are based on fixed-gain, cascade controllers.

In the preliminary design phases of a project, the concepts and options must be benchmarked rapidly. During these early phases, emphasis is generally put on power sizing under mass, envelope, reliability and thermal constraints [8–10]. Although the natural dynamics of the EMA power part is sometimes addressed, control is never considered in a realistic way. This puts a high penalty on the preliminary design process for two main reasons:

- The sizing of EMAs is highly dependent on the mission profile (time history of position and force at actuator/load interface), which affects mechanical, magnetic and thermal stresses. It involves two sizing loops because the motor sizing depends on the motor design itself (rotor inertia and mean and maximal temperatures of the windings). A simple second-order representation model of the closed-loop performance is generally used to translate the mission profile from the load to the motor shaft levels. This method ignores how the controllers will solicit the EMA in practice.
- Although the power sizing ensures sufficient power capability, there is no early validation that the choices made are consistent with the specified closed-loop performance.

When the control is addressed in more detail, the well-established approach consists in using a bottom–up process [11,12]: the current loop is first addressed, and then the speed loop is considered. The position loop is rarely addressed in the literature because it is not present in many electric drives that aim to control speed (e.g., electric vehicles, fan or pump drives). For each internal loop, the bottom-up process allocates a flat-top target bandwidth that is related to the position loop specified dynamics. Unfortunately, this blind allocation deprives the control designer of a realistic and quantified view of the effective contribution of an inner loop to the stability and rapidity of its upper loops.

The research work that is reported hereafter has been driven by these considerations. It puts emphasis on the design and implementation of a top–down process that serves as a straightforward preliminary control design of a cascade position controller from top-level specifications. This work was driven by two major constraints:

- Linking formally, in a noncausal manner, the control and digital implementation parameters to the EMA dynamic specification and top-level design parameters;
- Avoiding the use of unrealistic linear control models of phenomena by verifying a posteriori the control robustness to unmodelled dynamics and nonlinearities, with resort to high-fidelity virtual tests.

Section 1 introduces the context. Section 2 details the proposed process and its implementation. The soundness of the proposal is shown in Section 3, which reports the control design validation through virtual testing. Section 4 provides important elements of discussion. The Appendices A and B merge all major resources that are used to generate the proposed preliminary control design process.

2. Top-Down Controller Design

Given the specified dynamic performance of the position loop, the proposed process outputs the proportional and integral control gains that are defined sequentially for the position, speed and current loops. Additionally, it provides the sampling frequencies for the digital implementation of the controllers, the target dynamics of the measurement chains and some values of interest for analysis purposes.

This common architecture of a cascade position controller, Figure 1, takes the benefit of the current and speed measurements that are needed to implement the brushless motor control so as to feed the controller back with measured state variables.

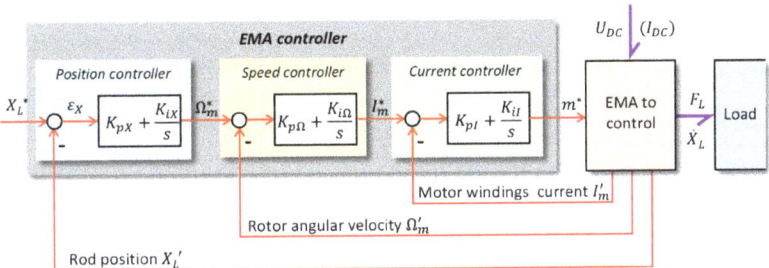

Figure 1. Cascade control of EMAs (notations are detailed in the Appendices A and B, full arrows and half arrows highlight the signal and power flows, respectively).

The current controller computes the duty cycle setpoints for the motor power drive. Current sensors and their conditioning provide the required feedback signals. The speed controller determines the current setpoints according to the power-operating domain of the motor. The speed feedback signal is commonly acquired from a resolver sensor and its resolver-to-digital converter (RDC), which measures the relative motion between the motor rotor and the stator. This chain also provides the position and speed signals for the field-oriented control (FOC) and back electromotive force (BEMF) compensation. The position controller determines the motor speed setpoint. The position feedback signal is commonly provided by a linear variometer differential transformer (LVDT), which measures the relative position between the EMA rod and the housing. In addition to the three loops, a force loop is sometimes required to meet the specific requirements related to the force limitation or rejection of dynamic loads [13].

2.1. Step 1: Design of the Position Controller and Specification of the Speed Loop Dynamics

The power architecture of an aerospace EMA typically involves a three-phase inverter, which is supplied by the DC-link and drives a brushless motor of the permanent magnet synchronous machine (PMSM) type. The motor shaft power is transmitted to the driven load through a mechanical reducer (a nut-screw system in the most common direct-drive, linear EMA design). In the following, the PMSM is considered as its DC motor equivalent and the inverter is assumed to be perfect. Figure 2 displays the linear control model of the EMA that is used in the proposed process.

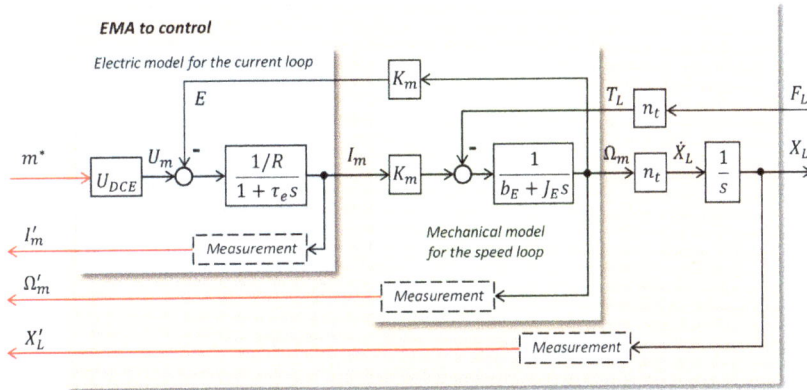

Figure 2. Block-diagram representation of the equivalent linear control model of the EMA to control (see Appendices A and B for details).

The first step of the process deals with the position loop. It uses as inputs the dynamic requirements of the position control, either in the frequency domain (f_3 frequency for -3 dB magnitude or f_{45} for $45°$ phase lag) or in the time domain (settling time t_{sX}), the design margin parameter DM and the transmission ratio n_t of the EMA. As a result, it provides the proportional gain K_{pX} of the position controller and specifies the speed loop dynamics for the second step and the minimal sampling frequency of the position controller. It also outputs additional performance indicators, in particular the angular frequency for phase margin ω_{PMX}, which is used to specify the sampling frequency of the position controller.

According to the author's experience, using an integral action in the position controller is not welcome for several reasons. First, the rejection of disturbances is quite low because the I gain is hardly limited by stability considerations. Second, many nonlinear effects (e.g., friction, compliance, backlash, measurement noise, quantisation) combine to produce a low-frequency limit cycle in the presence of the I action. The magnitude of the limit cycle is linked to the minimal position step that can be produced at the rod output. Therefore, it is not affected by any change in the I gain, which only acts on the frequency of this limit cycle. This explains why it is preferred to keep the position controller purely proportional, as seen in Equation (1), however with output limitation.

$$\Omega_m^* = K_{pX}(X_L^* - X_L) \tag{1}$$

In the absence of friction or backlash (or compliance), the EMA internal mechanical transmission between the motor shaft and the EMA rod links the rotor and rod mechanical power variables by:

$$\begin{cases} sX_L = n_t \Omega_m \\ T_L = n_t F_L \end{cases} \tag{2}$$

and:

$$n_t = l/2\pi N \tag{3}$$

with l as the nut-screw lead and N as the reduction ratio of the intermediate gear.

2.1.1. Performance of the Position Loop with I-P Speed Controller

As given in Table A2, the speed loop behaves as a second-order system versus the speed demand and the rate of external load. When a first-order, low-pass filter of time constant $\tau_\Omega = K_{p\Omega}/K_{i\Omega}$ is applied to the speed demand, the controller becomes of the I-P type, and the speed loop transfer is given by:

$$\Omega_m = \frac{\Omega_m^* - \frac{1}{K_m K_{i\Omega}} s T_L}{1 + \frac{2\zeta_\Omega}{\omega_{n\Omega}} s + \frac{1}{\omega_{n\Omega}^2} s^2} \tag{4}$$

The pole of the feedforward filter compensates the zero introduced by the speed P-I controller in the pursuit transfer function, as shown in Figure 3.

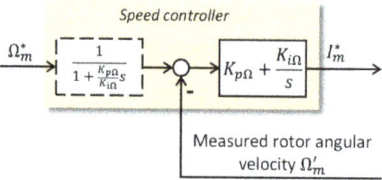

Figure 3. I-P speed controller obtained by filtering the speed setpoint (continuous-time domain).

One can implement the speed controller in the I-P form because this makes the open-loop position transfer simpler. This therefore enables the closed-loop position transfer to be expressed formally in a canonical form. In this case, the open-loop position transfer for the pure proportional control of gain K_{pX} and perfect position measurement ($X'_L = X_L$) becomes, in the nonsaturated domain:

$$X_L = \frac{K_{lX}(X_L^* - X_L) - K_{XF}sF_L}{s\left[1 + \frac{2\zeta_\Omega}{\omega_{n\Omega}}s + \frac{1}{\omega_{n\Omega}^2}s^2\right]} \quad (5)$$

where F_L is the external load applied at the EMA rod.

The position loop gain K_{lX} is linked to the position proportional gain K_{pX} and the EMA transmission factor by:

$$K_{lX} = K_{pX} n_t \quad (6)$$

while the dynamic compliance of the position control is given by:

$$K_{XF} = \frac{n_t}{K_m K_{i\Omega} K_{pX}} \quad (7)$$

which becomes:

$$K_{XF} = \frac{n_t}{K_{pX} J_E \omega_{n\Omega}^2} \quad (8)$$

It is only linked to the motor torque constant, the EMA transmission factor and the integral control gain of the speed loop. Therefore, the K_{pX} proportional position control gain depends only on the speed loop target dynamics $\omega_{n\Omega}$ given the EMA-specified dynamics and design parameters (K_m, J_E, n_t).

The open-loop transfer function for position pursuit, Equation (5), combines a pure gain (K_{lX}) with integral and second-order dynamics (ζ_Ω, $\omega_{n\Omega}$). It is therefore welcome to link the closed-loop performance to these parameters in a dimensionless manner by introducing the dimensionless angular frequency $\overline{\omega} = \omega/\omega_{n\Omega}$, which gives:

$$X_L/(X_L^* - X_L) = X_L/\varepsilon_X = \frac{\overline{K}_{lX}}{s(1 + 2\zeta_\Omega \overline{\omega} s + \overline{\omega}^2 s^2)} \quad (9)$$

where $\overline{K}_{lX} = K_{lX}/\omega_{n\Omega}$ is the dimensionless loop gain of the position loop.

The key enabler of the proposed process is the chart that is generated once numerically. It calculates, e.g., using a control toolbox, the position closed-loop performance indicators as a function of the two parameters $\overline{K}_{lX}$ and ζ_Ω, which maximises a given constrained objective. Figure 4 displays the chart obtained to secure the fastest closed-loop response to a step position demand (minimal settling time) without overshoot. The data are generated with 1% accuracy. Particular attention is paid to the $-45°$ phase lag requirement because it is a major one regarding the stability of the upper aircraft flight control loops.

Figure 4a displays the links among the dimensionless loop gain $\overline{K}_{lX}$, the phase margin and the dimensionless settling time $\bar{t}_s = t_{sX} \cdot \omega_{n\Omega}$ for a given value of the damping factor ζ_Ω. The best compromise between stability and rapidity is found for $\zeta_\Omega = 0.54$. Figure 4b summarises the closed-loop performance indicators expressed in the frequency domain. All the values are dimensionless, with reference to $\omega_{n\Omega}$. Again, the best bandwidth is obtained when ζ_Ω is close to 0.5. Figure 4c shows that the frequency for the phase margin varies in the range of 1 to 1.25 times the closed-loop bandwidth, while the phase margin is always greater than 65° (Figure 4a). On its side, Figure 4d confirms that for low values of the loop gain, the closed-loop system is equivalent to a first-order system of time constant $1/\overline{K}_{lX}$. However, when the loop gain increases, the stability is affected by the closed-loop imaginary poles. The greatest dimensionless bandwidth at $-45°$ phase is 0.287. It is obtained for $\overline{K}_{lX} = 0.58$, while the shortest dimensionless settling time of 5.89 is achieved for $\overline{K}_{lx} = 0.54$. Although $\zeta_\Omega = 0.54$ minimises the settling time, such a damping generates 13% overshoot for the speed loop. Setting $\zeta_\Omega = 1$ removes this overshoot. It is therefore welcome in the presence of backlash, and it still provides a good compromise for position loop stability and rapidity. However, it requires much faster speed (and current) loop dynamics than the first choices for a given dynamics of the position loop.

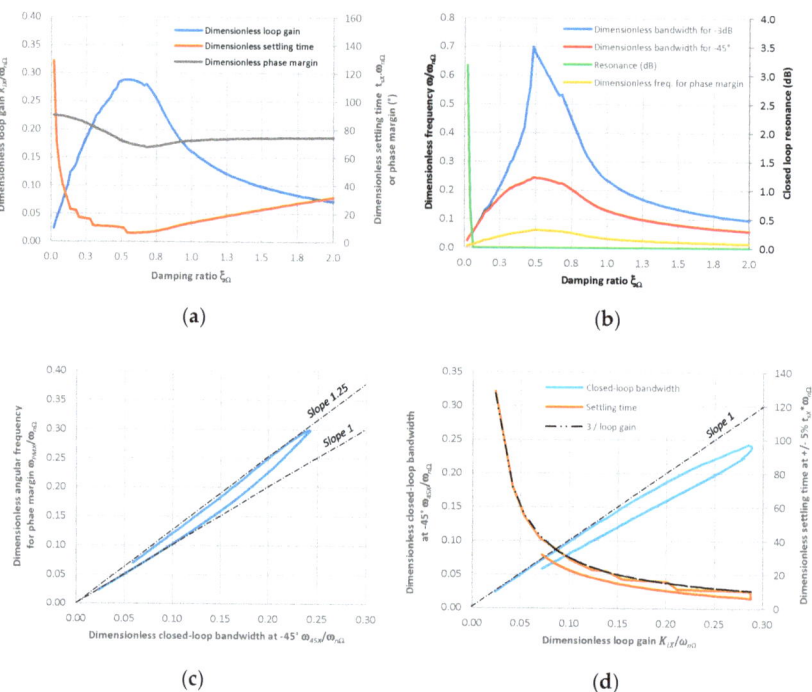

Figure 4. Performance indicators for position control for the fastest closed-loop step response without overshoot using dimensionless variables: (**a**) loop gain, settling time and phase margin; (**b**) performance indicators in the frequency domain; (**c**) frequency for the phase margin; and (**d**) closed-loop rapidity versus loop gain.

In the implemented approach, the control design parameter is ξ_Ω. The data plotted in Figure 4a or Figure 4b are first used to determine $\omega_{n\Omega}$ and then K_{IX} given the specified dynamics of the position control. This enables the K_{pX} P gain of the position controller to be calculated using Equation (6) from the EMA transmission factor n_t.

2.1.2. Performance of the Position Loop with P-I Speed Controller

When the low-pass filter of Figure 3 is not implemented, a zero remains in the pursuit transfer function Ω_m / Ω_m^*. This P-I implementation of the speed controller also has its merits. As it does not introduce any lowpass filtering of the speed setpoint that is generated by the position controller, it decreases the tracking error. The presence of the zero that remains in the pursuit transfer function of the closed-loop position, however, tends to introduce overshoot in the position step response. Nonetheless, it does not affect the K_{XF} parameter, which quantifies the load position sensitivity to the rate of external load.

In this case, the performance chart is generated to obtain the smallest response time of the position loop, ensuring that all closed-loop poles are stable and purely real. This helps to avoid back and forth motion in the presence of backlash and limits the overshoot generated by the zero of the speed loop. The main data of this chart are presented graphically in Figure 5.

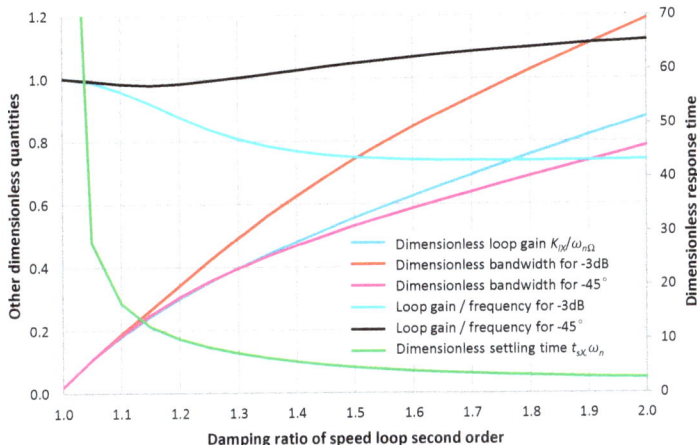

Figure 5. Performance indicators of position control with the P-I implementation of the speed controller giving the fastest response time under a purely real negative closed-loop poles constraint.

It can be remarked that the constraint imposed on the closed-loop poles cannot be met when the damping ratio ξ_Ω is lower than the unity. For values greater than 1.3, this constraint significantly impacts the speed loop natural frequency compared with the I-P implementation under the null overshoot constraint. For instance, when $\xi_\Omega = 1.3$:

- The speed loop natural frequency is $\omega_{n\Omega} = \omega_3/0.492 = \omega_{45}/0.396$ in the present case (Figure 5), while it was $\omega_{n\Omega} = \omega_3/0.1601 = \omega_{45}/0.0937$ formerly (Figure 4), so the present case is much less demanding in terms of speed loop (and consequently current loop) dynamics; and
- The loop gain becomes $K_{lX} = \omega_3/0.808 = \omega_{45}/1.002$ (Figure 5), while it was $K_{lX} = \omega_3/0.723 = \omega_{45}/1.236$ formerly (Figure 4), so the present case is disadvantageous concerning the rejection of disturbances, as shown by Equation (8).

2.1.3. Digital Implementation of the Position Controller

In this paper, the controllers are set in the continuous-time domain, using transfer functions as control models. Although this choice puts aside any advanced controller that does not exist in the continuous-time domain, it keeps a direct link with the physics through canonical parameters (time constants, damping factors and natural frequencies). Once designed, the controller is discretised for digital implementation. The phase lag introduced by filtering, sampling and computation is actively managed to specify the sampling frequencies.

When seen from the continuous-time domain, the zero-order sampling and hold function performed at the sampling frequency f_s in the digital implementation of the controller is equivalent to a pure delay of $\Delta_s = 1/2f_s$. At frequency f, it introduces a phase of $\varphi(°) = -180 f/f_s$. In a closed-loop system, this delay is, with rare exceptions, detrimental to the closed-loop stability. This is why it is important to select the sampling frequency consistently with the target dynamics of the considered closed-loop system. There are a few practical known recommendations to make this decision [14]:

- There must be at least 7 to 15 samples in the rise time of the system response to a step input; or
- The sampling frequency must be at least 15 to 25 times the closed-loop bandwidth.

However, these general rules are not directly driven by the stability of the loop under consideration. This is why the author prefers the following more direct approach that can be expressed as follows. In total, the digital control introduces at the frequency f a parasitic phase lag (phase lag stands for the opposite value of phase) $\varphi_{d(°)}$. It comes from the sampling delay plus the phase lag of the antialiasing filter and the Δ_c time spent for conversions and processing. If it is assumed that the antialiasing filtering is achieved with a Butterworth second-order low-pass filter of cut-off frequency of $f_s/2$, then the phase lag is given by:

$$\varphi_{d(°)} = \frac{180}{\pi}\left[\pi/n_s + atan\left\{\frac{2.8/n_s}{1-(2/n_s)^2}\right\} + 2\pi f_s \Delta_c/n_s\right] \qquad (10)$$

where:

$$n_s = f_s/f \qquad (11)$$

For outer loops having a low bandwidth, the delay Δ_c is generally negligible in comparison with other contributors. Depending on the implementation of the digital control, it may, however, be significant for the most inner (e.g., current) loops. If Δ_c is neglected, $\varphi_{d(°)}$ already reaches 80.5° when n_s drops to 4.35. Above this value, the phase lag of the antialiasing filter varies almost linearly versus frequency, and $\varphi_{d(°)}$ can be approximated by:

$$\varphi_{d(°)} \cong 180\,(1+2.8/\pi)/n_s \cong 340.4/\,n_s \qquad (12)$$

with less than 2.9% error of underestimation (a 360 factor, instead of 340.4, corresponds to the phase lag produced by a full sampling period delay). Antialiasing contributes to almost half the total phase lag, while the magnitude effect of the low-pass filter remains below ±0.003 dB. Of course, the $\varphi_{d(°)}$, as shown in Equation (12), can be adapted to the current context, e.g., for the antialiasing filter or if the processing time becomes the major source of phase lag. If necessary, Equation (12) can be modified to include the phase lag introduced by the sensor and measurement chain.

These results provide a straightforward means to quantify (or specify) the reduction of the open-loop phase margin given the digital implementation of a controller that has been designed in the continuous time domain. For example, if this contribution (including the antialiasing filter) must not exceed 10° parasitic phase lag, Equation (12) indicates that the sampling frequency must be at least 34 times the frequency at which the phase margin is determined. This is a really huge value.

If the dynamics of the position measurement can be neglected, the minimal sampling frequency of the position controller can be specified using Equation (12). This option has been anticipated when building the performance charts, Figures 4 and 5, which explicitly provide the angular frequency $\omega_{PMX} = 2\pi f_{PMX}$, at which the phase margin of the position control is determined when parasitic phase lags are not considered.

Using, e.g., Equation (12), the sampling frequency f_{sX} of the position controller must satisfy the constraint:

$$f_{sX} \geq 340.4\, f_{PMX}/\varphi_{dX(°)} \qquad (13)$$

where $\varphi_{dX(°)}$ is the allocated parasitic phase lag introduced by the digital implementation of the position controller.

Notes

- The phase lag introduced by the position measurement is not considered. Although it is generally negligible, this assumption must be verified (when the measurement chain is known), ensured by relevant specification (when the measurement chain is to be defined) or removed by adding the position measurement dynamics in Equation (13).
- For LVDTs position sensors, the demodulation filter is welcome to avoid any frequency aliasing.

2.2. Step 2: Design of the Speed Controller and Specification of the Current Loop Dynamics

2.2.1. Viscous Friction vs. Real Friction

A pure viscous friction coefficient of coefficient b_E is most of the time considered in the accounts dealing with setting the speed controllers of electric drives [15,16]. However, real friction is far different from pure viscous friction (where the friction force is proportional to the velocity). This is particularly true for motion control when the actuator drives a variable load at variable speed with frequent speed reversals. In this case, the friction force mainly depends on load, temperature, and, in a much lesser amount, relative speed [17]. This is clearly illustrated by the examples given in Figure 6.

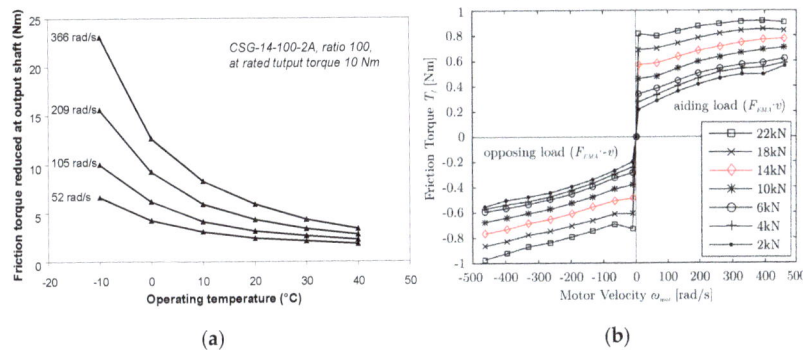

Figure 6. Friction force in EMAs: (**a**) the influence of temperature on friction torque for a geared EMA at a rated output torque, calculated from harmonic drive efficiency data (size 14, ratio 100) [17]; (**b**) the influence of speed and load at room temperature (demonstration EMA sized for the Airbus A320 aileron–gear drive, ball screw, no preload) [18].

This figure clearly shows that a pure viscous friction model totally fails to reproduce real friction. For linear control design, it is therefore preferred to consider friction as an unmodelled effect. This requires the controller to be robust enough against it. In this work, this robustness is assessed a posteriori, either by simulation (when validated models are available), from partial real tests or through former capitalised experience. This approach is not only applied to friction but also to backlash and compliance, whether they concern the EMA itself or the kinematics linking the EMA to the driven load. It works particularly well in the field of aerospace, e.g., for flight controls or landing gears actuation. Indeed, for such applications, the natural dynamics generated by the combination of moving bodies' inertance and the backlash/compliance of the mechanical transmissions is far greater than the specified bandwidth of the actuator position control.

2.2.2. Setting the Speed Loop Controller

The first step of the proposed process has specified the second-order dynamics of the speed loop: the undamped natural frequency $\omega_{n\Omega}$ and the damping factor ξ_Ω. These target values are used as inputs in Table A2 to obtain the proportional ($K_{p\Omega}$) and integral ($K_{i\Omega}$) gains of the speed controller from the total equivalent reflected inertia at the motor rotor J_E and the EMA motor electromagnetic constant K_m:

$$K_{i\Omega} = J_E \omega_{n\Omega}^2 / K_m \quad (14)$$

$$K_{p\Omega} = 2 J_E \xi_\Omega \omega_{n\Omega} / K_m \quad (15)$$

These settings are directly linked to the specified position loop dynamics. It is interesting to remark that the time constant of the P-I speed controller,

$$\tau_\Omega = K_{p\Omega} / K_{i\Omega} = 2\xi_\Omega / \omega_{n\Omega} \quad (16)$$

is only linked to the dynamics specified for the speed loop, determined in Step 1, once the damping factor ξ_Ω is chosen. It is therefore independent of the EMA parameters.

2.2.3. Digital Implementation of the Speed Controller

The sampling frequency $f_{s\Omega}$ for the digital implementation of the speed controller is specified in the same manner as for the position loop:

$$f_{s\Omega} \leq 340.4 \frac{f_{PM\Omega}}{\varphi_{d\Omega(°)}} \tag{17}$$

This is constrained by the frequency $f_{PM\Omega} = \omega_{PM\Omega}/2\pi$, at which the phase margin of the speed loop is determined, and by the parasitic phase lag $\varphi_{d\Omega(°)}$ introduced by the digital implementation of the speed controller.

Note

The motor speed measurement can generate significant phase lag. Allocating the accepted phase lag for motor angle measurement can add another constraint to specify the dynamics of the rotor speed/angle measurement chain.

2.2.4. Specification of the Current Loop Dynamics

Step 2 is also used to specify the current loop dynamics. Again, the objective is to limit the parasitic phase lag that the current loop introduces into the speed loop or, in other words, to ensure the validity of the results summarised in Table A2. This is achieved as follows.

It can be shown that the frequency at which the phase margin of the speed loop is given by:

$$\omega_{PM\Omega} = \omega_{n\Omega} \sqrt{2\zeta_\Omega^2 + \sqrt{1 + 4\zeta_\Omega^4}} \tag{18}$$

If the current controller is set as usual, its P-I time constant is made equal to that of the motor windings, leading to:

$$\tau_{CI} = K_{pI}/K_{iI} = L/R \tag{19}$$

In this case, the current loop behaves as a first-order lag of time constant:

$$\tau_{lI} = L/K_{pI}U_{DCE} \tag{20}$$

Thus, the dynamics of the current loop is specified by limiting the parasitic phase lag φ_I that it introduces in the speed loop at the $\omega_{PM\Omega}$ angular frequency at which the speed loop phase margin is determined:

$$\tau_{lI} \leq tan(\varphi_I)/\omega_{PM\Omega} \tag{21}$$

It is worth remarking that this constraint does not involve any EMA design parameter.

2.3. Step 3. Design of the Current Controller

2.3.1. Setting the P-I Controller of the Current Loop

The proportional and integral gains of the current loop controller are set according to Appendix A, given the following two constraints:

$$K_{pI} \geq L/U_{DCE}\tau_{lI} \tag{22}$$

$$K_{iI} = R K_{pI}/L \tag{23}$$

Note

These equations involve quantities related to the EMA design (L, R, U_{DCE}), which can vary significantly during the EMA operation and consequently alter the performance of the current loop. To make the EMA sufficiently robust, the setting of the current controller gains must consider the worst conditions and their effect on rapidity and stability.

2.3.2. Digital Implementation of the Current Controller

According to Appendix A, when the P-I time constant of the current controller compensates the electric time constant of the motor, the open-loop transfer function becomes a pure integrator of gain $K_{iI} U_{DCE}/R$. In the presence of pure parasitic delays, the angular frequency ω_{PMI}, at which the phase margin of the current loop is defined, is given by:

$$\omega_{PMI} = 2\pi f_{PMI} = K_{iI} U_{DCE}/R \tag{24}$$

This frequency can be used to specify the sampling frequency f_{sI} for the digital implementation of the current controller. Given the high dynamics required for the current loop, it may be important to consider not only sampling and antialiasing but also additional effects that can limit the allowable controller gains by alteration of the closed-loop stability: time spent for computation and conversions, and dynamics of the currents measuring chain.

All these effects increase the open-loop phase lag. Thus, they can be merged to consider their negative contribution to the phase margin globally. In the very common case, the dynamics of the current measurement chain is negligible compared with that introduced by the various delays. However, a simple conservative option consists of considering that the overall delay is equal to a full sampling period, giving the constraint:

$$f_{sI} \geq 360 \frac{f_{PMI}}{\varphi_{dI}(°)} \tag{25}$$

2.4. Synthesis of the Top-Down Process

All these results can be represented graphically to summarise the interdependencies among the parameters involved in the design of the EMA position control. This is achieved using the diagram shown in Figure 7.

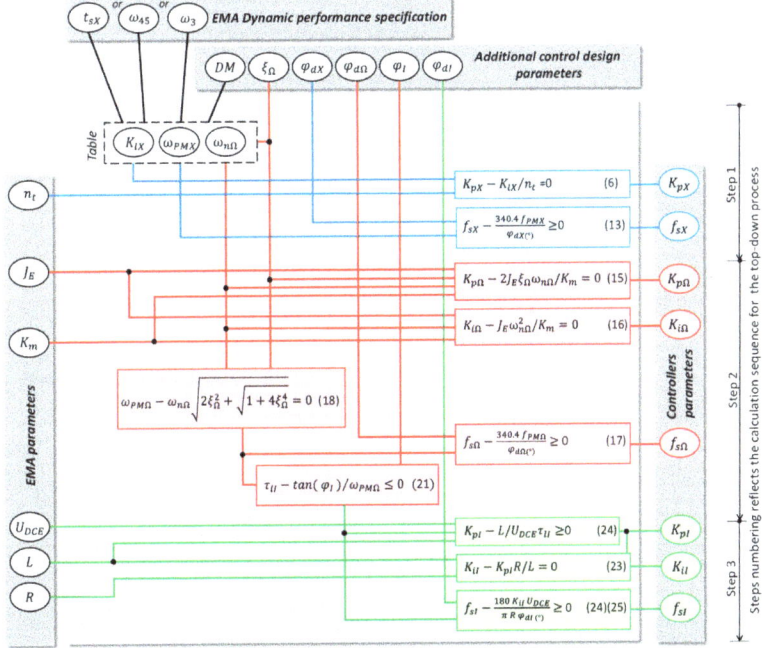

Figure 7. Noncausal representation of the interdependence among parameters for the design of the EMA position control.

The blue, red and green blocks use the equations related to the position, speed and current controllers, respectively. A noncausal representation is preferred (nonoriented signal lines) because this enables the calculation causality to be adapted to the current context. This possibility is particularly attractive, e.g., for EMA preliminary sizing when the control hardware is imposed. When read from the top down, the data flow implements the proposed top-down process, where each sequential step from 1 to 3 is dedicated to the setting of a given controller and the specification of the next inner controller. The eight controller parameters (right) are computed given the performance specification and control design choices (top), given the main EMA design parameters (left). When the process is combined with preliminary sizing and optimisation during the early phases of a project, the J_E, L and R parameters can be obtained from estimation models, for example, using scaling laws or metamodels [8], from the main design parameters n_t, K_m and U_{DCE}.

3. Illustrative Example

The example of a wingtip, direct-drive, linear flight control actuator used for regional aircraft [19] is used to illustrate the proposed control design. The process is validated through the simulation of an accurate lumped-parameter model of the actuator (control and electromechanical units) and the driven load, which was developed in former studies [20,21].

3.1. Virtual Prototype

The modelled and unmodelled phenomena are summarised in Table 1. The high-fidelity model is implemented and simulated in the Simcenter-AMEsim (2020.1, Imagine, Roanne, France) environment. It involves 75 state variables, no implicit variable and +200 parameters. Iron losses and magnetic saturation at the motor are not modelled as they are not significant in this application. Any energy loss is made sensitive to temperature, enabling isothermal simulations to be run for various operating temperatures. Given the dynamics in presence and the sampling/switching frequencies, a 1 s simulation with integration accuracy of 10^{-7} typically takes a 290 s CPU on a 64-bit personal computer (Intel Core I7-8550U CPU at 1.8 GHz).

Table 1. Model used for the virtual validation of the controller design.

	Modelled	Not Modelled
DC link *	Diode and capacitance Braking resistance, chopper and its control	Parasitic serial and parallel resistances or capacitance
Three-phase inverter	3 legs, 6 transistors Conduction and switching losses	
Three-phase PMSM	Motor constant Windings resistance and inductance Temperature effects on motor constant and windings resistance	Cyclic inductance Magnetic saturation Iron losses
Mechanical transmission *	Mechanical transformation (nut-screw) Inertia of rotating and mass of translating assemblies Rotational and translational friction with true sticking and effects of speed, load and temperature Transmission compliance and backlash (in translational domain) End stops	Moving body Side loads
Kinematics to load	Three-bar mechanism (variable lever arm) Transmission compliance and backlash	Friction and side loads at eye or hinge joints
Sensors *	Gain, range, dynamics, demodulation, antialiasing, sampling, quantisation, saturation and noise	Offset and thermal drift Hysteresis and nonlinearity
Controller	Discrete, with saturation and antiwindup, time for processing BEMF compensation if used, FOC using dq0 model Limitation of speed, current and voltage demands (according to motor operating range)	
PWM *	Symmetrical triangle carrier, sampling, saturation Timing and synchronisation with current loop	
Thermal *	All energy losses made temperature-dependent and generating heat	Thermal transients

* See [22,23].

3.2. Virtual Validation

The EMA position controller is virtually validated using a bottom–up incremental approach that follows the real validation process, i.e., the integration branch of the V-model of product lifecycle [3]. The operation of every (simulated) element of the EMA (motor, inverter, measurement chains and mechanical transmission) has been virtually validated, as should be done with partial tests for the real elements. The current loop is validated first, followed by the speed loop and finally the position loop.

The controllers have been designed with the following allocation of the parasitic phase lag because of digital implementation: $5°$ for the position loop, $10° + 10°$ for the speed loop and $20°$ for the current loop.

The loops are excited to assess both pursuit and rejection performances on the same response plot. As numerical simulation naturally provides time responses, a demand step is applied first, followed by a disturbance step. To make the virtual validation realistic, a random noise is introduced on each measured quantity, typically very few percent of the maximal or rated values. The time responses given in this section have been plotted using the realistic magnitudes that were identified during real tests of the power and signal electronics: 4% of the maximal RMS phase current, 3% of the rated rotor speed, $5°$ for the rotor angle and 6% of LVDT secondary voltage magnitude. Particular attention is also paid to the effect of nonlinearities and unmodelled dynamics on the performance expected from the linear continuous control model. In this attempt, the responses are analysed for various step magnitudes. High magnitude leads to saturations as a result of power and signal limitations. Medium magnitude generally enables the EMA to operate far from hard nonlinear effects and static imperfections. Very low magnitude points to the influence of static imperfection such as quantisation, breakaway friction and backlash.

The dimensionless responses are presented to provide on a single figure the demand, the response of the linear continuous control model and the response of the high-fidelity, nonlinear, high-order model. In the responses provided for the high-fidelity model, the EMA is assumed to operate at room temperature. The excitation magnitudes are referred to the rated values and to the noise magnitude (before the antialiasing filter). The time values are hidden for confidentiality. However, it can be mentioned that the time ranges of Figures 8–10 are in the ratio 1:8:100, respectively, to indicate the relative dynamics of the current, speed and position loops.

3.2.1. Current Loop

To obtain the current loop responses, the speed and position loops are opened. The motor is tested without connection to the nut-screw while externally imposing the rotor angular velocity. A current step demand (I_m^* or I_q^*) is applied first with the rotor blocked, followed by a motor speed step disturbance Ω_m. The response of the linear continuous control model is obtained from the last transfer function of Table A1. The responses are displayed in Figure 8.

This figure elicits the following comments:

- The responses of the controlled virtual prototype globally agree well with the responses expected from the linear model and control strategy.
- As anticipated, stability is degraded by sampling and antialiasing but remains acceptable given the active management of this effect in the control design process and the $10°$ φ_{dI} allocation.
- Figure 8a shows the influence of the current and speed measurement noises. Although the current demand is only twice the peak noise of the current measurement prior to the antialiasing filter, the current response remains globally stable.
- Under medium-magnitude excitations (Figure 8b), the relative importance of noises on response decreases.
- Under high-current and high-speed excitations (Figure 8c), the current response to the speed disturbance is affected by a significant ripple. As explained in [24], this effect comes from the tracking error of the rotor position measurement used by the dq0

transforms, although this is very fast, which generates an alias I_d current proportionally to the I_q current demand.

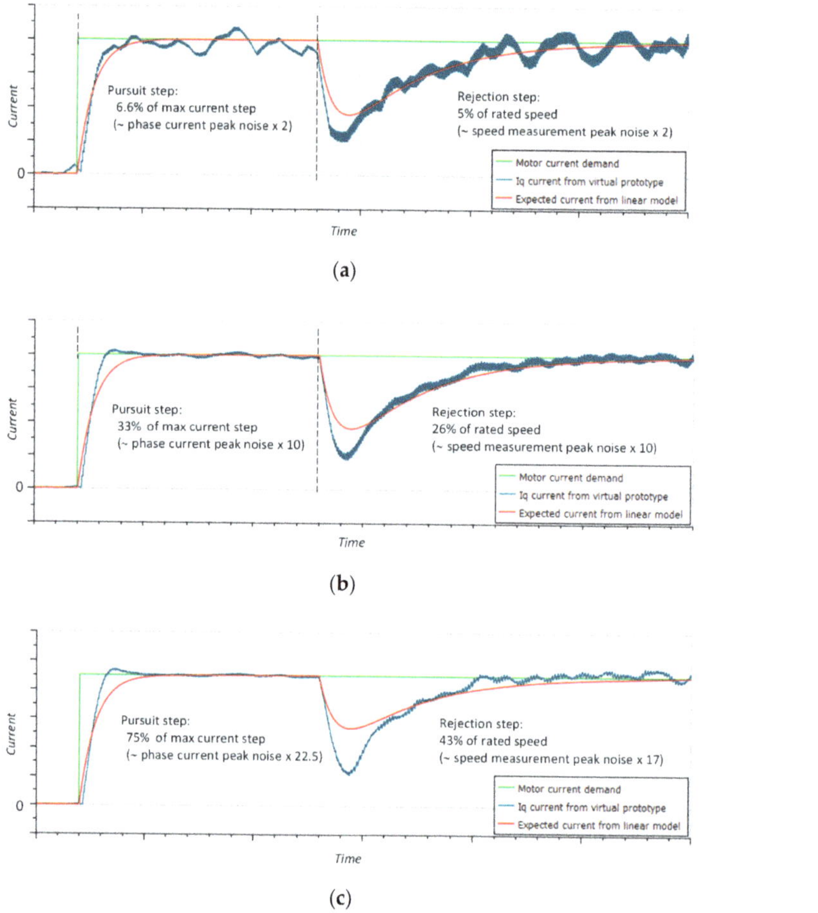

Figure 8. Time response of the current loop for step excitations: (**a**) small-magnitude excitations, noise effects magnified; (**b**) medium-magnitude excitations, operation close to linear; and (**c**) large-magnitude excitations, close to internal saturation.

3.2.2. Speed Loop

For the speed loop test, the position loop is opened and the translating part of the nut-screw is removed. A rotor speed step demand Ω_m^* is applied first for a free rotor shaft, followed by a step disturbance torque T_L applied to the rotating part of the nut-screw. The response of the continuous linear control model is obtained from the last transfer function of Table A2. The most relevant time responses are displayed in Figure 9, which elicits the following comments:

- Once again, the responses of the controlled virtual prototype globally agree well with the response expected from the linear model and control strategy.
- Even when the excitation magnitudes are only a few times the measurement noise before the antialiasing filter (Figure 9a), the expected dynamics and average response are still satisfactory.

- There is very little difference between the simulated and expected responses for medium-excitation magnitudes when they do not lead to saturation (Figure 9b).
- For high magnitudes of excitations (Figure 9c), the current and the voltage demand saturate for a long time during transients (Figure 9d). This makes the EMA operate temporarily in an open loop. At the end of the saturating phases, the normal control is recovered with high rapidity and stability. The absence of an excessive overshoot or limit cycle proves the correct setting and efficiency of the antiwindup function of the P-I controllers, implemented using the back calculation and tracking scheme [25].

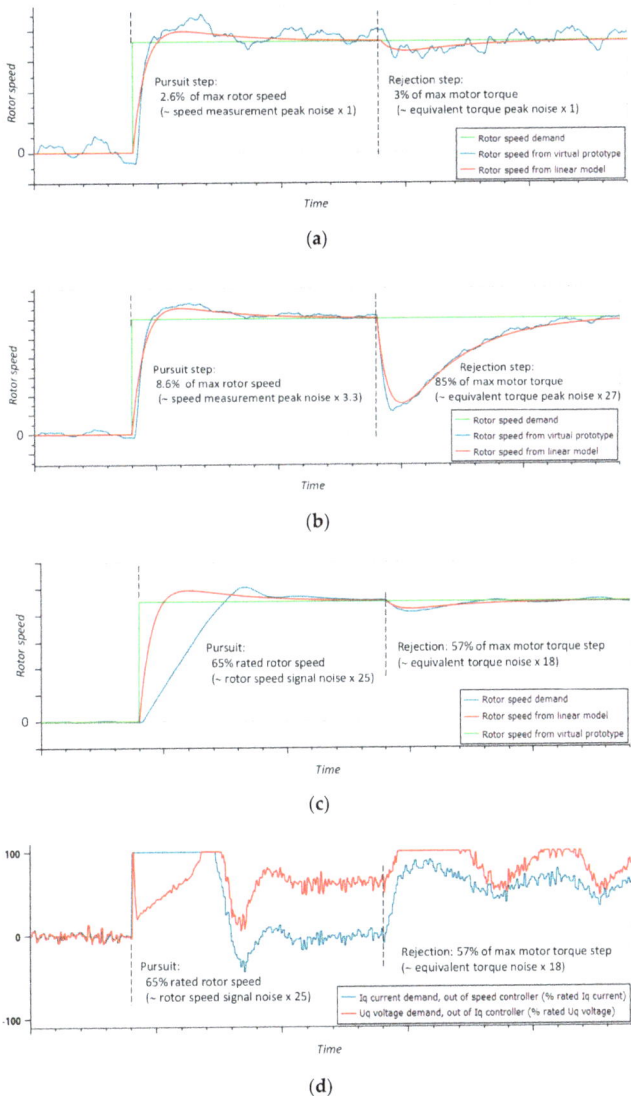

Figure 9. Time response of the speed loop for step excitations: (**a**) small-magnitude excitations, noise effects magnified; (**b**) medium-magnitude excitations, at the limit of saturation; (**c**) large-magnitude excitations, introducing long saturation of the current and modulation ratio; and (**d**) the speed and current controller outputs for large-magnitude excitations.

3.2.3. Position Loop

The simulations are run with all control loops active. According to the customer's specification for the validation of the EMA performance, a pure mass equivalent to the reflected mass of the driven load is attached to the EMA rod. The rod position step setpoint X_L^* is applied first from rest at the null position, without any external force.

To better highlight the combined effect of EMA friction and backlash, the external disturbance force F_L is then applied at the EMA rod, without any change in the position demand, as two opposite and consecutive steps. The response of the continuous linear control model is obtained by a simulation of the speed closed-loop model combined with Equations (5) and (6). The most relevant time responses are displayed in Figure 10:

- At a very low magnitude of rod position demand (twice the EMA internal backlash), the response is still smooth and close to that expected from the linear model (Figure 10a). Logically, the combination of Coulomb friction, backlash and I control (speed and current loops) generates a limit cycle, albeit with a very low magnitude (<10% of the backlash).
- Even in the presence of backlash, the rod force step is rejected with the same dynamics as that of the linear model (Figure 10b). For 100% force, the transient position error does not exceed 155% of the backlash or 10% of the nut-screw lead. This excellent capability of external force disturbance confirms the soundness of the proposed approach, which is intended to maximise the position loop gain for a given ζ_Ω.
- The position response under saturating excitations is illustrated by Figure 10c,d. The position demand is a pulse whose magnitude is just lower than the actuator stroke. A 100% rated rod force is applied to assess the position response under aiding and opposite loads. The influence of speed, current and voltage saturations clearly appears in Figure 10c, where the position response becomes unable to meet the expected dynamics. When the controllers leave the saturation domain, control is rapidly recovered in the linear domain with very few oscillations.

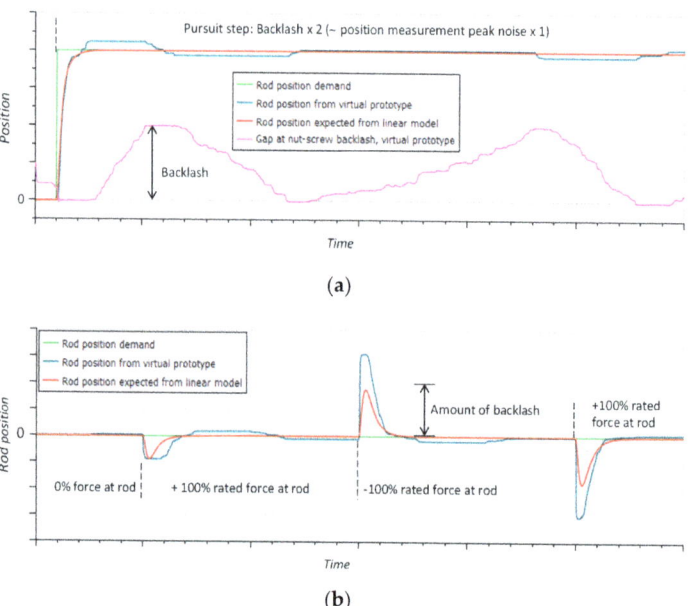

Figure 10. *Cont.*

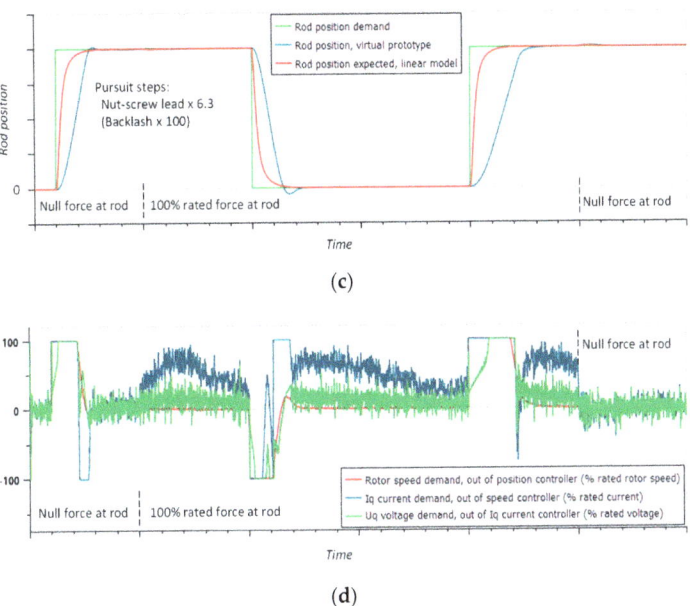

(c)

(d)

Figure 10. Time response of the position loop for step excitations: (**a**) small-magnitude excitations, noise effects magnified; (**b**) rejection of rod force disturbance, null position demand; (**c**) large-magnitude excitations, introducing long saturation of the current and modulation ratio; and (**d**) controller outputs for large-magnitude excitations.

4. Discussion

Within the nonsaturating domain of operation, two main nonlinear effects act as disturbances in the linear model that is used for control synthesis. It is important to relate the excellent robustness of the position control to these unmodelled effects and to their magnitude: in the reported validation, the EMA internal backlash is 6.3% of the nut-screw lead, while the load-independent friction force represents 3% of the EMA rated force. At the rated output force, this percentage rises to 12.4% under the contribution of the load-dependent friction. When they are expressed as their equivalent at the EMA rod level, the motor rotor inertia is 44 times greater than that of the driven load.

Besides this particular example, it is worth addressing more general comments:

As shown by Equation (8), the K_{XF} rod position sensitivity to the rate of external load is inversely proportional to $K_{pX}\omega_{n\Omega}^2$. Given the proposed control design process, the selection of ζ_Ω can offer a means to act on K_{XF} to meet a given pursuit dynamics.

- For very preliminary control design, it is advised to link the EMA internal parameters (e.g., inertia, nut-screw lead, motor windings resistance) to the EMA top-level specifications (maximal or rated speed and force at rod, along with reliability). This can, for example, be achieved using scaling laws, as proposed in [8]. Combining power and control preliminary designs would then enable more global and automated EMA design exploration and optimisation.
- Two examples have been provided in Section 2.1 to generate the performance charts. As these charts are precalculated once, other types of constraints can be applied and combined, without any need to change the process.
- Although the top-down process is driven by the overall control need, it is worth mentioning that it may output hardware and software specifications that cannot be met because of, for example, a lack of performance, maturity, availability or excessive cost. Therefore, precautions must be taken to check that each subspecification generated

can be met in practice. If not, the concerned subspecification has to be replaced by a constraint, and the data flow of the design process has to be revised accordingly. This is enabled by the noncausal representation used in Figure 7.

5. Conclusions

The design of position control of EMAs has been addressed for aerospace safety-critical applications. The focus has been placed on the proposition and implementation of a top-down process requiring very few input data for the design of cascade controllers, when certifiability constraints and design assurance levels welcome common control techniques. This work was primarily intended to enable control considerations to be added to the preliminary sizing phases in order to accelerate the development process. This objective was achieved by several contributions. The first one comes from the generation of charts that link the position control gain to the speed loop dynamics and damping targets, when a position control performance criterion is maximised. For a given loop, the second one consists in determining the control parameters, the numerical implementation and the specification of the following loop, in a formal and simple way which calls upon a minimal number of EMA parameters. The third contribution lies in the graphical representation that synthetises in a noncausal way the interdependencies between the control specifications, the EMA key design parameters, the control choices and the controller parameters. The proposed process has been virtually validated using a very detailed high-fidelity model of the EMA. The EMA responses derived from the virtual testbench have shown the efficiency of the proposed process, even in the presence of significant noise, saturation, friction and backlash, which are unmodelled in the linear models used for control synthesis.

Funding: The reported work was funded by the Clean Sky 2 European project ASTIB (JTI-CS2-2014-CPW01-REG-01-01). This project aims at supporting the improvement of the Technological Readiness Level for a number of significant equipment items that are considered of critical importance for the future Green Regional Aircraft (GRA).

Institutional Review Board Statement: Not applicable.

Informed Consent Statement: Not applicable.

Data Availability Statement: Not applicable.

Acknowledgments: The author sincerely thanks the project partners of Work Package 2.4 for their constructive discussions during his development and implementation of EMA models and control laws.

Conflicts of Interest: The author decla es no conflict of interest.

Nomenclature

Acronyms

BEMF	Back ElectroMotive Force	PMSM	Permanent Magnet Synchronous Machine
EMA	ElectroMechanical Actuator	PWM	Pulse Width Modulation
DAL	Design Assurance Level	RMS	Root Mean Square
FOC	Field-Oriented Control	RDC	Resolver To Digital Converter
I-P	Integral-Proportional	SE	Systems Engineering
LVDT	Linear Variometer Differential Transformer	TRL	Technology Readyness Level
P-I	Proportional-Integral		

Notations

b	Viscous friction coefficient	R	Resistance
DM	Design margin	t	Time
E	Electromotive force	T	Torque
f	Frequency	U	Voltage

F	Force	X	Position
I	Current	θ	Angle
J	Moment of inertia	Δ	Delay
K	Gain	ε	Error
l	Nut-screw lead	φ	Phase
L	Inductance	ζ	Dimensionless damping ratio
m	Modulation ratio	τ	Time constant
n	Ratio	ω	Angular frequency
N	Gear reduction ratio	Ω	Angular velocity
s	Laplace variable		
Subscript			
a	Antialiasing	p	Proportional
c	Computation	PM	Phase margin
C	Controller	n	Natural, undamped
d	Digital, direct (in-phase)	q	Quadrature
DC	Direct current supply	r	Settling
e	Electric	s	Sampling, or specified
E	Equivalent	t	Transmission
i	Integral	T	Torque
I	Current	XF	Position–force
l	Loop	Ω	Angular velocity
L	Load	3	At −3 dB magnitude
m	Motor	45	At −45° phase
Superscript			
′	Modified or measured value		
*	Setpoint		
¯	Dimensionless		
°	Angle expressed in degree		

Appendix A. Current Loop

With reference to Figures 1 and 2, the elements involved in the motor current loop make a double-input, single-output dynamic system. The motor current I_m is the controlled variable that must follow the demand I_m^* (pursuit function) and reject the disturbance E (rejection function). The modelling and analysis of the current loop are summarised in Table A1.

The PMSM motor is assumed to be of three phases with star connection. Being controlled under the max torque per current (null direct current, $I_d^* = 0$) strategy, it is considered as its equivalent brushed DC machine [26]. The motor constant K_m stands for the torque constant K_T (Nm/A), where the current is the I_q quadrature current of the power conservative dq0 transform. The torque constant equals the motor BEMF constant K_E (Vs/rad) when it is defined using the root mean square (RMS) line-to-line voltage. In the linear operating range of the PWM, the maximal line-to-line RMS voltage U_{DCE} at the motor windings is defined from the DC-link supply voltage U_{DC} as:

$$U_{DCE} = \frac{\sqrt{3}}{2\sqrt{2}} U_{DC} = 0.612 \, U_{DC} \tag{A1}$$

Allowing the PWM to operate in the pseudo-linear range extends the 0.612 factor to $1/\sqrt{2} = 0.707$ [27].

The last part of the table displays the main performance indicators for the very common setting that fixes the P-I time constant τ_{CI} of the controller to the motor electric time constant τ_e. In this case, the zero introduced by the P-I current controller compensates exactly the pole corresponding to the motor electric time constant. Therefore, the pursuit dynamics is fixed by the K_{iI} integral control gain, while the BEMF disturbance E is rejected at order 1 (s factor at the numerator) instead of order 0. The BEMF rate is rejected with the first-order dynamics of the electric time constant τ_e, whose gain is fixed by the proportional control gain K_{pI}. The BEMF disturbance can be theoretically removed thanks to feed-forward or compensation schemes. They involve the motor electromagnetic constant K_E and the measurement (or the estimate) of the motor shaft angular velocity.

At this level, the gains of the P-I current controller only depend on two parameters (the resistance and inductance of the motor windings). They seem to be independent of the target dynamics of the current loop. However, although Table A1 does not explicitly show any limitation in these gains, several additional effects bound in practice the dynamics and accuracy of the current loop:

- The modulation ratio is bounded to $[-1; +1]$;
- Of course, the BEMF (disturbance) is correlated to the motor current (controlled variable) through the airgap torque and the motor shaft dynamics. However, this coupling can be neglected, except in very specific cases, for the current loop study. This is because this coupling appears at frequencies that are significantly lower than the current loop bandwidth.

Table A1. Current loop model and analysis in continuous time domain.

		Constitutive Equations	
Pulse width modulation	$U_m = m\, U_{DCE}$ $I_m = \frac{1}{m} I_{DCE}$	Variables: U_m motor voltage, I_{DCE} supply current, I_m motor current, m modulation ratio Parameters: U_{DCE} equivalent line to line voltage	
Motor windings electrical circuit	$U_m = E + RI_m + LsI_m$	Variables: Ω_m motor shaft angular velocity, s Laplace variable, E motor BEMF Parameters: R motor windings resistance, L motor windings inductance	
P-I current controller	$m = (K_{pI} + K_{iI}/s)(I_m^* - I_m')$	Parameters: K_{pI} current loop proportional gain, K_{iI} current loop integral gain Variables: I_m^* motor current setpoint, I_m' measured current (equals the real current I_m if the measurement is perfect)	
		Current Open-Loop Transfer	
	$I_m = \dfrac{\frac{K_{iI}U_{DCE}}{s}(1+\tau_{CI}s)(I_m^*-I_m) - E}{R(1+\tau_e s)}$	Motor electric time constant $\tau_e = L/R$ Current controller time constant $\tau_{CI} = K_{pI}/K_{iI}$	
		Current Closed-Loop Transfer (P Control Only)	
	$I_m = \dfrac{1}{R'} \dfrac{K_{pI} U_{DC} I_m^* - E}{1 + \tau_e' s}$	Apparent windings resistance $R' = R + K_{pI}U_{DCE}$ Apparent electric time constant $\tau_e' = L/(R + K_{pI}U_{DCE})$	
		Current Closed-Loop Transfer (P-I Control, $\tau_{CI}=\tau_e$)	
	$I_m = \dfrac{1}{1+\tau_{lI}s}\left(I_m^* - \dfrac{\tau_e/K_{pI}U_{DCE}}{(1+\tau_e s)} sE\right)$		
Static pursuit gain:		$I_m/I_m^* = 1$	
Tracking rejection gain:		$I_m/sE = \tau_e/K_{pI}U_{DCE}$	
Denominator time constant (pursuit):		$\tau_{lI} = L/K_{pI}U_{DCE}$	
Denominator time constant (rejection):		τ_e	
Controller setting for τ_{lI} target pursuit dynamics:		$K_{pI} = L/\tau_{lI}U_{DCE},\ K_{iI} = K_{pI}R/L$	

Appendix B. Speed Loop

With reference to Figures 1 and 2, the elements involved in the actuator speed loop make a double-input, single-output dynamic system. The motor shaft speed Ω_m is the controlled variable that must follow the demand Ω_m^* (pursuit function) and reject the disturbance torque T_L (rejection function). Table A2 summarises the simplified modelling and linear analysis of the speed loop in the continuous time domain. It is obtained under the following assumptions:

- The dynamics of the current loop is neglected because in the very general case, it is much greater than the speed loop dynamics.
- At the speed loop level, the BEMF disturbance that applies to the current loop has no effect, either because the BEMF is compensated or because the integral action of the current controller removes its effect much faster than the speed loop dynamics.
- All mechanical effects are considered as their overall equivalent, expressed at the motor rotor level.

- The backlash and mechanical compliance of the actuator are not considered.
- Friction is assumed to be purely viscous, making it linearly dependent on relative speed only (see Section 2.2.1 for the discussion).
- As for the current loop, the digital implementation of the controller, the sensors and their conditioning, and thermal effects (in particular on K_m through the magnet's sensitivity to temperature) are not considered.

The P-I speed controller makes the speed closed loop behave as generalised second-order dynamics. The $\omega_{n\Omega}$ natural undamped frequency is fixed by the $K_{i\Omega}$ integral control gain. Given this gain, the ξ_Ω dimensionless damping factor is set linearly by the proportional gain $K_{p\Omega}$. As for the current loop, the integral action of the controller removes the speed dependence on constant external loads, while the dependence on the load rate is directly proportional to the integral control gain.

Given the linear modelling assumptions, Table A2 indicates no limitation in setting the speed controller gains. They are only linked to three EMA parameters (equivalent inertia J_E, equivalent viscous friction b_E and motor torque constant K_m) and to the target second order (damping factor ξ_Ω and natural frequency $\omega_{n\Omega}$).

Table A2. Speed loop model and analysis in continuous-time domain.

	Constitutive Equations	
Perfect current control	$I_m \approx I_m^*$	Variables: I_m motor actual current, I_m^* current setpoint
PI speed controller	$I_m^* = \left(K_{p\Omega} + K_{i\Omega}/s\right)\left(\Omega_m^* - \Omega_m'\right)$	Parameters: $K_{p\Omega}$ speed loop proportional gain, $K_{i\Omega}$ speed loop integral gain. Variables: s Laplace variable, Ω_m^* rotor speed setpoint, Ω_m' measured rotor speed (equals real value Ω_m if measurement perfect)
Dynamics of the moving parts reflected at the rotor level	$J_E s \Omega_m = K_m I_m - b_E \Omega_m - T_L$	Parameters: J_E equivalent inertia reflected at the rotor, b_E equivalent viscous friction reflected at the rotor, K_m motor torque constant. Variables: T_L EMA equivalent load reflected at the motor rotor
	Speed Open-Loop Transfer	
$\Omega_m = \dfrac{K_m(K_{p\Omega}+K_{i\Omega}/s)(\Omega_m^* - \Omega_m) - T_L}{J_E s + b_E}$		Mechanical time constant $\tau_m = J_E/b_E$
	Speed Closed-Loop Transfer	
	$\Omega_m = \dfrac{(1+\tau_\Omega s)\Omega_m^* - \frac{1}{K_m K_{i\Omega}} s T_L}{1 + (\tau_\Omega + \frac{b_E}{K_m K_{i\Omega}})s + \frac{J_E}{K_m K_{i\Omega}} s^2} = \dfrac{(1+\tau_\Omega s)\Omega_m^* - \frac{1}{K_m K_{i\Omega}} s T_L}{1 + \frac{2\xi_\Omega}{\omega_{n\Omega}} s + \frac{1}{\omega_{n\Omega}^2} s^2}$	
Static pursuit gain: $\Omega_m/\Omega_m^* = 1$		Tracking rejection gain: $\Omega_m/sT_L = 1/K_m K_{i\Omega}$
Speed controller time constant: $\tau_\Omega = K_{p\Omega}/K_{i\Omega}$		($\tau_\Omega = 2\xi_\Omega/\omega_{n\Omega}$ if b_E is neglected)
Denominator natural frequency: $\omega_{n\Omega} = \sqrt{K_m K_{i\Omega}/J_E}$		Denominator damping factor: $\xi_\Omega = \dfrac{K_m K_{p\Omega} + b_E}{2\sqrt{K_m K_{i\Omega} J_E}}$
Controller gains for target second order: $K_{i\Omega} = J_E \omega_{n\Omega}^2/K_m$, $K_{p\Omega} = (2J_E \xi_\Omega \omega_{n\Omega} - b_E)/K_m$		

References

1. Gonzalez, C.M. Engineering Refresher: The Basics and Benefits of Electromechanical Actuators. Available online: https://cdn.baseplatform.io/files/base/ebm/machinedesign/document/2019/04/machinedesign_15522_electromechanical.pdf (accessed on 8 February 2022).
2. O'dwyer, A. *Handbook of PI and PID Controller Tuning Rules*, 3rd ed.; Imperial College Press: London, UK, 2009.
3. *Incose Systems Engineering Handbook: A Guide for System Life Cycle Processes and Activities*, 4th ed.; Wiley-Blackwell: San Diego, CA, USA, 2015.
4. Ali, Z.A.; Han, Z. Maneuvering Control of Hexrotor UAV Equipped With a Cable-Driven Gripper. *IEEE Access* **2021**, *9*, 65308–65318. [CrossRef]
5. Fadel, M. Position Control for Laws for Electromechanical Actuator. In Proceedings of the 2005 International Conference on Electrical Machines and Systems, Nanjing, China, 27–29 September 2005; pp. 1708–1713. [CrossRef]
6. Zhang, M.; Li, Q. A Compound Scheme Based on Improved ADRC and Nonlinear Compensation for Electromechanical Actuator. *Actuators* **2022**, *11*, 93. [CrossRef]

7. Khanh, N.D.; Kuznetsov, V.E.; Lukichev, A.N.; Chung, P.T. Adaptive Control of Electromechanical Actuator Taking into Account Nonlinear Factors Based on Exo-model. In Proceedings of the 2022 Conference of Russian Young Researchers in Electrical and Electronic Engineering (ElConRus), St. Petersburg, Russia, 25–28 January 2022; pp. 803–807. [CrossRef]
8. Budinger, M. Preliminary design and sizing of actuation systems. In *Mechanical Engineering [physics.class-ph]*; UPS Toulouse: Toulouse, France, 2014.
9. Wu, S.; Bo, Y.; Jiao, Z.; Shang, X. Preliminary design and multi-objective optimization of electro-hydrostatic actuator. *Proc. Inst. Mech. Eng. Part G J. Aerosp. Eng.* **2010**, *231*, 1258–1268. [CrossRef]
10. Vaculik, S.A. A Framework for Electromechanical Actuator Design. Ph.D. Thesis, The University of Texas at Austin, Austin, TX, USA, May 2008. Available online: https://repositories.lib.utexas.edu/bitstream/handle/2152/18161/vaculiks84501.pdf?sequence=2&isAllowed=y (accessed on 8 February 2022).
11. Krishnan, R. *Permanent Magnet Synchronous and Brushless DC Motor Drives*; Taylor and Francis Group: Boca Raton, FL, USA, 2010. [CrossRef]
12. Sang-Hoon, K. *Electric Motor Control*; Elsevier: Amsterdam, The Netherlands, 2017. [CrossRef]
13. Dee, G.; Vanthuyne, T.; Alexandre, P. An electrical thrust vector control system with dynamic force feedback. In Proceedings of the 3rd International Conference on Recent Advances in Aerospace Actuation Systems and Components, Toulouse, France, 13–15 June 2007; pp. 75–79.
14. Landau, I.D. *Identification et Commande des Systèmes*; Hermes: Paris, France, 1988.
15. Krishnan, R. *Electric Motor Drives: Modeling, Analysis, and Control*; Prentice Hall: Upper Saddle River, NJ, USA, 2001.
16. Crowder, R. *Electric Drives and Electromechanical Systems Applications and Control*, 2nd ed.; Elsevier: Oxford, UK, 2020.
17. Maré, J.-C. Friction modelling and simulation at system level—Considerations to load and temperature effects. *IMechE Part I J. Syst. Control. Eng.* **2015**, *229*, 27–48. [CrossRef]
18. Arriola, D. Model-based Design and Fault-tolerant Control of an Actively Redundant Electromechanical Flight Control Actuation System. Ph.D. Thesis, Technical University Hamburg, Hamburg, Germany, 2019.
19. Dimino, I.; Gallorini, F.; Palmieri, M.; Pispola, G. Electromechanical Actuation for Morphing Winglets. *Actuators* **2019**, *8*, 42. [CrossRef]
20. Wang, L. Force Equalization for Active/Active Redundant Actuation System Involving Servo-Hydraulic and Electro-Mechanical Technologies. Ph.D. Thesis, INSA, Toulouse, France, December 2012. Available online: http://www.theses.fr/2012ISAT0038/document (accessed on 8 February 2022).
21. Karam, W. Générateurs de Forces Statiques et Dynamiques à Haute Puissance en Technologie Électromagnétique. Ph.D. Thesis, INSA, Toulouse, France, November 2007. Available online: http://eprint.insa-toulouse.fr/archive/00000182/http://www.theses.fr/2007ISAT0035/document (accessed on 8 February 2022).
22. Fu, J.; Maré, J.-C.; Fu, Y. Modelling and simulation of flight control electromechanical actuators with special focus on model architecting, multidisciplinary effects and power flows. *Chin. J. Aeronaut.* **2017**, *30*, 47–65. [CrossRef]
23. Fu, J.; Maré, J.-C.; Yu, L.; Fu, Y. Multi-level virtual prototyping of electromechanical actuation system for more electric aircraft. *Chin. J. Aeronaut.* **2018**, *31*, 889–913. [CrossRef]
24. Maré, J.-C. Recent Experiences in Extending the Scope of EMA Simulation for Virtual V&V. In Proceedings of the SAE A-6B3 Panel, Spring Meeting, Virtual, 28 April 2021; Available online: https://www.sae.org/works/meetingminuteResources.do?comtID=TEAA6&resourceID=867983 (accessed on 8 February 2022).
25. Åström, K.J. Advanced Control Methods: Survey and Assessment of Possibilities. In *Advanced Control in Computer Integrated Manufacturing: A Control Engineer's View*; Morris, H.M., Kompass, E.J., Williams, T.J., Eds.; Purdue University: West Lafayette, IN, USA, 1987.
26. Maré, J.-C. Practical Considerations in the Modelling and Simulation of Electromechanical Actuators. *Actuators* **2020**, *9*, 94. [CrossRef]
27. Mohan, N.; Undeland, T.M.; Robbins, W.P. *Power Electronics: Converters, Applications and Design*, 1st ed.; Wiley: Hoboken, NJ, USA, 1989.

MDPI
St. Alban-Anlage 66
4052 Basel
Switzerland
Tel. +41 61 683 77 34
Fax +41 61 302 89 18
www.mdpi.com

Aerospace Editorial Office
E-mail: aerospace@mdpi.com
www.mdpi.com/journal/aerospace

www.ingramcontent.com/pod-product-compliance
Lightning Source LLC
LaVergne TN
LVHW070142100526
838202LV00015B/1879